**ACPL ITEM
DISCARDED**

5-15-73

Man's Impact
on Terrestrial
and Oceanic Ecosystems

The MIT Press
Cambridge, Massachusetts,
and London, England

**Man's Impact
on Terrestrial
and Oceanic Ecosystems**

Edited by
William H. Matthews
Frederick E. Smith
Edward D. Goldberg

Copyright © 1971 by
The Massachusetts Institute of Technology

This book was designed by the MIT Press Design Department.
It was set in Linotype and Monotype Baskerville
by The Colonial Press Inc.
Printed on Mohawk Neotext Offset
by Semline, Inc.
and bound in Fictionette FNV 3453
by The Colonial Press Inc.
in the United States of America.

All rights reserved. No part of this book may be reproduced in any form or by any means, electronic or mechanical, including photocopying, recording, or by any information storage and retrieval system, without permission in writing from the publisher.

Library of Congress Cataloging in Publication Data

Matthews, William Henry, 1942–   comp.
   Man's impact on terrestrial and oceanic ecosystems. Includes bibliographies.
   1. Pollution. 2. Man—Influence on nature. 3. Ecology. I. Smith, Frederick E., joint comp. II. Goldberg, Edward D., joint comp. III. Title.
TD174.M39   574.5   79–160055

ISBN 0-262-13077-7

To Carroll L. Wilson

The Director of the Study
of Critical Environmental
Problems

Contents

Preface	x
Contributors	xiii

Part I
General Ecological Effects of Man's Activities — 1

1
Work Group on Ecological Effects (Abridged) — 4
SCEP Report

2
Subtraction by Multiplication: Population, Technology, and the Diminished Man — 33
F. H. Bormann

3
Effects of Pollution on the Structure and Physiology of Ecosystems — 47
G. M. Woodwell

4
Population, Natural Resources, and Biological Effects of Pollution of Estuaries and Coastal Waters — 59
Bostwick H. Ketchum

5
Health Effects Research Program of the National Air Pollution Control Administration — 80
Vaun A. Newill, Gory J. Love, F. Gordon Hueter, and Robert J. M. Horton

Part II
Pollution and Terrestrial Ecosystems — 99

6
Air Pollution and Plants — 101
H. E. Heggestad

7
Air Pollution and Trees — 116
George H. Hepting

8
Pollution and Range Ecosystems — 130
Dixie R. Smith

9
A Synopsis of the Pesticide Problem (Abridged) — 144
N. W. Moore

Part III
Climatic Change and Terrestrial Ecosystems   173

10
Climatic Effects of Man's Activities   174
Summary of SCEP Report

11
Potential Effects of Global Temperature Change on Agriculture   184
Sherwood B. Idso

12
Potential Effects of Global Atmospheric Conditions on Forest Ecosystems   192
Karl F. Wenger, Carl E. Ostrom, Philip R. Larson, and Thomas D. Rudolph

13
Climate and Forest Diseases   203
George H. Hepting

Part IV
Pollution and Oceanic Ecosystems   227

14
Runoff from Agricultural Land as a Potential Source of Chemical, Sediment, and Waste Pollutants   230
Lloyd L. Harrold

15
Runoff from Forest Lands   240
Howard W. Lull

16
Waste-Solid Disposal in Coastal Waters of North America   252
M. Grant Gross

17
Chemical Invasion of Ocean by Man   261
Edward D. Goldberg

18
Chlorinated Hydrocarbons in the Marine Environment   275
SCEP Task Force

19
Ocean Pollution by Petroleum Hydrocarbons 297
Roger Revelle, Edward Wenk, Bostwick H. Ketchum,
and Edward R. Corino

20
Phosphorus and Eutrophication 319
Excerpt from SCEP Work Group on Ecological Effects

Part V
Measurements and Monitoring 325

21
Work Group on Monitoring (Abridged) 327
SCEP Report

22
Global Biological Monitoring 351
Dale W. Jenkins

23
Identification of Globally Distributed Wastes in the Marine Environment 371
Edward D. Goldberg and M. Grant Gross

24
Proposal for a Base-Line Sampling Program 377
Edward D. Goldberg, Geirmundur Arnason,
M. Grant Gross, Frank G. Lowman, and Joseph L. Reid

25
International Environmental Monitoring Programs 392
Robert Citron

Part VI
Modeling: A Tool for Understanding and Management 429

26
Estuary Modeling 430
Geirmundur Arnason

27
General Circulation Patterns in the World Ocean 448
Joseph L. Reid

28
Hydrodynamic Modeling of Ocean Systems 460
Kirk Bryan

Part VII
Some Implications of Change 467
29
Implications of Change and Remedial Action 469
Summary of SCEP Report
30
Expectations of the Decision Maker 473
Richard A. Carpenter
31
Residuals Management 477
Walter O. Spofford, Jr.
32
Phosphates, Heavy Metals, and DDT: Pollution Control Costs and Implications 489
John F. Brown, Jr.
33
DDT: The United States and the Developing Countries 499
Rita F. Taubenfeld

Index 519

Preface

Over the past few years, the concept of the earth as a "spaceship" has provided many people with an awareness of the finite resources and the complex natural relationships on which man depends for his survival. These realizations have been accompanied by concerns about the impacts that man's activities are having on the global environment. Some concerned individuals, including well-known scientists, have warned of both imminent and potential global catastrophes.

Theories and speculations of the global effects of pollution have included assertions that the buildup of CO_2 from fossil-fuel combustion might warm up the planet and cause the polar ice to melt, thus raising the sea level several hundred feet and submerging coastal cities. Equally foreboding has been the warning of the possibility that particles emitted into the air from industrial, energy, and transportation processes might prevent some sunlight from reaching the earth's surface, thus lowering global temperature and beginning a new ice age. Demands to ban DDT have been increasing steadily as its effects on the reproductive capabilities of birds have been determined, and as evidence is found of its accumulation in other species including man. Serious questions have been raised about the effects on terrestrial and oceanic ecosystems of systematically discharging into the environment air pollutants and such toxic materials as heavy metals, oil, and radioactive substances; or of nutrients such as phosphorus that can overenrich lakes and coastal areas.

These and related global environmental problems were considered in depth by the one-month, interdisciplinary Study of Critical Environmental Problems (SCEP) conducted during July 1970. This Study, which was sponsored by the Massachusetts Institute of Technology, produced a report entitled *Man's Impact on the Global Environment,* which was published by The M.I.T. Press in October 1970. That report presents an assessment of the existing state of scientific knowledge on several global problems and contains specific recommendations for action which would reduce the harmful effects of pollution or would provide the information required to understand more adequately the impact of man on the global environment.

The Director of SCEP was Professor Carroll L. Wilson, and the Associate Director was Professor William H. Matthews, both of M.I.T. Chairmen of major SCEP Work Groups included Drs. Frederick E. Smith, Harvard University; Edward D. Goldberg, Scripps Institution of Oceanography; William W. Kellogg, National Center for Atmospheric Research; and G. D. Robinson, Center for the Environment and Man, Inc. These group leaders with Dr. Matthews have compiled this volume and a second volume, *Man's Impact on the Climate,* from the background materials prepared for SCEP, working papers written during the Study, and a few selected articles that have been previously published.

These two volumes are intended to provide more detailed scientific and technical information on global environmental problems than could adequately be summarized in the SCEP Report. It is hoped that these volumes, which reproduce, supplement, and complement material in the SCEP Report, will serve as useful reference works to researchers and students in the many disciplines involved in solving global environmental problems.

This volume contains 33 papers of varying length, scope, depth, complexity, and style. Some are general reviews; others are detailed analyses of one problem. Some were prepared during the Study, and others have been polished for previous publication. Despite these variations, we believe that this volume represents a unique collection of ideas, facts, techniques, wisdom, and judgment.

Such a collection could have been compiled only through the cooperation of several federal agencies with responsibilities in these areas, particularly those of the U.S. Department of Agriculture, and through the active participation in SCEP of scientists and professionals from over a dozen universities and research institutions. To all these and many others, we owe a sincere thanks. We also express our deep appreciation for the extraordinary assistance of Miss Ada Demb of M.I.T. in converting this collection of papers into a publishable manuscript.

The subject of this volume is the impact of man's activities on terrestrial and oceanic ecosystems. The book is divided into seven parts, each preceded by an introduction to the papers in that section. These introductions serve to alert the reader to the content of the papers and to relate the papers to one another.

Part I provides a broad semitechnical overview of the general nature of ecological and biological problems that result from population growth and technology utilization. The concepts discussed in the SCEP Work Group report and the other papers are fundamental to an understanding of the ecological context of the many, more specific pollution problems that must ultimately be solved.

Parts II and III deal with the direct and indirect effects of pollution on large terrestrial ecosystems. The direct effects are the results of specific pollutants acting directly on living organisms, while the indirect effects are much more subtle and result from changes that man may cause in the climate of the planet. The potential for disrupting these ecosystems that have evolved over centuries is clearly evident from these papers.

The introduction of pollutants into the marine environment and the effects of specific pollutants on oceanic ecosystems are the subjects treated in the seven papers in Part IV. In each paper, the same theme emerges with disturbing regularity—man knows very little about the effects his activities may have on the marine environment. In order to obtain the information required for protecting the earth's ecosystems, measurement and monitoring programs must be expanded and supplemented. Recommendations on how to proceed in this area and a review of present activities are presented in Part V. Another important tool for gaining additional knowledge and for effective environmental management is mathematical modeling of complex estuarine and ocean systems. The state of the art of such models is reviewed in Part VI.

The final section of this volume, Part VII, contains several papers that illustrate the complex social and political issues which must be addressed if decisions that could avert potential disasters are ever to be made and implemented. These are issues that cannot simply be resolved by "more research"—the scientist and his fellow citizens in other fields must confront them together. We hope that this volume will contribute to those efforts.

William H. Matthews
Frederick E. Smith
Edward D. Goldberg

Contributors

Geirmundur Arnason
CENTER FOR THE ENVIRONMENT
AND MAN, INC.

F. H. Bormann
YALE UNIVERSITY

John F. Brown, Jr., Manager
LIFE SCIENCES BRANCH
GENERAL ELECTRIC RESEARCH
AND DEVELOPMENT CENTER

Kirk Bryan
GEOPHYSICAL FLUID DYNAMICS
LABORATORY
NATIONAL OCEANIC AND
ATMOSPHERIC ADMINISTRATION

Philip A. Butler
BUREAU OF COMMERCIAL
FISHERIES
U.S. FISH AND WILDLIFE SERVICE

Richard A. Carpenter, Chief
ENVIRONMENTAL POLICY
DIVISION
CONGRESSIONAL RESEARCH
SERVICE
LIBRARY OF CONGRESS

Robert Citron, Director
CENTER FOR SHORT-LIVED
PHENOMENA
THE SMITHSONIAN INSTITUTION

Edward R. Corino
ESSO RESEARCH AND
ENGINEERING COMPANY

Gifford Ewing
WOODS HOLE OCEANOGRAPHIC
INSTITUTION

Edward D. Goldberg
SCRIPPS INSTITUTION OF
OCEANOGRAPHY

M. Grant Gross
MARINE SCIENCES RESEARCH
CENTER
STATE UNIVERSITY OF NEW YORK
AT STONY BROOK

Lloyd L. Harrold
AGRICULTURAL RESEARCH
SERVICE
U.S. DEPARTMENT OF
AGRICULTURE, AND OHIO STATE
UNIVERSITY

H. E. Heggestad
AGRICULTURAL RESEARCH
SERVICE
U.S. DEPARTMENT OF
AGRICULTURE

George H. Hepting
FOREST SERVICE
U.S. DEPARTMENT OF
AGRICULTURE, AND
NORTH CAROLINA STATE
UNIVERSITY

Robert J. M. Horton
DIVISION OF HEALTH EFFECTS
RESEARCH
AIR POLLUTION CONTROL OFFICE
ENVIRONMENTAL PROTECTION
AGENCY

F. Gordon Hueter, Associate
Director
DIVISION OF HEALTH EFFECTS
RESEARCH
AIR POLLUTION CONTROL OFFICE
ENVIRONMENTAL PROTECTION
AGENCY

Sherwood B. Idso
AGRICULTURAL RESEARCH
SERVICE
U.S. DEPARTMENT OF
AGRICULTURE

Merton Ingham, Chief
OCEANOGRAPHY UNIT
U.S. COAST GUARD

Dale W. Jenkins
DIRECTOR OF ECOLOGY
THE SMITHSONIAN INSTITUTION

Philip C. Kearny
AGRICULTURAL RESEARCH SERVICE
U.S. DEPARTMENT OF AGRICULTURE

Bostwick H. Ketchum, Associate Director
WOODS HOLE OCEANOGRAPHIC INSTITUTION

Philip R. Larson
FOREST SERVICE
U.S. DEPARTMENT OF AGRICULTURE

Gory J. Love, Assistant Director
DIVISION OF HEALTH EFFECTS RESEARCH
AIR POLLUTION CONTROL OFFICE
ENVIRONMENTAL PROTECTION AGENCY

Frank G. Lowman, Head
RADIOECOLOGY DIVISION
PUERTO RICO NUCLEAR CENTER

Howard W. Lull
FOREST SERVICE
U.S. DEPARTMENT OF AGRICULTURE

Paul Meier
UNIVERSITY OF CHICAGO

David W. Menzel
WOODS HOLE OCEANOGRAPHIC INSTITUTION

N. W. Moore
MONK'S WOOD EXPERIMENTAL STATION
HUNTINGTON, ENGLAND

Vaun A. Newill, Director
DIVISION OF HEALTH EFFECTS RESEARCH
AIR POLLUTION CONTROL OFFICE
ENVIRONMENTAL PROTECTION AGENCY

Carl E. Ostrom
FOREST SERVICE
U.S. DEPARTMENT OF AGRICULTURE

Joseph L. Reid
SCRIPPS INSTITUTION OF OCEANOGRAPHY

Roger Revelle, Director
CENTER FOR POPULATION STUDIES
HARVARD UNIVERSITY

Robert Risebrough
INSTITUTE OF MARINE RESOURCES
UNIVERSITY OF CALIFORNIA AT BERKELEY

Thomas D. Rudolph
FOREST SERVICE
U.S. DEPARTMENT OF AGRICULTURE

Dixie R. Smith
FOREST SERVICE
U.S. DEPARTMENT OF AGRICULTURE

Walter O. Spofford, Jr.
QUALITY OF THE ENVIRONMENT PROGRAM
RESOURCES FOR THE FUTURE, INC.

Lucille F. Stickel
PATUXENT WILDLIFE RESEARCH CENTER
DEPARTMENT OF THE INTERIOR

Rita F. Taubenfeld
INSTITUTE OF AERO-SPACE LAW
SOUTHERN METHODIST UNIVERSITY

Karl F. Wenger
FOREST SERVICE
U.S. DEPARTMENT OF AGRICULTURE

Edward Wenk, Jr.
UNIVERSITY OF WASHINGTON

G. M. Woodwell
BROOKHAVEN NATIONAL LABORATORY

Man's Impact
on Terrestrial
and Oceanic Ecosystems

Part I   General Ecological Effects
                           of Man's Activities

The first paper in this section and several papers in later sections are products of the Study of Critical Environmental Problems (SCEP), which was sponsored by the Massachusetts Institute of Technology and conducted during July 1970 at Williamstown, Massachusetts. The papers are reprinted from the SCEP Report, *Man's Impact on the Global Environment,* which was published by The M.I.T. Press shortly after the Study.

SCEP was a one-month interdisciplinary examination of the global climatic and ecological effects of man's activities. The disciplines represented by the more than fifty participants included meteorology, oceanography, ecology, chemistry, physics, biology, geology, engineering, economics, social sciences, and law. The participants were drawn from seventeen universities, thirteen federal agencies, three national laboratories, and eleven nonprofit and industrial corporations. The Study was supported professionally and/or financially by twelve federal agencies, four foundations, six industrial corporations, a research institution, and an academic institution.

The SCEP Report contains a summary of the major findings and recommendations of the Study and also the reports of the seven SCEP Work Groups. These Work Group reports were developed through intensive, full-time discussion and study by the group members. The report of the Work Group on Ecological Effects which appears in this section is the product of four weeks of intensive effort. Several sections of the report have been deleted here because they presented summaries of specific areas that will be treated in much more detail in other papers of this volume. The group was chaired by Frederick E. Smith, Harvard University, and the participants were Geirmundur Arnason, Center for the Environment and Man, Inc.; Edward D. Goldberg, Scripps Institution of Oceanography; M. Grant Gross, State University of New York; Arthur Hasler, University of Wisconsin; Bruce B. Hanshaw, U.S. Geological Survey; J. B. Hilmon, U.S. Forest Service; Dale W. Jenkins, The Smithsonian Institution; Philip C. Kearny, Agricultural Research Service; Frank Lowman, Puerto Rico Nuclear Center; Jerry S. Olson, Oak Ridge National Laboratory; Joseph Reid, Scripps Institution of Oceanography; Clarence Tarzwell, Federal Water Quality Administration; Edward

Wenk, University of Washington; George Woodwell, Brookhaven National Laboratory. Stephen Burbank and Robert Stoller of Harvard Law School were rapporteurs.

Major themes of the report of the SCEP Work Group on Ecological Effects concern the interactions among man, technology, and environment and the constraints surrounding future development. The second paper of this series, which was recently prepared by Dr. F. H. Bormann, President of the Ecological Society of America, brings together in a forceful, succinct fashion the problems generated by population growth and technology. Although the problem of population growth was not considered in depth by SCEP, it was regarded as fundamental in the solution of global environmental problems. This paper also complements the SCEP Work Group report through comments on the distribution of resources, the problems of developing and developed nations, and ultimate questions of life-style.

In studying pollution problems, it is extremely important—though exceedingly difficult—to focus attention on the general effects of pollution on ecosystems rather than only on analysis of the effects of particular pollutants. Dr. G. M. Woodwell provided many insights during SCEP in this critical area. His paper on these general effects was published in *Science* shortly before the Study and is reproduced in this volume.

Coastal and estuarine regions are complex ecosystems that are increasingly being overtaxed by the growth of population, industry, and agriculture. Even more significant than increased use is the problem of fundamental conflicts in the fisheries and recreational demands on these waters and their use as disposal areas for the waste products of our civilization. Dr. Ketchum's paper reviews the present state of our understanding about these relationships and then presents two thought-provoking matrices on the effects of human uses on estuarine quality and on the compatibility of present and potential human uses of estuaries.

In delimiting the problem areas to be considered by SCEP, the Steering Committee that planned the Study during the winter and spring of 1969–1970 decided that the effects of pollution on human health would not be studied. It was felt that this complex area could not be treated adequately within the scope and scale of the envisioned study of global climatic and ecological problems. Yet, human health is an area of critical importance, and man is indeed a major component in the ecology of the planet. In order to provide this perspective to those at SCEP, Dr. Vaun A. Newill of the National Air Pollution Control Administration, NAPCA (then an agency of the Department of Health, Education, and Welfare and now in the Environmental Protection Agency), presented to the participants early in July a paper on the NAPCA health effects research program. This paper concludes the section on the general ecological effects of man's activities.

1
Work Group on Ecological Effects (Abridged) — SCEP Report

Introduction
This report focuses on three major areas of ecological concern in the biosphere:* (1) man's overall effect on terrestrial vegetation and terrestrial ecosystems; (2) his cumulative effect on the estuaries and coastal oceans that support most of the world's fisheries; and (3) problems relating to the biology of the oceans. In all three areas a lack of critical information severely handicapped problem evaluation. Nevertheless, major environmental problems have been identified that exist now or seem certain to exist soon.

Only a small part of the biosphere is managed by man, yet he relies upon all of it for the regulation of climate, composition of the atmosphere, supplies of clean water, and many other services. In this sense the entire biopshere is used. The increasing scope and intensity of such use, compounded by practices whose side effects have spread throughout the globe, threaten the capacity of the biosphere to continue its normal functions. Man may suddenly be faced with far greater management obligations than he is willing or able to undertake. Changes in human action are urgently needed, both to ameliorate our impact upon the environment and to develop new management capabilities.

The Problem
The paragraphs that follow summarize information on man's total impact, general changes in ecosystems, and the consequences of such changes for man. These are the general arguments in support of the contention that relations between man and biosphere are approaching a stage of crisis.

Present Scale of Human Activity
Man is now using about 41 percent of the total land surface (Table 1.1). More than half of the remainder is not usable because it is too cold, frozen, or mountainous. The most optimistic report

---

Reprinted from Study of Critical Environmental Problems (SCEP), 1970. *Man's Impact on the Global Environment* (Cambridge, Massachusetts: The M.I.T. Press), pp. 114–126, 160–166.

* The biosphere of the earth is composed of all life-forms and of that portion of air, land, and water in which life exists. The following is the scientific definition of biosphere to which this Work Group subscribed: a zone of the earth intersecting with the lower atmosphere, hydrosphere, and upper lithosphere.

Table 1.1 Present and Potential Uses of the Land Surface of the Planet (Percent of total area)

Use	Present	Potential
Croplands	11	24
Rangelands	20	28
Managed forests	10	15
Reserves (80% forest)	26	0
Not usable	33	33
Total Land	100	100

Source: President's Science Advisory Committee (PSAC), 1967.

Table 1.2 Past and Projected World Land-Use Changes (Percent of total land area)

Use	1950–1968	1968–2000
Cropland	+1.5	+0.4
Rangeland	+5.2	−1.3
Total forest	−0.2	−1.3

Sources: Food and Agriculture Organization (FAO), 1951, 1958, 1969.

on land use that has appeared in recent times (President's Science Advisory Committee [PSAC] Report, 1967) estimates that as much as two-thirds of the total land surface is usable. These figures can be taken as the upper limit. If an estimate is weighted for the different intensities of different land uses, we appear to be using about half of the earth's land resources at the present time (Table 1.1).

The Food and Agriculture Organization of the United Nations (FAO) projects very little exploitation of the remaining potential (Table 1.2). The land already in use tends to be the best for such use, and intensification of management will probably be chosen rather than expansion into marginal lands. Expansion has ended in the developed nations and, by 1985, is expected to be ended in most of the world. Thus, except for an expansion in the utilization of forest land, use is expected to become stabilized. The difference between the FAO projections and the potential estimated in the PSAC Report is only a factor of 2, which is not large in comparison with problems of population growth over the next century.

The present scale of human activity can also be estimated for a number of specific man-induced processes in comparison with natural rates of geological and ecological processes. H. J. M. Bowen has compared global mining activities with estimates of the annual discharge of materials through rivers into the oceans. Such natural rates are well suited to an analysis of pollution problems in the coastal oceans (Bowen, 1966). Twelve cases for which the man-induced rate is as large or larger than the natural rate are shown in Table 1.3. With a 5 percent annual growth increment in the mining industries, many more materials will soon fall into this category (Table 1.4). Although data on some natural rates may be erroneous and therefore lead to some incorrect inclusions and omissions, the table nonetheless gives a reasonable overview.

These comparisons show that at least some of our actions are large enough to alter the distribution of materials in the biosphere. Whether these changes are problems depends upon the toxicity of the material, its distribution in space and time, and its persistence in ecological systems. Better estimates of natural

Table 1.3 Man-Induced Rates of Mobilization of Materials Which Exceed Geological Rates as Estimated in Annual River Discharge to the Oceans (Thousands of metric tons per year)

Element	Geological Rates[a] (In Rivers)	Man-Induced Rates[b] (Mining)
Iron	25,000	319,000
Nitrogen	8,500	9,800 (consumption)
Manganese	440	1,600
Copper	375	4,460
Zinc	370	3,930
Nickel	300	358
Lead	180	2,330
Phosphorus	180	6,500 (consumption)
Molybdenum	13	57
Silver	5	7
Mercury	3	7
Tin	1.5	166
Antimony	1.3	40

Sources:
[a] Bowen, 1966.
[b] United Nations, *Statistical Yearbook*, 1967. Data for mining except where noted.

Table 1.4 World Average Annual Rates of Increase for the Period 1951–1966 for Selected Aspects of Human Activity[a]

	Percentage[b]
Agricultural production	3
Industry based on farm products	6
Mineral production (including fuel)	5
Industry based on mineral products	9
Construction and transportation	6
Commerce	5

Source: Digested from United Nations, *Statistical Yearbook*, 1967.
[a] Data are for the world excluding Mainland China.
[b] Rates are in constant dollars.

processes, and of man-induced rates that include waste contributions from other industries, would be very useful.

If total production rates of materials are small in comparison with natural rates, problems may be locally intensive but are not expected to be large on a global scale. Where technological exploitation or production rates are high, data on loss rates to the environment are needed. In the absence of such data it is impossible to ascertain the effects of technology on the environment. For example, an industry with a high recycling rate and an accumulating inventory in products may have a small output of waste.

Growth in Human Demands

The world population has been increasing at about 2 percent per year, a rate that produces a doubling every 35 years (United Nations, *Statistical Yearbook,* 1967). Population alone, however, does not reflect our total demand upon the environment. As our numbers increase, the technology needed to support people must increase even more. This arises in part from the increasing complexity of activities associated with larger populations and in part from the increasing difficulty of achieving higher and higher rates of production of goods. The average rates of annual increase of relevant activities for the last fifteen years are shown in Table 1.4.

The problem of achieving higher rates of production is well defined in agriculture, where increasing yields are obtained only

Table 1.5 World Average Rates of Increase for the Period 1951–1966 for Selected Aspects of Human Activity Related to Food Production

	Percentage[a]
Food	34
Tractors	63
Phosphates	75
Nitrates	146
Pesticides	300

Source: Digested from United Nations, *Statistical Yearbook*, 1967.
[a] Rates in constant dollars.

Table 1.6 Pesticide Usage and Agricultural Yields in Selected World Areas

Area or Nation	Pesticide Use		Yield	
	Grams per hectare	Rank	Kilograms per hectare	Rank
Japan	10,790	1	5,480	1
Europe	1,870	2	3,430	2
United States	1,490	3	2,600	3
Latin America	220	4	1,970	4
Oceania	198	5	1,570	5
India	149	6	820	7
Africa	127	7	1,210	6

Source: FAO, *Production Yearbook*, 1963.

from still greater inputs. Table 1.5 shows these increases over a fifteen-year period.

Finally, Table 1.6, from the FAO *Production Yearbook* for 1963, shows how the quantity of pesticide application relates to increasing food supplies. Note that Japan, with twice the U.S. yield of food per acre, uses ten times as much pesticide, while we, with just over twice the yield of Africa, use eleven times as much pesticide.

Such a high ecological cost for increased food production may be the result of a deeply entrenched agricultural system, or it may be unavoidable. In either case it poses a very difficult specific problem associated with population growth.

A measure of *ecological demand* is needed that summarizes man's impact upon the environment. This would include both the utilization of resources and the disposal of wastes and would reflect the intensity of interaction between the biosphere and the aggregate of the man-made world. One such statistic is the Gross Domestic Product (minus services) which is compiled by the United Nations (Table 3 in United Nations, *Statistical Yearbook, 1968*). Since 1950 this index has been increasing between 5 and 6 percent per year, doubling in $13\frac{1}{2}$ years, which appears to be a reasonable estimate for the current rate of increase in ecological demand. If this should continue, then, in the time taken for the human population to double, our total ecological demand would increase sixfold.

More than anything else, such a rapid rate of growth suggests why environmental problems seem to have erupted so suddenly, why the future will surely bring more problems than the present, and why slowness to respond may be disastrous.

Global Ecological Effects

The significant aspect of human action is man's total impact on ecological systems, not the particular contributions that arise from specific pollutants. Interaction among pollutants is more often present than absent. Furthermore, the total effect of a large number of minor pollutants may be as great as that of one major pollutant. Thus, the total pollution burden may be impossible to estimate except by direct observation of its overall effect on ecosystems.

Recognition of this overall effect is essential for defining the real pollution problem. However, the effects and ultimate costs of specific pollutants must also be documented if effective control is to become possible. Of the pollutants that pose threats to ecological systems, this Work Group chose to concentrate on four classes of materials released by man into the environment. Case studies for these are presented in the next section of this Work Group report.

SELECTIVE IMPAIRMENT OF PREDATORS

Among animals, predators (those that eat other animals) are generally more sensitive to environmental stress than herbivores (those that eat plants). Furthermore, in aquatic systems the top-level predators (which eat other predators, for example, most game fish) are the most sensitive of all. This has been found for stressors such as oxygen deficiency, thermal stress, toxic materials, and pesticides. It is as general in terrestrial as in aquatic systems, and among insects and other invertebrates as well as among vertebrates.

Examples of this phenomenon are numerous. Overenrichment by sewage waste and fertilizer runoff of fresh waters, or pollution with industrial wastes, leads to the rapid loss of trout, salmon, pike, and bass. Spraying crops for insect pests has inadvertently killed off many predaceous mites, resulting in outbreaks of herbivorous mites that obviously suffered less. Forest spraying has similarly "released" populations of scale insects after heavy damage to their wasp enemies.

In addition to suffering direct damage from pollutants, predators also suffer from the damage done lower in the food web to the herbivores and plants. In effect such pollutants compete with the predators, their damage resulting in a loss of food to the predators. Both in nature and in computer models, severe damage to lower levels in the food chain usually leads to the extinction of the predator before its food is extinct.

Persistent pesticides reveal a third mechanism by which predators may suffer more than their prey. Such fat-soluble pesticides as DDT are concentrated as they pass from one feeding level to the next. In the course of digestion a predator retains rather than eliminates the DDT content of its prey. The more it eats, the more DDT is accumulates. The process results in especially high concentrations of toxins in predaceous terrestrial vertebrates.

EXCESSIVE DAMAGE AT ONE BRIEF STAGE IN LIFE CYCLES

The need for more thorough toxicity studies is presented later in this report. The frequency with which damage is brief but intense, hitting one particular stage without necessarily harming older or younger stages, makes cause-and-effect studies very difficult. Usually some phase of the reproductive cycle is involved. Oyster larvae are wiped out by pollution with copper that poses no danger to established oysters. DDT can kill fish larvae at the moment of yolk absorption without seeming to have harmed the parent that put the pesticide into the egg. The final effect is that entire populations can disappear, and the apparent health of the older individuals may lead to a lack of corrective action until it is too late.

This cause of local extinction is less likely to be selective among herbivores and predators than the causes of damage already discussed.

GREATER INSTABILITY IN ECOSYSTEM REGULATION

Terrestrial ecosystems of landscape size usually contain hundreds of species of plants and thousands of species of animals, but the herbivores rarely consume more than 5 to 15 percent of the vegetation.

Occasionally the set of processes that generally keeps herbivores in check fails, and they become numerous enough to damage the foliage appreciably. Even so, it is rare for such outbreaks to be severe. Ecosystem regulation is so general that exceptions are objects of intense study. The tundra-lemming system in the Arctic and the conifer-spruce budworm system in northern forests appear to be two examples of systems in which strong fluctuations are inherent (Morris and Miller, 1954; Pitelka, Tomick, and Treichel, 1955).

The normal degree of control is best summarized by pointing out that for each species of terrestrial plant there are about 100 species of animals capable of eating it, yet most of the time, most of the plant production falls to the ground uneaten. For many insects able to defoliate trees the population density rarely rises as high as one per tree.

Nevertheless, this stability is fragile enough to be easily destroyed. The list of disturbances that lead to population outbreaks (almost always of herbivores) includes almost anything that removes a number of species or impairs the health or numbers of predators. Forests are familiar with the consequences that may follow cleaning out underbrush, selectively removing tree species, or growing single-aged one-species stands (see, for example, Davis, 1954).

The two general effects discussed strongly affect the frequency and severity of population explosions in ecosystems. The loss of species from excessive damage leaves a system less able to function or to recover from dysfunction, while selective damage to predators can precipitate population outbreaks.

ATTRITION OF ECOSYSTEMS

The most general effect of pollutants on ecosystems combines all of the effects discussed previously with additional direct depression of plant or animal vigor. This leads to overall deterioration of the system, characterized by instability and species loss. The result is a system of few species, generally weeds, pollution-tolerant in the sense of being able to support high death rates. The only further step is annihilation, a stage that so far has occurred only in areas of extreme pollution.

Many lakes and urban centers have severely deteriorated ecosystems. Less severe deteriorations occur more commonly, often as temporary afflictions in ecosystems that otherwise manage to survive intact. Once an ecosystem is severely damaged and becomes unattractive, its death is usually considered an improvement by the people who live within it.

This general problem is labeled "attrition" because it lacks discrete steps of change. Stability is lost more and more frequently, noxious organisms become more common, and the aesthetic aspects of waters and countryside become less pleasing. This process has already occurred many times in local areas. If it were to happen gradually on a global scale, it might be much less noticeable, since there would be no surrounding healthy ecosystems against which to measure such slow change. Each succeeding generation would accept the status quo as "natural."

The gradual decline in ecosystem function brings with it a decline in services for man. When at the same time nature becomes unappealing to man, the decision is easily reached to dispense with the natural systems (for example, filling in a polluted lake). The real costs of dispensing with more and more natural functions need to be appraised.

Environmental Services

It is a mark of our time, and a signal of the degree to which man is ecologically disconnected, that the benefits of nature need to be enumerated. More important, however, is the need to evaluate each service in terms of the cost of replacing it or the cost of doing without it (including future costs that may result from the loss of additional services). Such an evaluation cannot be performed here, but a listing of a few benefits with some indication of their value can be attempted.

PEST CONTROL

At least 99 percent of the potential plant pests in the world are controlled by natural means. Man attempts to control only a few thousand, most of them on crops (Hoffman and Henderson, 1967).

The total use of pesticides has not yet been sufficient to destroy the numerous herbivore-predator systems that operate so well in natural vegetation, although the list of crops that have lost their natural means of control increases steadily.

Forest growth is slowed by pest outbreaks, a cost that has long been appreciated in forestry, leading to major research programs on forest pests. The loss of value in parks and suburban areas is primarily one of aesthetics, perhaps best estimated by the price that people are willing to pay for a tree. If the pest problem is ignored and annual defoliation accepted as a norm, the ecosystem will respond by producing a scrubby kind of vegetation dominated by annual weeds.

INSECT POLLINATION

A few groups of insects pollinate most of the vegetables, fruits, berries, and flowers. The unintended killing of pollinators with insecticides has often occurred on a local scale, usually noticed only when a desired crop fails to set. Farmers have learned to be careful.

The problem arises anew in the context of general vegetation rather than particular crops. The current rate of increase in the use of insecticides, the large amount of acreage involved, and the slowly increasing broad-scale applications to large areas combine as a threat to the continued function of insect pollinators.

The cost of pollination under human management is so great that the production of dependent crops would decrease greatly. Most wildflowers and whole sections of natural flora would disappear. Except for a few favored crops, man is certainly unable to replace this service; the cost of irreversible loss will have to be placed against the apparent advantage of any massive insecticide program.

FISHERIES

Although aquaculture produces important sources of fish protein, most of the harvest comes from fisheries where production is entirely natural. Marine fishery yields have been increasing at 5 percent per year. Although basic productivity may remain high in lakes that are overenriched (Lake Erie is still the most productive of the Great Lakes), the kinds of fish caught are usually the least desirable on the market.

The present threat to marine fisheries lies not so much in

the danger of polluting the grounds where the fish are caught but in the pollution of inshore areas where many species reproduce. These nurseries are being dredged, filled, and polluted with both nutrients and heavy metals. The value of commercial yields becomes a cost to be considered in evaluating policy for estuaries and coastal oceans.

CLIMATE REGULATION

Natural vegetation air-conditions the landscape and also serves as windbreaks. The day-night regimes of temperature, humidity, and wind on barren land indicate the role of vegetation in the human environment. If the human population expands to the degree that the attrition of ecosystems leads to poor vegetation, this free service will not be well performed. The cost of air-conditioning the landscape or the cost of living with an unpleasant or even unbearable outdoor climate will ultimately have to be considered.

SOIL RETENTION

Vegetation holds the soil on the land. Regulating the total volume of our soils, which decrease slowly in croplands and increase slowly under vegetation, is another asset provided free in nature.

FLOOD CONTROL

Flood frequency and intensity vary greatly with the presence of vegetation along waterways. Vegetation distributes the flow of water evenly. By contrast, flash floods and mudflows are common in the barren gullies of the southwest deserts.

SOIL FORMATION

Soil is a mixture of plant debris and weathered rock, formed through the joint action of bacteria, fungi, worms, soil mites, and insects. These agents comprise about 40 percent of the total biomass of terrestrial animals. Without these biological actions, plant debris disintegrates very slowly, and the soil fails to develop water-absorbent, ion-exchange properties. Larger insects also play a vital role in the processes of decay that recycle materials by breaking down branches and tree trunks.

CYCLING OF MATTER

The cycling of the elements in organisms depends upon green plants as the basic energy source, combined action of animals and microbes for decomposition, and specific groups of plants for several specialized steps involving nitrogen and sulfur transfer. Without these three major components, material would accumulate in one part of the cycle and ecosystems would deteriorate.

COMPOSITION OF THE ATMOSPHERE

Much of the oxygen, carbon dioxide, and nitrogen in the air and dissolved in the oceans passes through organisms in ecological cycles. These cycles, interacting with the geological cycle of erosion and deposition, lead to the regulation of atmospheric composition on a geological time scale. If natural life were largely destroyed, these functions could be taken over by regulating industries. However, we would have to balance all of our own actions, for example, reducing carbon dioxide as fast as it is made in combustion.

The amounts of nitrate in land and water, and the amounts of methane and ammonia in the air and water, are regulated primarily by microbes. The loss of this function would coincide with the ultimate destruction of life.

Evaluation of Current Status

Man does not yet threaten to annihilate natural life on this planet. Nevertheless, his present actions have a considerable impact on ecosystems, and his future actions and numbers will certainly have even more. The critical issue is the danger that we may curtail an environmental service without being able to carry the loss or we may irreversibly lose a service that we cannot live comfortably without.

In general, the expected losses from present impacts do not exceed our capacity to carry the burden; this leads us to the conclusion that an intractable crisis does not now seem to exist. Our growth rate, however, is frightening. The impact of two, four, or eight times the present ecological demand will certainly incur greater losses in the environment. If the process of change were gradual, the present ecological advantage that is reflected in our 5 to 6 percent annual growth would taper off in the face of decreased environmental services, and growth would be correspondingly slowed. Instead, the risk is very great that we shall overshoot in our environmental demands (as some ecologists claim we have already done), leading to cumulative collapse of our civilization.

It seems obvious that before the end of the century we must accomplish basic changes in our relations with ourselves and with nature. If this is to be done, we must begin now. A change system with a time lag of ten years can be disastrously ineffectual in a growth system that doubles in less than fifteen years.

Information Needs

Introduction

The need for better information on sources, effects, and distributions of pollutants is documented in the subsequent case studies. Indeed, it can be argued that the single greatest contributor to environmental pollution is ignorance. Furthermore, data sufficient to raise an alarm are not adequate for designing and operating a management system. It is in the nature of any study of the environment to seek effective means whereby the information needed now may be expeditiously obtained and at the same time to devise a system that will furnish such information at regular intervals in the future. This section presents a set of specific recommendations which, if implemented, would provide information needed to improve management decisions and research designs.

Information on Distribution

Of the two proposals that follow, the first suggests simplified methods for the collection of data pertinent to an analysis of ecosystem function, while the second reflects the great gaps in our knowledge of the present concentrations of potentially toxic pollutants in the marine environment. The design of more ambitious programs in these two areas of study is presented in subsequent sections. Hence, the title of this subsection is intended to suggest not only a gap in our knowledge of the location of various pollutants in, and their impact on, the biosphere but also different approaches to gathering that information.

ECOLOGICAL SYSTEMS

A considerable body of data exists on ecological systems. If more of it had been collected in an organized, systematic fashion, and if more of it were critically aimed at measuring ecosystem function, we would not now suffer so severe a handicap in environmental evaluation. These deficiencies can be remedied on a local scale without waiting for organized national and international programs. Moreover, inexpensive but critical data, systematically

17 Work Group on Ecological Effects

collected in a number of places, would become useful in regional and global evaluation.

Specific suggestions follow for the collection of data that are useful in the diagnosis of ecosystem function. An important characteristic of these methods is that data can be collected in different ways at different places using various species or species groups. The major constraints are (1) that the method chosen be used consistently in any one place, and (2) that measurements be made systematically over time. The following list is suggestive, not exhaustive, of the variety of measures to be considered, and it is organized according to the global ecological effects described earlier.

DETECTING DAMAGE TO PREDATORS

(1) In local and regional populations of land and marine birds, especially for species that have recognized juvenile forms, the reproduction and survival of the young to maturity (animal recruitment) need to be counted. (2) In commercial and game-fish populations that normally contain many year classes, catches should be sampled to estimate recruitment of the youngest-year classes. (3) Native herbivorous insects should be raised and set out at appropriate life stages to attract insect enemies. Species should be used that are not pests and are themselves so sparse that their enemies are virtually impossible to find. Samples of eggs can be observed for the frequency of egg parasites, and insect larvae can be set out to attract parasites of later growth stages. Moths of the silkworm family exist throughout the temperate and tropical zones but rarely become numerous. Aphids are another group that should be used intensively.

DETECTING LOSS OF SPECIES

(1) Plankton samples of algae can be sorted to species and counted, a task that would provide an estimate of diversity and species abundance. Zooplankton are easier to sort to species and are just as informative. Both should be done, but the minute plants and animals should be collected in nets with different mesh sizes. (2) Samples of soil are easily examined for diversity among the soil animals. With an overhead light as a heat source, the soil in a funnel, and water or alcohol in a beaker below, hundreds to thousands of animals crawl downward and fall into the beaker. (3)

Inverted light traps collect hundreds of individuals and dozens of species of flying insects. Records over time give good approximations of the abundance and diversity of species that are attracted to light.

MISCELLANEOUS MEASURES

(1) Tree leaves, especially at the moment of leaf fall, can be examined for the relative area removed or damaged. The latter averages around 6 percent in a healthy forest ecosystem and may prove to be a very sensitive measure of the functional success of natural control of leaf-eating insects. (2) Oxygen depletion in deep water, especially in August, provides an estimate of the accumulated burden of organic matter that has sunk from upper levels during the summer. Useful where the water is stratified (warm above, cold below, or fresh above, salty below), low oxygen levels are a signal of overenrichment and danger to game fish. (3) The growth of trees (tree rings) can be followed annually as well as backward in time. Enough estimates to obtain a good local average will reveal trends in growth that are moderately pronounced. The latter becomes a measure of annual forest production.

SAMPLING THE MARINE ENVIRONMENT

Data are available about the emissions of various pollutants into the atmosphere, the rivers, the estuaries, and the ocean through rivers and coastal sewage outfalls. We have some information concerning the resulting concentrations on land and in fresh water, some about the resulting concentrations and effects of pollutants in estuaries and lagoons, but very little about the coastal waters. Open ocean concentrations of the various pollutant substances introduced by man have not yet been measured.

We can show that many pollutants are available to the oceans through various routes, such as the atmosphere, runoff, dumping, and ships. If we knew the form in which a particular substance reached the ocean (gas, solute, particle, liquid), it might be possible to calculate its concentration and distribution in various areas and depths of the ocean, provided that it did not undergo any biological or chemical transformation. However, it is precisely because these substances do take part in the chemical and biological cycling in the ocean that they are of serious concern. Because the manner of cycling of these substances in the open ocean is not yet understood, we cannot make accurate estimates of their distributions, concentrations, and effects.

Analysis of a small number of samples collected from the open ocean would give immediate clues as to the amounts of various substances the ocean has received, the paths through which they have entered, how long they persist, how they are transformed, how they are concentrated, and what their effects are now and are likely to be in the future.

It seems worthwhile to carry out a brief (1-year) and simple sampling program of the marine environment. Enough is known of the atmospheric circulation over the ocean, the water circulation, including convergences, divergences, and upwelling, about the areas of high and low primary productivity and biomass concentration, and about the major fish populations, so that a coarse array of samples (at most 1,000) would be useful in the definition of the important pollution problems in the oceans.

Of these samples (biomass, sediment, water, surface film, air), the most important will be those taken from the biomass (fish and plankton). For, since the biomass may concentrate many of these substances from the water, pollutants may be most easily detected there.

Many of these samples, especially fish and plankton, can be obtained from collections already at hand or from those being compiled by existing programs. Some samples, such as oil slicks, can be obtained through minor additions to present programs. Others will be more difficult and expensive.

The choice of materials to be assayed as potential threats to marine ecosystems may be sought on the basis of the following categories:

1. *The minerals or solid phases that result from the combustion of fossil fuels and industrial activities.* The production of energy through the burning of coal, oil, and natural gases is accompanied by the release of such gases as carbon dioxide, sulfur dioxide, and nitrogen oxides, as well as by the atmospheric entry of many of the constituents of the fuel itself (vanadium, nickel, and copper). A measure of fossil-fuel combustion may be sought in concentrations of carbon (soot, cokey balls) in wind dust loads, rains, and lake and oceanic sediments. The elemental carbon may

act as a most important tracer of the paths of materials introduced to the atmosphere and the oceans as a result of energy production.

2. *Elements introduced to the atmosphere or rivers in amounts that are of the same order of magnitude as those brought to the world's rivers by natural weathering processes.* Cadmium, mercury, nickel, lead, and vanadium appear to be five heavy metals in this category.

3. *Synthetic organic chemicals, produced in large quantities, which can interfere with the metabolic processes of marine organisms.* Initially, one can consider such materials as the dry-cleaning solvents, perchlorethylene and trichlorethylene, freon, gasoline, and the chlorinated biphenyls.

Information on Effects

The protection and enhancement of our water resources require not only a continuing measure of what is being added to them but also a thorough knowledge of those concentrations of wastes and materials that are not harmful even with continuous exposure. The latter is essential for the definition and detection of pollution, for the determination of the objectives and goals of waste treatment and pollution abatement, and for the establishment of water quality standards.

Toxicity studies conducted over the past 40 years have not supplied data essential for the setting of definitive water quality standards. Little is known of the requirements of the most sensitive species, thus the results do not indicate safe levels for the biota. To meet this problem, well-coordinated, thoroughly planned, and adequately funded national and international programs are essential.

The toxicity of acute doses of many pollutants is known for humans and, in some cases, for lower organisms in both terrestrial and aquatic ecosystems. Although the long-range effects of continued exposure to some toxic materials are also known for man, these data are almost nonexistent for other organisms.

Plants and animals that inhabit the waters of the world are in intimate contact with their environment. There is a phase boundary between organism and water across which toxic substances (especially the soluble and colloidal forms) transfer efficiently and directly. Because aquatic organisms may easily accumulate added pollutants throughout their lifetimes, the long-range effects of continued additions of these materials must be determined soon if man desires to maintain or to improve the life-supporting and aesthetic qualities of the world in which he lives.

The development of information essential for the setting of water-quality standards for the protection of aquatic life resources requires the determination of (1) toxicants that may be added in significant quantities to the hydrosphere; (2) the important species of the aquatic biosphere, based on their economic or recreational values, their importance in the maintenance of food chains, ecosystem balance, or in the production of oxygen or food for man; and (3) maximum concentrations of wastes or materials that are not harmful with continuous exposure.

The selection of test organisms must be given careful consideration, because there is great variation in the toxicity of the same materials to different species and life stages and of different materials to the same species. In many aquatic organisms the early life stages, including the egg stage, are particularly sensitive to toxic materials or to other environmental stresses, and representatives of this group should be included in a test program. Other organisms, especially benthic forms and planktonic filter feeders, show marked propensities for accumulating precipitated and coprecipitated materials and thus may accumulate sufficient toxic particles to cause their own deaths or to pass concentrated amounts of the pollutants along the food chain to higher trophic levels. Representatives of this group should be tested.

Some adult organisms with high metabolic rates are especially susceptible to stresses from increased pressure and salinity or from decreased pH and oxygen levels. Trout and tunas are representatives of this group. The phytoplankton are important as primary producers (and concentrators) and in helping to maintain the oxygen-carbon dioxide balance in water. These plants should receive high priority in the test program. Finally, some pollutants, including DDT and other fat-soluble materials, are concentrated with increased trophic levels. Sublethal toxicity studies of DDT, aldrin, toxaphene, and petrochemical components should thus be conducted with carnivorous fish, sharks, and porpoises.

The results of laboratory tests must be corroborated in the field to determine their adequacy under natural conditions, where the additional stresses of parasites, disease, predators, competitors, and other environmental factors are operative. Studies must also be made of the effects of concentrations of wastes in bottom sediment and of the relation between the concentration of a material in the water, in bottom sediment, in the bodies of organisms, and in passage through the food chain.

Even with the use of shortcuts by which long-term studies are carried out through two generations with only the most sensitive species, considerable time will elapse before needed data are compiled for all important wastes. In the meantime, all available data should be used to set tentative standards.

From the international viewpoint, the persistent pollutants are of prime importance. When water quality requirements are determined for these materials, they will in general be applicable worldwide, making it possible to set international water quality standards for open ocean areas. Because of differences in areal biota, standards will vary for estuaries and coastal waters; however, it is believed that there will be sufficient similarity to render beneficial the exchange of data on an international basis.

Information on Sources

Man now produces more than a million different kinds of products, both as waste and as final products that eventually end up as waste. Certain of these are easily identified as threatening to the environment; for some, production and use figures are given in this report. The problem is how to maintain an adequate surveillance of all major pollutants. While a complete documentation of the required activities is not feasible, the following steps will greatly improve our ability to recognize new problems sooner.

SURVEYS OF NEW OR RAPIDLY EXPANDING ACTIVITIES

Especially in industry, innovation and change should come under the scrutiny of a data-gathering-and-evaluating group.

TENTATIVE IDENTIFICATION OF POTENTIAL THREATS

All new products and activities similar to existing products and activities that are known to be deleterious to the environment should be noted for special study. For example, these products would include all of the chlorinated hydrocarbons. Major land changes and plans to build new industries, especially along water, should also be included. The total number of categories to be assessed will be large but very much smaller than the totality of change.

IMMEDIATE RESEARCH ON ENVIRONMENTAL EFFECTS

For new wastes or products that will enter the environment, research should focus on (1) toxicity or other biological action at low and chronic levels, (2) persistence in the environment, and (3) the time response of the potential target.

PRIORITY FOR INTENSIVE STUDY

Depending upon the results of this research, upon the estimated time for recovery should damage occur, and upon current and projected rates of release, a high priority for intensive study can be given. Additionally or alternatively, the burden of proving the material safe could be placed on the producer, with production prohibited until such proof is satisfactory.

General Considerations of Ecology and Life-Style

In addition to acting immediately on many specific items, such as those recommended in the preceding section, man must soon solve some very basic problems that adversely affect many of his activities. A few are described here which require solutions that will demand far-reaching changes in some human values and activities.

Population Growth

For some time, and indefinitely into the future, additional numbers of people can be accommodated only by a more intensive use of land and consumption of a larger share of the earth's resources. This prognosis implies less room for other forms of life. Sooner or later, we must decide how much life to displace. Population will increase whether we decide to eliminate all other natural life or whether we try to preserve a biosphere that still provides us with most of our major needs. At the current rate of population increase such a biosphere may not last for more than 100 years.

Ecological Growth

An indefinitely rising material standard of living has nearly the same effect on the biosphere as an indefinitely rising population. As with population, this problem must also be solved sooner or later. A shift involving increased emphasis upon spiritual, aesthetic, and intellectual components in a standard of living remains the only way that standards could improve indefinitely. At present, the Gross Domestic Product (an approximation for the Gross National Product) for nations and its homologue for individuals stand as the most generally used criteria of status. A basic change in values in which increase in material wealth is not so highly rated must accompany any solution to the problem of ecological growth.

Pesticide Addiction

Realization that the use of pesticides increases the need to continue their use is not new, nor is the awareness that the constant use of pesticides creates new pests. For many of our crops on which pesticide use is heavy, the number of pests requiring control increases through time. In a very real sense, new herbivorous insects find shelter among our crops where their predator enemies cannot survive.

Fifty years ago most insect pests were exotic species, accidently imported to a country lacking their natural enemies. More recently many of the pests, including especially the mites, leafrolling insects, and a variety of aphids and scale insects, have been indigenous.

Thus pesticides not only create the demand for future use (addiction), they also create the demand to use more pesticide more often (habituation). Our agricultural system is already heavily locked into this process, and it is now spreading to the developing countries. It is also spreading into forest management. Pesticides are becoming increasingly "necessary" in more and more places.

Before the entire biosphere is "hooked" on pesticides, an alternative means of coping with pests should be developed. At the very least this will include crop breeding for resistance, changes in crop patterns that may not be so well suited to mechanization, massive developments in the techniques of managing predators, parasites, and diseases for use in control, relaxed acceptance values for blemished fruit, and a general acceptance of low levels of insects as a situation we do not have to fight.

Communication and Education

As our technology becomes increasingly powerful, it is increasingly urgent that we educate our children (and ourselves) to the dangers of misusing the environment. This can be done both in the schools as a part of the regular curriculum and in special programs designed to familiarize the student with the functions of natural and man-altered ecosystems. For the latter, "ecological classrooms" can be set aside to be used by the public to give meaning to its growing awareness of the environment. Similar areas are needed by scientists so that base lines can be established against which to measure change.

The present surge of activity in the classrooms, in the mass media, and in various park systems attests to the general recognition of these needs. In many cases, however, and especially in the schools, a critical lack of teaching material, experience, and expertise has forced many individuals to improvise programs whose quality and soundness are less than satisfactory. A much greater effort is needed to bring professional competence into the development of these programs.

No change in life-style will occur unless the need for change is broadly recognized and accepted. It remains the responsibility of those who study environmental problems and understand their dimensions to convey this need to the public. It is also their responsibility to determine that information organized in a form that can be understood by the public accurately reflects the evidence and scientific arguments taken from their professions. Several existing organizations, appropriately expanded, can aid greatly in this process of bringing information to the public and to decision makers. Imagination and energy are required for the development of additional means for increasing enlightened public discussion of critical environmental problems.

Appendix: Carbon Cycle in the Biosphere*

About half of the carbon dioxide released into the atmosphere by the burning of fossil fuel has disappeared. The two large carbon "pools" that interact with the pool in the atmosphere are the vegetation of land surfaces and the carbonate-bicarbonate system in the oceans (together with living matter in the oceans).

---

* This appendix presents the findings and recommendations of a SCEP group whose members were J. B. Hilmon, C. D. Keeling, Lester Machta, Jerry S. Olson, Roger Revelle, Walter O. Spofford, and Fred Smith. Jon Machta of the University of Michigan assisted this group in helping to define the components to be put into a computer model.

The data presented here are not scientifically firm; numbers frequently are based on educated guesses. Nevertheless, this presentation of these numbers is valuable in demonstrating an interdisciplinary analytical technique and is included as an illustrative approach to the problem of measuring the various carbon pools.

Slightly differing estimates from several sources (FAO Yearbooks, 1951–1969; Ryther, 1969; Olson, 1970; Whittaker and Likens, unpublished) were combined into a single set producing the summary given on Table 1.A.1. Although the accuracy of these estimates is poor, they are the best that are available.

For the purpose of analyzing the dynamics of interaction between the vegetation and the atmosphere, the organic carbon pools were divided into "short-lived" and "long-lived" components. The former includes leaves, litter, short-lived animals, and most algae, while the latter includes wood, large roots, upper soil humus, and dead organic matter in the oceans (see Table 1.A.2). While these estimates are speculative, they represent an approach to data analysis that usefully interrelates ecology and meteorology. Better estimates must await new research and model building that may be generated from this analysis.

The results suggest that the organic carbon pool, with a mean residence time of the order of 60 years (terrestrial wood, humus, and so forth), is about twice the size of the atmospheric pool and

Table 1.A.1 Organic Carbon and Its Rates of Production[a]
(Living and dead, excluding incipient fossil deposits)

Reservoir	Area[b] 10^6 km^2	Organic Carbon Pool[b] 10^9 metric tons	Production/Year[b] 10^9 metric tons
Forest and woodland	48	1,012	36
Grassland and tundra	38	314	9
Desert and semidesert	32	59	3
Wetlands	2	30	2
Glaciers and barren	15	0	0
Agricultural	15	165	6
Total terrestrial	150	1,580	56
Oceanic	361	703	22
Burning fossil fuel (1970)			4
Atmospheric pool		683	

[a] Resistant humus and other material with decay rates of 0.001 per year or less have been omitted. Production is "net primary production," that is, production from photosynthesis minus plant respiration; it represents the yield to animals and decomposers.
[b] The numbers shown here are intermediate values from the several sources listed in the text. Although these sources present similar estimates, their combined accuracy is not regarded as high; procedures for obtaining global estimates for characteristics of vegetation are still primitive.

Table 1.A.2 Subdivision of Organic Carbon into Materials with Short or Long Residence Times

	Organic Carbon Pool		Mean Residence Time (years)	
	Short	Long	Short	Long
Reservoir				
Forest and woodland	62	950	3.1	59
Grassland and tundra	11	303	2.6	63
Desert and semidesert	4	55	2.2	39
Wetlands	1	29	1	36
Agricultural	4	161	1	80
Total terrestrial	82	1,498	2.3	62
Oceanic	2	701	0.1	701

Mean residence of carbon in the atmosphere: 8.8 years.
Note: The numbers shown here are not data but intelligent guesses of the way the estimates from Table 1.A.1 can be subdivided. Definitions of "short" and "long" residence times are given in the text.

could easily have absorbed a significant portion of the carbon dioxide that has left the atmosphere. Nevertheless, the residence time is short enough so that decomposition would soon be returning larger amounts to the atmosphere.

Recommendations

A number of recommendations emerged from this effort, derived to a considerable extent from the fresh views we obtained through interdisciplinary discussion.

1. We recommend that an organized information system, such as that being started under the International Biological Program, be developed and maintained to eliminate uncertainties in comparing techniques and results from scattered sources.

2. We recommend that differences in experience among teams of experts of varying backgrounds be used to test whether their calculations can be checked out and improved. Japanese, Scandinavian, Australian, Central European, and Inter-American groups are already developing the necessary team approaches.

3. We recommend that radiocarbon and other techniques for determining age, together with studies of decomposition rates, be developed for the various components of dead organic matter. Total soil profile determinations of organic matter are exceed-

ingly scarce. In addition, the total amount of living and dead material is not known, and its residence time is also largely unknown.

4. We recommend that more detailed analysis be given to second-stage modeling. While a two-component (short versus long-lived) approach permits the development of initial models of the interaction between the atmosphere and vegetation, both fractions are parts of a broad spectrum of materials of various longevities.

5. We recommend that the effects of deforestation be given serious study. The conversion of forests to grassland and cropland and the harvesting of forests both greatly reduce the pool of organic carbon in the terrestrial system. Since the rise of agriculture, man has removed an amount of carbon that appears to be about half the size of the atmospheric pool and added it to the air as carbon dioxide. A historic analysis of periods of intense deforestation and of climatic variations may give empirical insight into the effect of carbon dioxide on the climate. Similarly, removal of the great tropical forests could have an impact on the dynamics of the carbon cycle at least as serious as the burning of fossil fuels.

References

Acree, F., Berova, M., and Bowman, M. C., 1963. Codistillation of DDT with water, *Journal of Agriculture and Food Chemistry, 11*.

American Chemical Society (ACS), 1969. *Cleaning our Environment: The Chemical Basis for Action* (Washington, D.C.: ACS).

Anderson, D. W., Hickey, J. J., Risebrough, R. W., Hughes, D. F., and Christensen, R. E., 1969. Significance of chlorinated hydrocarbon residues to breeding pelicans and cormorants, *Canadian Field Naturalist, 83*.

Bailey, T. E., and Hannum, J. R., 1967. Distribution of pesticides in California, *Proceedings of the Sanitary Engineering Division of the American Society of Civil Engineers, 93*.

Bitman, J., Cecil, H. C., and Fries, G. F., 1970. DDT–Induced inhibition of avian shell gland carbonic anhydrase: a mechanism for thin eggshells, *Science, 168*.

Blumer, M., 1969a. Oil pollution of the ocean, *Oil on the Sea*, edited by D. P. Hoult (New York: Plenum Press).

Blumer, M., 1969b. Oil pollution of the ocean, *Oceanus, 15*.

Blumer, M., Souze, M. G., and Sass, J., 1970. Hydrocarbon pollution of edible shellfish by an oil spill, Woods Hole Oceanographic Institute reference 70–71 (unpublished).

Bowen, H. J. M., 1966. *Trace Elements in Biochemistry* (London and New York: Academic Press).

Bowman, M. C., Acree, F., and Corbett, J., 1960. Insecticide solubility, solubility of carbon-14 DDT in water, *Journal of Agriculture and Food Chemistry, 8*.

Butler, P. A., 1964. Commercial fishing investigations in effects of pesticides on fish and wildlife, *Fish and Wildlife Service Circular, 226*.

Butler, P. A., 1966. Pesticides in the marine environment, *Journal of Applied Ecology, 3* (Supplement).

Butler, P. A., 1967. Pesticide residues in estuarine mollusks, *National Symposium on Estuarine Pollution* (Stanford University).

Butler, P. A., 1969. Monitoring pesticide pollution, *Bioscience, 19*.

Carnaghan, R. B. A., and Blaxland, J. D., 1957. Toxic effect of certain seed dressings on wild and game birds, *Veterinary Record, 69*.

Chemical Economics Handbook, 1969. (Menlo Park, California: Stanford Research Institute.)

Congressional Record, July 30, 1970 (Washington, D.C.: U.S. Government Printing Office).

Conney, A. H., 1967. Pharmacological implications of microsomal enzyme induction, *Pharmacological Review, 19*.

Cooke, R. F., 1969. Oil transportation by sea, *Oil on the Sea*, edited by D. P. Hoult (New York: Plenum Press).

Cronin, L. E., 1967. The role of man in estuarine processes, *Estuaries*, edited by G. H. Lauff (Washington, D.C.: American Association for the Advancement of Science).

Davis, K. P., 1954. *American Forest Management* (New York: McGraw-Hill).

Federal Water Pollution Control Administration (FWPCA), 1968. *Cost of Clean Water*, Vol. III (Washington, D.C.: U.S. Government Printing Office).

Federal Water Pollution Control Administration (FWPCA), 1970. *National Estuarine Inventory* (Washington, D.C.: U.S. Government Printing Office).

Food and Agriculture Organization (FAO), 1951, 1958, 1963, 1968. *Production Yearbook* (Rome: FAO).

Food and Agriculture Organization (FAO), 1951, 1958, 1969. *Yearbook on Food and Agricultural Statistics*, 1950, 1957, 1968 (Rome: FAO).

Goldberg, E. D., 1970. Atmospheric transport, background paper prepared for SCEP (unpublished).

Gress, F. Reproductive success of the brown pelicans on Anacapa Island in 1970, *Transactions of the San Diego Natural History Society*, forthcoming.

Grolleau, G., and Giban, J., 1966. Toxicity of certain seed dressings to game birds and theoretical risks of poisoning, *Journal of Applied Ecology, 3*.

Hampson, G. R., and Sanders, H. L., 1969. Local oil spill, *Oceanus, 15*.

Hasler, A. D., 1969. Cultural eutrophication is reversible, *Bioscience, 19*.

Heath, R. G., Spann, J. W., and Kreitzen, J. F., 1969. Marked DDE impairment of mallard reproduction in controlled studies, *Nature, 224*.

Herman, S. G., Kirven, M. N., and Risebrough, R. W., 1970. Pollutants and raptor populations in California (unpublished).

Hickey, J. J., and Anderson, D. W., 1968. Chlorinated hydrocarbons and eggshell changes in raptorial and fish eating birds, *Science, 162*.

Hoffman and Henderson, 1967. The fight against insects, *Yearbook of Agriculture* (Washington, D.C.: U.S. Department of Agriculture).

Holme, R. A., 1969. Effects of *Torrey Canyon* pollution on marine life, *Oil on the Sea*, edited by D. P. Hoult (New York: Plenum Press).

Holmes, R. W., 1969. The Santa Barbara oil spill, *Oil on the Sea*, edited by D. P. Hoult (New York: Plenum Press).

Irukayama, K., 1967. The pollution of Minimata Bay and Minimata disease, *Advances in Water Pollution Research, 3*.

Jefferies, D. J., 1969. Induction of apparent hyperthyroidism in birds fed DDT, *Nature, 222*.

Jefferies, D. J., and French, M. C., 1969. Avian thyroid: effect of p,p'-DDT on size and activity, *Science, 166*.

Jefferies, D. J., and Prestt, I., 1966. Post-mortems of peregrines and lannens with particular reference to organochlorine residues, *British Birds, 59*.

Jehl, J. Shell thinning in eggs of the brown pelicans of western Baja California, *Transactions of the San Diego Natural History Society*, forthcoming.

Ketchum, B. H., 1969. Eutrophication of estuaries, *Eutrophication: Causes, Consequences, Correctives*, edited by G. Rohlich (Washington, D.C.: National Academy of Sciences).

Lofroth, G., and Duffy, M. E., 1969. Birds give warning, *Environment, 11*.

McHugh, J. L., 1967. Estuarine nekton, *Estuaries*, edited by G. H. Lauff (Washington, D.C.: American Association for the Advancement of Science).

Manigold, D. B., and Schulze, J. A., 1969. Pesticides in selected western streams —a progress report, *Pesticides Monitoring Journal, 3*.

Matsumura, F., and Patil, K. C., 1969. Adenosine Triphosphatase sensitive to DDT in synapses of rat brains, *Science, 166*.

Menzel, D. W., Anderson, J., and Randtke, A., 1970a. Marine phytoplankton vary in their response to chlorinated hydrocarbons, *Science, 167*.

Menzel, D. W., Anderson, J., and Randtke, A., 1970b. The susceptibility of two species of zooplankton to DDT (unpublished).

Morris, R. F., and Miller, C. A., 1954. The development of a life table for the spruce budworm, *Canadian Journal of Zoology, 32*.

Muto, S., and Suzuki, T., 1967. Analytical results of residual mercury in the Japanese storks, *ciconia ciconia boyciana swinhoe*, which died at Okama and Toyooka regions, *Japanese Journal of Applied Entomology and Zoology*.

Nace, R. L., 1967. Water resources: a global problem with local roots, *Environmental Science and Technology, 1*.

Nash, R. G., and Beal, M. L., 1970. Chlorinated hydrocarbon insecticides: root uptake versus vapor contamination of soybean foliage, *Science, 168*.

National Air Pollution Control Administration (NAPCA), 1970. *Nationwide Inventory of Air Pollutant Emissions* (Raleigh, North Carolina: NAPCA).

Nimmo, D. R., Wilson, A. J., Jr., and Blackman, R. R., 1970. Localization of DDT in the body organs of pink and white shrimp, forthcoming.

Olson, J. S., 1970. Carbon cycle and temperate woodlands, *Analysis of Temperate Forest Ecosystems*, edited by D. E. Reichle (New York: Springer-Verlag).

Peakall, D. B., 1970. p,p'-DDT: effect on calcium metabolism and concentration of estrodol in the blood, *Science, 168*.

Peterle, T. J., 1969. DDT in Antarctic snow, *Nature, 224*.

Pitelka, F. A., Tomick, P. Q., and Treichel, G. W., 1955. Ecological relations of jaegers and owls as lemming predators near Barrow, Alaska, *Ecological Monograph, 25*.

President's Science Advisory Committee (PSAC), 1967. *The World Food Problem* (Washington, D.C.: U.S. Government Printing Office).

Report of the Secretary's Commission on Pesticides and Their Relationship to Environmental Health, 1969. (Washington, D.C.: U.S. Department of Health, Education, and Welfare, U.S. Government Printing Office).

Ricker, W. E., 1969. Food from the sea, *Resources and Man* (San Francisco: W. H. Freeman and Co.).

Risebrough, R. W., and Anderson, D. W. Pollutants and shell thinning in the brown pelicans, *Transactions of the San Diego Natural History Society*, forthcoming.

Risebrough, R. W., and Coulter, M. C. Chlorinated hydrocarbons and shell thinning in the ashy petrel, *Oceanochoma Lamochroa* (unpublished).

Risebrough, R. W., Davis, J. D., and Anderson, D. W. Effects of various chlorinated hydrocarbons on birds, forthcoming.

Risebrough, R. W., Huggett, R. J., Griffin, J. J., and Goldberg, E. D., 1968. Pesticides: transatlantic movement in the northeast trades, *Science, 159*.

Risebrough, R. W., Menzel, P. B., Martin, D. J., and Olcott, H. S. DDT Residues in pacific marine fish, *Pesticides Monitoring Journal*, forthcoming.

Risebrough, R. W., Sibley, F. C., and Kirven, M. N. Reproductive success of the brown pelicans on Anacapa Island in 1969, *Transactions of the San Diego Natural History Society*, forthcoming.

Rohlich, G., ed., 1969. Eutrophication: Causes, Consequences, Correctives (Washington, D.C.: National Academy of Sciences).

Ryther, J., 1969. Photosynthesis and fish production in the sea, *Science, 166*.

Schreiber, R. W. Pollutants and shell-thinning of eggs of brown pelicans in Florida, *Transactions of the San Diego Natural History Society*, forthcoming.

Schwartz, H. L., Kosyreff, W., Surks, M., and Oppenheimer, J. H., 1969. Increased deiodination of L-thyroxine and L-trilodothyronine by liver microsomes from rats treated with phenobarbital, *Nature, 221*.

Seba, D., and Cochrane, E., 1969. Surface slicks as a concentrator of pesticides, *Pesticide Monitoring Journal*.

Skou, J. C., 1965. Enzymatic basis for active transport of Na^+ and K^+ across cell membrane, *Physiological Reviews, 45*.

Stockinger, H. E., 1963 Mercury Hg^{297}, *Industrial Hygiene and Toxicology*, Vol. II, edited by F. A. Patty (New York: Interscience).

Sverdrup, H. V., Johnson, M. W., and Fleming, R. H., 1942. *The Oceans* (New York: Prentice-Hall).

Tarrant, K. B., and Tatton, J., 1968. Organo-pesticides in rainwater in the British Isles, *Nature, 219*.

United Nations. *Statistical Yearbook,* 1967 (New York: Statistical Office of the United Nations), 1968.

Verduin, J., 1966. Eutrophication and agriculture (paper presented at American Association for the Advancement of Science Symposium, Washington, D.C.).

Vollenweider, 1969. Eutrophication (Paris: OECD Mimeograph Report).

Weibel, S. R., 1969. Urban drainage as a factor in eutrophication, *Eutrophication: Causes, Consequences, Correctives,* edited by G. Rohlich (Washington, D.C.: National Academy of Sciences).

Whittaker, R. H., and Likens, G. E. *1961 Woodland Forest Working Group of International Biological Program* (unpublished).

Wolman, A. A., and Wilson, A. J., Jr., 1970. Occurrence of pesticides in whales, *Pesticide Monitoring Journal, 4.*

Wurster, C. F., 1968. DDT reduces photosynthesis by marine phytoplankton, *Science, 159.*

Yates, M. L., Holswade, W., and Higer, A. L., 1970. Pesticide residues in hydrobiological environments, *Water, Air and Waste Chemistry Section of the American Chemical Society Abstracts.*

2
Subtraction by Multiplication: Population, Technology, and the Diminished Man

F. H. Bormann

Almost overnight, as it were, the population question has pushed, shoved, and elbowed its way onto the political stage in America. There is, however, much confusion as to how much of the world is affected.

Many believe that the population explosion is a primary concern of the poor, undeveloped nations of the earth. This conclusion is fallacious. It stems from a lack of appreciation of ecologic interrelationships that ultimately govern human success and bind nation states to world membership whether they like it or not.

The population bomb threatens all of us, but the nature of the threat is quite different for poor countries than it is for rich countries. For poor countries, mass starvation looms as the principal problem. Food supply is also a problem for rich countries but only when viewed in the context of ecological food chains. For rich countries, population growth is coupled with the growth of technology and the two together constitute a more fundamental threat to man than the simpler problem of food supply. This threat is directed toward the whole human life support system, the environment. As a result of the population-technology interaction, the quality of life in many rich countries has already been seriously diminished, there is widespread environmental decay with its inescapable effects on the dignity of man, the maintenance of law and order has become a serious problem, and the possibility now looms that the very potential of the earth to support human life will be seriously reduced. In the following paragraphs, I will explore the implications of the population explosion, first for the poor underdeveloped countries, then for the rich developed nations, and conclude with some thoughts on political priorities I believe imperative if we are to stem the tide which is running strongly against the human race.

For the poor countries, the primary problem is one of enough food production to keep pace with burgeoning population. These nations, with two-thirds of the world's people, have the highest rates of population growth and generally speaking a smaller pro-

Reprinted with permission from the May 1970 issue of the *Yale Alumni Magazine,* copyright by Yale Alumni Publications, Inc.

portional share of the world's productive lands (1). Various experts (1, 2, 3) predict that famine will run rampant through these countries within a few decades.

More optimistically, the Green Revolutionists (those trying to develop modern agricultural technology in poor countries) claim that famines will be averted by the introduction of new varieties of crops, fertilizers, pesticides, and new farming techniques (4, 5). In some countries, notably the Philippines, India, Pakistan, and Turkey, the Green Revolutionists can point to remarkable local successes in raising crop yields. However, the development of new varieties of plants and new farming techniques alone are not guaranteed to solve food problems of underdeveloped nations (6, 7). Very difficult economic, sociologic, political and ecologic problems must be overcome before any poor nation can make the fundamental change in its agriculture as demanded by the Green Revolutionists. Price and credit systems must be developed that will maintain the incentive of local farmers to grow the new varieties. The new fertilizer-sensitive varieties will require millions of tons of fertilizer. This means substantial investments for new fertilizer plants and the development of transport and distribution facilities to get fertilizers to the fields and crops to the market. Pesticides are an intimate part of the "Green Revolution"; thousands of tons will be needed. New chemical industries will have to be developed or hard-to-come-by foreign exchange will have to be spent. Increasing dependence on pesticides coupled with massive plantings of relatively genetically uniform varieties poses the threat of increasingly difficult pest problems and the need of large and sophisticated agricultural research apparatus to cope with them. Exhaustive research on ecological systems tells us that the effects of an alteration of any part of a system will be transmitted to all other parts of the system or to interconnected systems (8, 9). With this in mind, we must ask what secondary ecological problems will arise from the massive use of fertilizers and pesticides in the poor countries of the earth (10)? Will increased yields in grain or rice be bought at the expense of decreased yields in fresh water and oceanic fish resulting from water pollution? Will poor nations begin to contribute, more massively to world pollution problems? These and other problems, coupled with the increasing instability of some national governments, such as the Indian government (11), and the penurious attitude of rich countries toward massive foreign aid brings sobering thoughts to mind when advocates of the "Green Revolution" tell us not to worry. In this regard, it is interesting to note that in 1970 the U.S. Congress voted the smallest foreign aid appropriation in the program's history (12).

Even if the "Green Revolution" lives up to highest expectations, in the face of a population growth rate of 2.5% in the poor countries it seems likely the best the "Green Revolution" can do is buy a few years of time. This becomes evident when one considers that the Food and Agriculture Organization of the U.N. estimates that by the year 2000 overall food supplies will have to be increased 160% for Africa, 240% for Latin America, and 300% for the Far East in order to provide a minimum adequate diet for all (13).

We must also bear in mind that pressures generated by massive food shortages or by their anticipation may be expressed in ways other than malnutrition or starvation. For example, what role does the exploding population of Red China play in the increasingly dangerous border clashes with the Soviet Union? To a nation anticipating a population of 1.5 billion people in 2015, the relatively empty spaces of Soviet Asia might be worth the risk of nuclear war. On a smaller scale, overpopulation must be credited as a major factor in the vicious border war between El Salvador and Honduras.

At the present time the population bomb is not a threat to the food resources of the rich technologic nations. These nations which constitute about a third of the world's people are seemingly well insulated against increased food demands of their own growing populations. First the good fortune of the lottery of history has dealt the nations of northern Europe, North America, the southern tips of South America and Africa, and Australia a large share of the world's best agricultural land (1). Second, these nations have made massive investments in the development of new varieties of crops, fertilizers, herbicides and pesticides, agricultural machinery, storage techniques, and marketing facilities; not to mention superb research and educational facilities (1, 14). The

United States alone spends approximately one billion dollars a year on research and development in agriculture (14). Modern agricultural technology has had remarkable success in raising yields throughout temperate areas. Grain output in developed countries increased about 50% between 1938–60. All of this increase came from rising per acre yields. The area in grain in 1960 was about the same as in 1938 (15).

In the race to capture the resources of the oceans, some wealthy technologic countries have developed superb fishing fleets —notably the U.S.S.R., Japan, and Norway. These and other fleets, mostly from rich nations, fish the oceans of the world and more or less guarantee that a large share of the resources of the sea will continue to go to the well-to-do nations of the world. Of the present oceanic catch, only 17% is used as food by poor nations, the rest, 83%, goes to the developed nations (16).

This brief discussion suggests that for the next few decades, at least, additional food supplies will not be a problem for the developed nations.

However, there is an important ecological consideration that casts food and population statistics from developed and underdeveloped countries in a different light.

Green plants form the base of all food chains. Thus, plant-eating animals are dependent on green plants for food. In turn, plant-eating animals may be used as food by flesh-eating animals. Ecological studies of wild nature tell us that the transfer of food from green plants to plant-eating animals to flesh-eating animals is far from efficient. As a rule of thumb, ecologists figure an efficiency of about 10%. Thus it takes about a hundred pounds of green plant tissue to produce about ten pounds of plant-eating animal and about ten pounds of plant-eating animal to produce about one pound of flesh-eating animal.

Man can fill either the role of a plant-eating animal or a flesh-eating animal. In poor countries, man behaves primarily as a plant-eating animal and consumes relatively few animal products. In rich countries man behaves more like a flesh-eating animal and consumes considerable quantities of animal products such as meat, milk, or eggs. In terms of production of green plants or plant products, it takes much more green plant production to support

Population, Technology, and the Diminished Man

a citizen of a developed country than a citizen of a poor country. One estimate (1) suggests that one American uses about six times the green plant production that one present-day citizen of India does.

This factor should be taken into account when figuring the effect of population increases on the world's supply of green food plants. The expected population increase in developed countries by the year 2000 is about 480 million (17). Since each new individual will require approximately six times the green plant tissue required by a citizen of a poor country, these 480 million additional persons will require additional supplies of green plants approximately equal to those required by the 2.8 billion additional persons expected in the poor countries by the year 2000 (17). In other words, population growth in the developed countries will have about the same effect on world food supplies as population growth in the undeveloped countries. This places rich nations like the United States and the Soviet Union in an awkward position when they advocate population control abroad but fail to regulate their own populations at home.

The chronic famine and starvation facing poor nations is not the only result of unrestricted population growth or perhaps even the most important result. Unrestricted population growth in the developed nations coupled with an even more rapid growth of technology is rapidly eroding the environment and reducing the quality of life and is threatening the very nature of life on this earth.

How does the interaction between population and technology work? In the well-to-do developed nations, on a per capita basis, citizens consume not only more food, but more TV sets, cars, gasoline, coal, lumber, iron, water, cloth, synthetics, medicines, detergents, pesticides, herbicides, food additives, fertilizers, and machinery. They live in larger houses, with heating and cooling, they have larger armies and navies, more roads, and airports. In other words, the average citizen of a rich nation uses a vastly greater quantity of the world's natural resources than does a citizen of a poor country. Using 1967 motor fuel consumption as a measure, the average American uses 250 times more than the average Indian (18). Even more startling is the estimate that although we in

America are but 6% of the total population of the world, we use about 50% of the principal minerals (iron, copper, lead, zinc) extracted from the world (19).

I should hasten to add that I do not advocate that we adopt the standards of present-day India or China, rather I want to emphasize that it takes many more acres of land, gallons of water, pounds of steel, cubic yards of air to supply the material wants of a citizen of a rich country than it does to supply the needs of a citizen of a poor country.

Not only do the developed countries consume vast quantities of natural resources, but they produce vast quantities of waste products that are voided into the environment. The United States each year discards 48 billion cans and 28 billion bottles and jars —1600 lbs. of solid waste per person (20). In 1965, 437,000 tons of pesticides were manufactured, presumably most of this finds its way into the environment (21). In addition to 200 million cubic yards of sewage sludge produced per year by the human population, there are one billion cubic yards of animal wastes, much of which must now be handled as sewage (22). U.S. chimneys belch 100,000 tons of sulfur dioxide per day while autos emit 230,000 tons of carbon monoxide (20). By 1970 we will add seven million tons of nitrogen fertilizers to our fields, lawns, and forests (23), a substantial proportion of which will be contributed to the pollution of ground water, streams, lakes, and rivers. The U.S. Food and Drug Administration estimates that we are now exposing ourselves and our environment to over a half-million different chemicals, all of which must eventually be imposed on the earth environment (24).

The effects of individual men are greatly multiplied by our bulldozers, our jet planes, our capacity to produce an almost infinite variety of chemical products, our ability to move vast quantities of things from place to place. Thus, modern technologic man has a huge power to alter the face of the earth. Utilization of our environment for disposal of wastes is not necessarily a crime; for man to live, some alteration and change of his environment must be expected. But little did we realize, even thirty years ago, the enormous effect technologic man could have on individual countries not to mention the earth as a whole.

It has come as something of a shock to learn that we have

almost destroyed many desirable ecological aspects of Lake Erie; that industrial, agricultural, and sewage pollution is a serious problem in most of our major river systems and lakes; that we can maintain an almost permanent pall of air pollution over our major cities; that the haphazard expansion of our towns, cities, and roads can use prodigious amounts of land, indiscriminately consuming open space and destroying amenities; that we can create vast congestion that can constrict human movement over land or through the air; that we can create widespread noise that is especially disturbing to the serenity of life. The very existence of these problems tells us that technologic man, endowed with vast power by his machines, has not yet learned to control and direct that power to achieve greater human dignity. Rather we are racing toward what Archibald MacLeish calls the diminishment of man (25).

We may now ask what will be the effect of rising populations in the rich countries on the world's resources and on contamination of the world's environment? For example, in the U.S., we expect our population to double in 63 years (17). Does that mean that in 63 years we will be consuming approximately double the amount of world resources we now consume and that we will be producing approximately double the amount of pollutants? The answer is a resounding no! For while our population grows at a rate of about 1%, our technology is growing at far greater rates; for example, production of electrical power is growing at about 5%, production of trash is growing at 4%, use of nitrogen fertilizer is increasing at 4%, and the annual growth rate of industrial production 5%. Although I do not have estimates, we can assume that many other aspects of our economy such as consumption of space, including farmlands and open space, or production of noise, thermal pollutants, air and water pollutants are growing at rates exceeding the growth of our population. If we assumed that our total effect on the environment grew by 4%, roughly the rate at which our GNP is growing (26), then in 63 years our effect on the world environment would be about 12 times greater than it now is. This is astounding for it is about six times greater than the environmental effect we would expect by population growth alone. This means that we would have to improve pollution abatement about 1200 percent in the next 63 years merely to maintain the unacceptable environmental conditions we have today. There is evidence that with continued unrestricted growth of population and technology, even with the most sophisticated pollution abatement, we simply will not be able to control the cancerous growth of pollution and the deterioration of the environment. In the State of New York and in Los Angeles where vigorous, well-informed, and massive economic attacks have been made on water and air pollution, respectively, it is now conceded that current programs have failed to clean the environment, and, in fact, have failed to stem the growth of pollution (27, 28, 29). The environmental program outlined by President Nixon in his State of the Union message although long on rhetoric gives little concrete evidence that the executive branch is ready for a crash program on pollution. The President's projected 10 billion dollar expenditure for sewage treatment plants is well below the 25 billion dollars proposed by Senator Muskie and the 30 to 40 billion considered necessary by a bipartisan group of congressmen. On a per capita basis, the Federal expenditure requested by President Nixon is far below existing water pollution expenditures in New York State. The portent of our growing failure to contain pollution coupled with a sluggish attitude in the White House is ominous indeed.

To return to the discussion of the meaning of population growth, I believe it is clear that per capita effect of the population growth in rich countries, coupled as it is to technology, has a vastly greater impact on our world's environment than does per capita effect of the growth of population in poor countries.

I have reserved one of the major effects of man until last. Man, rich or poor, has tended to ignore the fact that he is utterly dependent on the biosphere for his existence. Within the biosphere, or the thin outer covering of the earth, a vast web of interacting processes and organisms form biogeochemical cycles and food chains that maintain life on this earth as we know it. Although there are many examples of individual countries ruined by poor environmental policies, until a very few decades ago, most of us thought that the overall functioning of the biosphere was immune to human disturbance and that human error was limited to individual countries. The first rude shock of awakening came with the atomic bomb, when radioactivity from an explosion in one country spread over the face of the earth. Now evidence is accumulating that we are altering the biosphere in other ways. The amount of carbon dioxide in our atmosphere has risen about 14% since the beginning of the nineteenth century (31). Scientists worry that rising levels of carbon dioxide will slow down the escape of heat from our atmosphere into space. They foresee the possibility of a hotter earth with consequent melting of polar ice caps. This could lead to a rising sea level, as much as 400 feet, and inundation of coastal areas throughout the world.

Atmospheric turbidity or dustiness has equally great potential for altering world temperatures. Estimates indicate that since 1930, there has been a rapid increase in atmospheric dustiness quite apart from contributions of volcanic activity (32). This trend may be subtly coupled with the population explosion not only because of increasing industrial pollution, but because of agricultural activities primarily in undeveloped countries. In the last three or four decades, it seems probable that wind erosion has greatly increased as millions of acres of marginal land have been opened to agricultural use (1, 3) or as slash and burn agricultural systems, utilized on 14 million square miles of land, have broken down under population pressures (33, 34). Responsible scientists now speculate that increased dustiness of the world's atmosphere has had a greater effect in altering world temperature than has the increase in carbon dioxide!

Technologic man has now spread DDT over the entire surface of the world. It has been recovered from the fat of Antarctic seals and penguins, from fish all over the high seas, and from ice in Alaskan glaciers (24). DDT is concentrated in fat as it is passed along the world's food chains. Widespread decline in populations of predatory birds is reported and there is a strong probability that DDT or other chlorinated hydrocarbons are responsible. Many fish are reported to contain chlorinated hydrocarbons. The species hardest hit are predatory species at the end of the food chains. These species help to regulate populations of other species lower down the food chains. Lessening the effect of predators may result in outbreaks of new pest species, and increased instability of the world ecosystem (35).

My major point here is that man has great power to alter the environment and that these alterations are not held within the borders of one nation (36). Many environmental effects are spread over the face of the globe by the great circulatory systems of the earth—air, water, and biological food chains. We know we have the power to drastically change the face of the earth by radioactivity generated in nuclear war, and we now suspect that the same results may be achieved by more subtle means such as alteration of carbon dioxide or dust levels in the atmosphere, circulation of DDT in the biosphere, or by some unanticipated effect resulting from one or several thousands of chemicals we dump into our environment. Nor can we forget incremental damage done to our earth. The list of damages grows longer with each passing year: Lake Erie, the polluted rivers of the world, the Baltic Sea, the growing destruction of the world's wetlands and estuaries, the worldwide increase in soil erosion, the smog-induced destruction of forests miles from the source of pollution, the increasing number of oil spills, the increasing destruction of parts of the continental shelves, and the thousands of less spectacular insults we inflict on the earth each day. If as a doctor viewing a patient, we carefully followed these events during the last three decades, we would probably conclude that the cancer of pollution had already metastasized over the skin of the earth!

The picture I have painted is not a bright one. Human population and technology are growing at an explosive rate. The effects of this growth may be expressed in malnutrition or famine, or in a more generalized deterioration of the environment and the quality of life. One thing is certain that the population problem is not an exclusive problem of the poor underdeveloped nations; it is equally a problem of the rich developed nations.

What can we in the United States do about this? The war in Viet Nam has taught us one major lesson—America alone cannot control and manage the world. It is obvious that neither can we control worldwide aspects of the population explosion and its accompanying environmental deterioration. Yet, rich America cannot live securely or happily as a member of a small group of fabulously wealthy nations in the midst of a world slum.

There are several areas of action that seem to me to be of primary importance.

Pollution problems clearly transcend international borders and thus require international solutions. Agreements between the Soviet Union and the United States are basic to the establishment of an international pattern. Yet, agreements affecting production costs and methods, international trade competition, and the growth of populations and economy are virtually impossible as long as we pursue the Viet Nam War and construct A.B.M.'s and the Russians continue to stir the Middle Eastern political pot. By continuing to pursue politics more appropriate to the 1950's the Soviet Union and the United States are fiddling while Rome burns. Obviously we need bold leadership, willing to take chances, if we are to save the biosphere.

On a smaller scale, the United States should stand against the spreading colonialism of the seas and for a more equitable division of the ocean's food resources among all nations. We should also do our very best to make the Green Revolution as effective as possible by appropriating the maximum possible economic, ecological, technologic, and social aid for the poor countries that request our help. We should recognize, however, that a more equitable distribution of ocean resources and the Green Revolution are not solutions but merely a means of buying time to allow nations to bring population and food supply into harmony.

At home we would do well to recognize that our real contribution to the world population problem is equal if not greater than that of the poor nations and that we, as well as they, must establish a population policy. To achieve this goal, we must be aware that much more is involved than merely setting some arbitrary optimum desirable population level. The connection between optimum population and the quality of each individual's life must be explored and understood before political support for this goal can be expected from the majority of our citizens. A first step toward the establishment of national population policy would be to revise our laws and provide appropriate medical facilities so that no unwanted children are born in the United States.

We must question one of our most cherished assumptions, the idea that we can survive only with a continuously growing economy. The ecological limits of the earth and the clear connection between unrestricted economic growth and rampant destruction of our environment tell us that continued addiction to Keynesian economics is a one-way ticket to world disaster. Coupled with optimum population size there must be a concept of optimum economic size.

President Nixon's State of the Union message, with its emphasis on environmental quality, may represent a cautious step toward these goals (30). However, the President's failure to mention population and his emphasis on continued and relatively unquestioned growth of the GNP lead one to assume that his environmental advisors do not recognize or have not dared to publicly explore the basic problems of population and technology growth underlying the grave crisis that exists in the world today.

At the root of population-technology question, there is a fundamental question of values. As individual citizens we must ask ourselves what it is all about, what do we want from life, what should be our relationship to our fellow men, for whom does society exist, does man's technologic brilliance free him from a fundamental dependence on nature? I think the great unrest among our youth today stems from the perception that we are failing to address ourselves in a serious way to these questions. It is imperative that vigorous discussion of these questions be pursued at all levels of society—it is the only hope if our democratic society is expected to make rational choices as to its future.

My belief is that our crisis of the environment is so all-pervasive and growing so rapidly that present political approaches must be regarded as temporary palliatives—patch and plaster tactics of yesterday. The real answer, if there is one, is in the development of a new kind of society. A society less devoted to the cultivation of individual consumerism and more devoted to larger social goals. A society in which we can regulate our own population, one where we will enjoy the fruits of technology but where our technology is in harmony with the maintenance of the biosphere, and where all citizens are guaranteed a decent environment and a maximum opportunity for individual fulfillment.

This is no small order of business. There are no ready answers to our problems; in fact, we do not fully know what the problems are. Nevertheless, of all the nations on the earth, America by virtue of its history of free discussion, its vast educational and technologic resources, and its historical concern about men is in the best position to rebuild society. If we are willing to try and if we succeed, we could provide the rest of the world with a new model of life and that in the long run could be our best contribution to humanity.

References

1. Borgstrom, G. The hungry planet. Collier-Macmillan, N.Y. 1967.

2. Ehrlich, P. R. The population bomb. Sierra Club, Ballantine, N.Y. 1968.

3. Paddock, W. and P. Paddock. Famine 1975! Little, Brown and Co., Boston. 1967.

4. Gaud, W. S. 1968. The emergence of the green revolution. Current. May.

5. Lelyveld, J. Green Revolution transforming Indian farming, but it has a long way to go. N.Y. Times, May 28, 1969.

6. Wharton, C. R. 1969. The Green Revolution: cornucopia or pandora's box. Foreign Affairs, Apr.

7. Ehrlich, P. R. and J. P. Holdren. 1969. Population and panaceas: a technologic perspective. BioScience 19: 1065–1071.

8. Bormann, F. H. and G. E. Likens. 1970. The watershed-ecosystem concept and studies of nutrient cycles. *In* The ecosystem concept and natural resource management. Ed. G. Van Dyne, Academic Press, Inc.

9. Likens, G. E. *et al.* 1970. Effects of forest cutting and herbicide treatment on nutrient budgets in the Hubbard Brook Watershed-Ecosystem. Ecol. Monog. Jan.

10. Carter, L. J. 1969. Development in the poor nations: how to avoid fouling the nest. Science 163: 1046–1048.

11. Lelyveld, J. Indian democracy, under continuing pressure, nears a test of survival. N.Y. Times, June 1, 1969.

12. New York Times. 1970. Feb. 27.

13. Six billions to feed. 1962. FAO World Food Problems No. 4, Rome.

14. Thurston, H. D. 1969. Tropical agriculture a key to the world food crises. BioScience 19: 29–34.

15. Voigt, G. K. Personal communication.

16. Borgstrom, G. 1970. The harvest of the seas: how fruitful and for whom? *In* Man's Struggle to Live with Himself: Issues in the Environmental Crises. H. Helfrich, Jr., Ed. Yale Univ. Press.

18. Times of India. Directory and Yearbook. Sham Lat., Ed. The Times of India Press. Bombay. Summer.

17. World Population Data Sheet. 1968. Population Reference Bureau. March.

19. Day, L. H. 1960. The American fertility cult. Columbia Univ. Forum. Summer.

20. The age of effluence. Time Magazine. May 10, 1968.

21. Nicholson, H. P. 1967. Pesticide pollution control. Science 158: 871–876.

22. Taiganides, E. P. 1967. The animal waste disposal problem. *In* Agriculture and the quality of our environment. Ed. N. C. Brady. AAAS Publ. 85.

23. Stewart, B. A., F. G. Viets, Jr. and G. L. Hutchinson. 1968. Agriculture's effect on nitrate pollution of groundwater. Jour. Soil and Water Conservation 23: 13–15.

24. Cole, L. C. 1970. Playing Russian roulette with biogeochemical cycles. *In* Man's Struggle to Live with Himself: Issues in the Environmental Crises. Ed. H. Helfrich, Jr., Yale Univ. Press.

25. MacLeish, A. The revolt of the diminished man. Sat. Rev. June 7, 1969.

26. Greenwald, D. The American economy. McGraw-Hill Economic Dept.

27. N.Y. Times. 1969. Oct. 25.

28. Water Newsletter. Vol. II, No. 23, Publ. by Water Information Center, Inc. Dec. 8, 1969.

29. Lear, J. Green light for the smogless car. Sat. Rev. Dec. 6, 1969.

30. New York Times. 1970. Jan. 23.

31. MacDonald, G. J. F. 1969. The modification of planet earth by man. Technology Review. Oct.–Nov.

32. News Release. 1968. AAAS 1968 National Meeting. Mimeographed.

33. Russel, W. M. S. 1968. The slash and burn technique. Natural History LXXVII, No. 3.

34. Netting, R. M. 1965. Heritage of survival, Kofyar terrace preserves soil and water. Nat. Hist. March 1965.

35. Niering, W. A. 1968. The effects of pesticides. BioScience 18: 869–875.

36. Woodwell, G. M. Toxic substances in ecological cycles. Sci. Am. March 1967.

3
Effects of Pollution on the Structure and Physiology of Ecosystems

G. M. Woodwell

The accumulation of various toxic substances in the biosphere is leading to complex changes in the structure and function of natural ecosystems. Although the changes are complex, they follow in aggregate patterns that are similar in many different ecosystems and are therefore broadly predictable. The patterns involve many changes but include especially simplification of the structure of both plant and animal communities, shifts in the ratio of gross production to total respiration, and loss of part or all of the inventory of nutrients. Despite the frequency with which various pollutants are causing such changes and the significance of the changes for all living systems (1), only a few studies show details of the pattern of change clearly. These are studies of the effects of ionizing radiation, of persistent pesticides, and of eutrophication. The effects of radiation will be used here to show the pattern of changes in terrestrial plant communities and to show similarities with the effects of fire, oxides of sulfur, and herbicides. Effects of such pollutants as pesticides on the animal community are less conspicuous but quite parallel, which shows that the ecological effects of pollution correspond very closely to the general "strategy of ecosystem development" outlined by Odum (1) and that they can be anticipated in considerable detail.

The problems caused by pollution are of interest from two viewpoints. Practical people—toxicologists, engineers, health physicists, public health officials, intensive users of the environment—consider pollution primarily as a direct hazard to man. Others, no less concerned for human welfare but with less pressing public responsibilities, recognize that toxicity to humans is but one aspect of the pollution problem, the other being a threat to the maintenance of a biosphere suitable for life as we know it. The first viewpoint leads to emphasis on human food chains; the second leads to emphasis on human welfare insofar as it depends on the integrity of the diverse ecosystems of the earth, the living systems that appear to have built and now maintain the biosphere.

The food-chain problem is by far the simpler; it is amenable at least in part to the pragmatic, narrowly compartmentalized solutions that industrialized societies are good at. The best example of the toxicological approach is in control of mutagens, particularly the radionuclides. These present a specific, direct hazard to man. They are much more important to man than to other organisms. A slightly enhanced rate of mutation is a serious danger to man, who has developed through medical science elaborate ways of preserving a high fraction of the genetic defects in the population; it is trivial to the rest of the biota, in which genetic defects may be eliminated through selection. This is an important fact about pollution hazards—toxic substances that are principally mutagenic are usually of far greater direct hazard to man than to the rest of the earth's biota and must be considered first from the standpoint of their movement to man through food webs or other mechanisms and to a much lesser extent from that of their effects on the ecosystem through which they move. We have erred, as shown below, in assuming that all toxic substances should be treated this way.

Pollutants that affect other components of the earth's biota as well as man present a far greater problem. Their effects are chronic and may be cumulative in contrast to the effects of short-lived disturbances that are repaired by succession. We ask what effects such pollutants have on the structure of natural ecosystems and on biological diversity and what these changes mean to physiology, especially to mineral cycling and the long-term potential for sustaining life.

Although experience with pollution of various types is extensive and growing rapidly, only a limited number of detailed case history studies provide convincing control data that deal with the structure of ecosystems. One of the clearest and most detailed series of experiments in recent years has been focused on the ecological effects of radiation. These studies are especially useful because they allow cause and effect to be related quantitatively at the ecosystem level, which is difficult to do in nature. The question arises, however, whether the results from studies of ionizing radiation, a factor that is not usually considered to have played an important role in recent evolution, have any general application. The answer, somewhat surprisingly to many biologists, seems to be that they do. The ecological effects of radiation follow patterns that

49   Effects of Pollution on Structure and Physiology of Ecosystems

are known from other types of disturbances. The studies of radiation, because of their specificity, provide useful clues for examination of effects of other types of pollution for which evidence is much more fragmentary.

The effects of chronic irradiation of a late successional oak-pine forest have been studied at Brookhaven National Laboratory in New York. After six months' exposure to chronic irradiation from a ^{137}Cs source, five well-defined zones of modification of vegetation had been established. They have become more pronounced through seven years of chronic irradiation (Figure 3.1).

Figure 3.1 The effects of chronic gamma radiation from a 9,500-curie ^{137}Cs source on a Long Island oak-pine forest nearly eight years after start of chronic irradiation
The pattern of change in the structure of the forest is similar to that observed along many other gradients, including gradients of moisture availability and of exposure to wind, salt spray, and pollutants such as sulfur dioxide. The five zones are explained in the text. The few successional species that have invaded the zones closest to the source appear most conspicuously as a ring at the inner edge of zone 2. These are species characteristic of disturbed areas such as the fire weed, *Erechtites hieracifolia*, and the sweet fern, *Comptonia peregrina*, among several others. (The successional changes over seven years are shown by comparison with a similar photograph that appeared as a cover of *Science* [16].)

The zones were:
1) A central devastated zone, where exposures were > 200 R/day and no higher plants survived, although certain mosses and lichens survived up to exposures > 1000 R/day.
2) A sedge zone, where *Carex pensylvanica* (2) survived and ultimately formed a continuous cover (> 150 R/day).
3) A shrub zone in which two species of *Vaccinium* and one of *Gaylussacia* survived, with *Quercus ilicifolia* toward the outer limit of the circle where exposures were lowest (> 40 R/day).
4) An oak zone, the pine having been eliminated (> 16 R/day).
5) Oak-pine forest, where exposures were < 2 R/day, and there was no obvious change in the number of species, although small changes in rates of growth were measurable at exposures as low as 1 R/day.

The effect was a systematic dissection of the forest, strata being removed layer by layer. Trees were eliminated at low exposures, then the taller shrubs (*Gaylussacia baccata*), then the lower shrubs (*Vaccinium* species), then the herbs, and finally the lichens and mosses. Within these groups, it was evident that under irradiation an upright form of growth was a disadvantage. The trees did vary —the pines (*Pinus rigida*) for instance were far more sensitive than the oaks without having a conspicuous tendency toward more upright growth, but all the trees were substantially more sensitive than the shrubs (3). Within the shrub zone, tall forms were more sensitive; even within the lichen populations, foliose and fruticose lichens proved more sensitive than crustose lichens (4).

The changes caused by chronic irradiation of herb communities in old fields show the same pattern—upright species are at a disadvantage. In one old field at Brookhaven, the frequency of low-growing plants increased along the gradient of increasing radiation intensity to 100 percent at > 1000 R/day (5). Comparison of the sensitivity of the herb field with that of the forest, by whatever criterion, clearly shows the field to be more resistant than the forest. The exposure reducing diversity to 50 percent in the first year was ~ 1000 R/day for the field and 160 R/day for the forest, a greater than fivefold difference in sensitivity (3).

The changes in these ecosystems under chronic irradiation are best summarized as changes in structure, although diversity, primary production, total respiration, and nutrient inventory are also involved. The changes are similar to the familiar ones along natural gradients of increasingly severe conditions, such as exposure on mountains, salt spray, and water availability. Along all these gradients the conspicuous change is a reduction of structure from forest toward communities dominated by certain shrubs, then, under more severe conditions, by certain herbs, and finally by low-growing plants, frequently mosses and lichens. Succession, insofar as it has played any role at all in the irradiated ecosystems, has simply reinforced this pattern, adding a very few hardy species and allowing expansion of the populations of more resistant indigenous species. The reasons for radiation's causing this pattern are still not clear (3, 6), but the pattern is a common one, not peculiar to ionizing radiation, despite the novelty of radiation exposures as high as these.

Its commonness is illustrated by the response to fire, one of the oldest and most important disruptions of nature. The oak-pine forests such as those on Long Island have, throughout their extensive range in eastern North America, been subject in recent times to repeated burning. The changes in physiognomy of the vegetation follow the above pattern very closely—the forest is replaced by communities of shrubs, especially bear oak (*Quercus ilicifolia*), *Gaylussacia,* and *Vaccinium* species. This change is equivalent to that caused by chronic exposure to 40 R/day or more. Buell and Cantlon (7), working on similar vegetation in New Jersey, showed that a further increase in the frequency of fires resulted in a differential reduction in taller shrubs first, and a substantial increase in the abundance of *Carex pensylvanica,* the same sedge now dominating the sedge zone of the irradiated forest. The parallel is detailed; radiation and repeated fires both reduce the structure of the forest in similar ways, favoring low-growing hardy species.

The similarity of response appears to extend to other vegetations as well. G. L. Miller, working with F. McCormick at the Savannah River Laboratory, has shown recently that the most radiation-resistant and fire-resistant species of twenty-year-old fields are annuals and perennials characteristic of disturbed places (8). An interesting sidelight of his study was the observation that the grass stage of long leaf pine (*Pinus palustris*), long considered a specific adaptation to the fires that maintain the southeastern savannahs, appears more resistant to radiation damage than the mature trees. At a total acute exposure of 2.1 kR (3 R/day), 85 percent of the grass-stage populations survived but only 55 percent of larger trees survived. Seasonal variation in sensitivity to radiation damage has been abundantly demonstrated (9), and it would not be surprising to find that this variation is related to the ecology of the species. Again it appears that the response to radiation is not unique.

The species surviving high radiation-exposure rates in the Brookhaven experiments are the ones commonly found in disturbed places, such as roadsides, gravel banks, and areas with nutrient-deficient or unstable soil. In the forest they include *Comptonia peregrina* (the sweet fern), a decumbent spiny *Rubus,* and the lichens, especially *Cladonia cristatella.* In the old field one of the most conspicuously resistant species was *Digitaria sanguinalis* (crab grass) among several other weedy species. Clearly these species are generalists in the sense that they survive a wide range of conditions, including exposure to high intensities of ionizing radiation—hardly a common experience in nature but apparently one that elicits a common response.

With this background one might predict that a similar pattern of devastation would result from such pollutants as oxides of sulfur released from smelting. The evidence is fragmentary, but Gorham and Gordon (10) found around the smelters in Sudbury, Ontario, a striking reduction in the numbers of species of higher plants along a gradient of 62 kilometers (39 miles). In different samples the number of species ranged from 19 to 31 at the more distant sites and dropped abruptly at 6.4 kilometers. At 1.6 kilometers, one of two randomly placed plots (20 by 2 meters) included only one species. They classified the damage in five categories, from "Not obvious" through "Moderate" to "Very severe." The tree canopy had been reduced or eliminated within 4.8 to 6.4 kilometers of the smelter, with only occasional sprouts of trees, seedlings, and successional herbs and shrubs remaining; this damage is equivalent to that produced by exposure to 40 R/day. The most resistant trees were, almost predictably to a botanist, red maple (*Acer rubrum*) and red oak (*Quercus rubra*). Other species surviving in the zones of "Severe" and "Very severe" damage included *Sambucus pubens, Polygonum cilinode, Comptonia peregrina,* and *Epilobium angustifolium* (fire weed). The most sensitive plants appeared to be *Pinus strobus* and *Vaccinium myrtilloides.* The pine was reported no closer than 25.6 kilometers (16 miles), where it was chlorotic.

This example confirms the pattern of the change—first a reduction of diversity of the forest by elimination of sensitive species; then elimination of the tree canopy and survival of resistant shrubs and herbs widely recognized as "seral" or successional species or "generalists."

The effects of herbicides, despite their hoped for specificity, fall into the same pattern, and it is no surprise that the extremely diverse forest canopies of Viet Nam when sprayed repeatedly with herbicides are replaced over large areas by dense stands of species of bamboo (11).

The mechanisms involved in producing this series of patterns in terrestrial ecosystems are not entirely clear. One mechanism that is almost certainly important is simply the ratio of gross production to respiration in different strata of the community. The size of trees has been shown to approach a limit set by the amount of surface area of stems and branches in proportion to the amount of leaf area (12). The apparent reason is that, as a tree expands in size, the fraction of its total surface devoted to bark, which makes a major contribution to the respiration, expands more rapidly than does the photosynthetic area. Any chronic disturbance has a high probability of damaging the capacity for photosynthesis without reducing appreciably the total amount of respiration: therefore, large plants are more vulnerable than species requiring less total respiration. Thus chronic disturbances of widely different types favor plants that are small in stature, and any disturbance that tends to increase the amount of respiration in proportion to photosynthesis will aggravate this shift.

The shift in the structure of terrestrial plant communities toward shrubs, herbs, or mosses and lichens involves changes in addition to those of structure and diversity. Simplification of the plant community involves also a reduction of the total standing crop of organic matter and a corresponding reduction in the total inventory of nutrient elements held within the system, a change that may have important long-term implications for the potential of the site to support life. The extent of such losses has been demonstrated recently by Bormann and his colleagues in the Hubbard Brook Forest in New Hampshire (13), where all of the trees in a watershed were cut, the cut material was left to decay, and the losses of nutrients were monitored in the run-off. Total nitrogen losses in the first year were equivalent to twice the amount cycled in the system during a normal year. With the rise of nitrate ion in the run-off, concentrations of calcium, magnesium, sodium, and potassium ions rose severalfold, which caused eutrophication and even pollution of the streams fed by this watershed. The soil had little capacity to retain the nutrients that were locked in the biota once the higher plants had been killed. The total losses are not yet known, but early evidence indicates that they will be a high fraction of the nutrient inventory, which will cause a large reduction in the potential of the site for supporting living systems as complex as that destroyed—until nutrients accumulate again. Sources are limited; the principal source is erosion of primary minerals.

When the extent of the loss of nutrients that accompanies a reduction in the structure of a plant community is recognized, it is not surprising to find depauperate vegetation in places subject to chronic disturbances. Extensive sections of central Long Island, for example, support a depauperate oak-pine forest in which the bear oak, *Quercus ilicifolia,* is the principal woody species. The cation content of an extremely dense stand of this common community, which has a biomass equivalent to that of the more diverse late successional forest that was burned much less recently and less intensively, would be about 60 percent that of the richer stand, despite the equivalence of standing crop. This means that the species, especially the bear oak, contain, and presumably require, lower concentrations of cations. This is an especially good example because the bear oak community is a long-lasting one in the fire succession and marks the transition from a high shrub community to forest. It has analogies elsewhere, such as the health balds of the Great Smoky Mountains and certain bamboo thickets in Southeast Asia.

The potential of a site for supporting life depends heavily on the pool of nutrients available through breakdown of primary minerals and through recycling in the living portion of the ecosystem. Reduction of the structure of the system drains these pools in whole or in part; it puts leaks in the system. Any chronic pollution that affects the structure of ecosystems, especially the plant community, starts leaks and reduces the potential of the site for recovery. Reduction of the structure of forests in Southeast Asia by herbicides has dumped the nutrient pools of these large statured and extremely diverse forests. The nutrients are carried to the streams, which turn green with the algae that the nutrients support. Tschirley (11), reporting his study of the effects of herbicides in Viet Nam, recorded "surprise" and "pleasure" that fishing had improved in treated areas. If the herbicides are not toxic to fish, there should be little surprise at improved catches of certain kinds of fish in heavily enriched waters adjacent to herbicide-treated forests. The bamboo thickets that replace the forests also reflect the drastically lowered potential of these sites to support living systems. The time it takes to reestablish a forest with the original diversity depends on the availability of nutrients, and is probably very long in most lateritic soils.

In generalizing about pollution, I have concentrated on some of the grossest changes in the plant communities of terrestrial ecosystems. The emphasis on plants is appropriate because plants dominate terrestrial ecosystems. But not all pollutants affect plants directly; some have their principal effects on heterotrophs. What changes in the structure of animal communities are caused by such broadly toxic materials as most pesticides?

The general pattern of loss of structure is quite similar, although the structure of the animal communities is more difficult to chart. The transfer of energy appears to be one good criterion of structure. Various studies suggest that 10 to 20 percent of the energy entering the plant community is transferred directly to the animal community through herbivores (14). Much of that energy, perhaps 50 percent or more, is used in respiration to support the herbivore population; some is transferred to the detritus food chain directly, and some, probably not more than 20 percent, is transferred to predators of the herbivores. In an evolutionarily and successionally mature community, this transfer of 10 to 20 percent per trophic level may occur two or three times to support carnivores, some highly specialized, such as certain eagles, hawks, and herons, others less specialized, such as gulls, ravens, rats, and people.

Changes in the plant community, such as its size, rate of energy fixation, and species, will affect the structure of the animal community as well. Introduction of a toxin specific for animals, such as a pesticide that is a generalized nerve toxin, will also topple the pyramid. Although the persistent pesticides are fat soluble and tend to accumulate in carnivores and reduce populations at the tops of food chains, they affect every trophic level, reducing reproductive capacity, almost certainly altering behavioral patterns, and disrupting the competitive relationships between species. Under these circumstances the highly specialized species, the obligate carnivores high in the trophic structure, are at a disadvantage because the food chain concentrates the toxin and, what is even more important, because the entire structure beneath them becomes unstable. Again the generalists or broad-niched species are favored, the gulls, rats, ravens, pigeons and, in a very narrow short-term sense, man. Thus, the pesticides favor the herbivores, the very organisms they were invented to control.

Biological evolution has divided the resources of any site among a large variety of users—species—which, taken together, confer on that site the properties of a closely integrated system capable of conserving a diversity of life. The system has structure; its populations exist with certain definable, quantitative relationships to one another; it fixes energy and releases it at a measurable rate; and it contains an inventory of nutrients that is accumulated and recirculated, not lost. The system is far from static; it is subject, on a time scale very long compared with a human lifespan, to a continuing augmentive change through evolution; on a shorter time scale, it is subject to succession toward a more stable state after any disturbance. The successional patterns are themselves a product of the evolution of life, providing for systematic recovery from any acute disturbance. Without a detailed discussion of the theory of ecology, one can say that biological evolution, following a pattern approximating that outlined above, has built the earth's ecosystems, and that these systems have been the dominant influence on the earth throughout the span of human existence. The structure of these systems is now being changed all over the world. We know enough about the structure and function of these systems to predict the broad outline of the effects of pollution on both land and water. We know that as far as our interests in the next decades are concerned, pollution operates on the time scale of succession, not of evolution, and we cannot look to evolution to cure this set of problems. The loss of structure involves a shift away from complex arrangements of specialized species toward the generalists; away from forest, toward hardy shrubs and herbs; away from those phytoplankton of the open ocean that Wurster (15) proved so very sensitive to DDT, toward those algae of the sewage plants that are unaffected by almost everything including DDT and most fish; away from diversity in birds, plants, and fish toward monotony; away from tight nutrient cycles toward very loose ones with terrestrial systems becoming depleted, and with aquatic systems becoming overloaded; away from stability toward instability especially with regard to sizes of populations of small, rapidly reproducing organisms such as insects and rodents that compete with man; away from a world that runs itself through a self-augmentive, slowly moving evolution, to one that requires constant tinkering to patch it up, a tinkering that is malignant in that each act of repair generates a need for further repairs to avert problems generated at compound interest.

This is the pattern, predictable in broad outline, aggravated by almost any pollutant. Once we recognize the pattern, we can begin to see the meaning of some of the changes occurring now in the earth's biota. We can see the demise of carnivorous birds and predict the demise of important fisheries. We can tell why, around industrial cities, hills that were once forested now are not; why each single species is important; and how the increase in the temperature of natural water bodies used to cool new reactors will, by augmenting respiration over photosynthesis, ultimately degrade the system and contribute to degradation of other interconnected ecosystems nearby. We can begin to speculate on where continued, exponential progress in this direction will lead: probably not to extinction—man will be around for a long time yet—but to a general degradation of the quality of life.

The solution? Fewer people, unpopular but increasing restrictions on technology (making it more and more expensive), and a concerted effort to tighten up human ecosystems to reduce their interactions with the rest of the earth on whose stability we all depend. This does not require foregoing nuclear energy; it requires that if we must dump heat, it should be dumped into civilization to enhance a respiration rate in a sewage plant or an agricultural ecosystem, not dumped outside of civilization to affect that fraction of the earth's biota that sustains the earth as we know it. The question of what fraction that might be remains as one of the great issues, still scarcely considered by the scientific community.

References

1. E. P. Odum, *Science* **164**, 262 (1969).

2. Plant nomenclature follows that of M. L. Fernald in *Gray's Manual of Botany* (American Book, New York, ed. 8, 1950).

3. G. M. Woodwell, *Science* **156**, 461 (1967); ——— and A. L. Rebuck, *Ecol. Monogr.* **37**, 53 (1967).

4. G. M. Woodwell and T. P. Gannutz, *Amer. J. Bot.* **54**, 1210 (1967).

5. ——— and J. K. Oosting, *Radiat. Bot.* **5**, 205 (1965).

6. ——— and R. H. Whittaker, *Quart. Rev. Biol.* **43**, 42 (1968).

7. M. F. Buell and J. E. Cantlon, *Ecology* **34**, 520 (1953).

8. G. L. Miller, thesis, Univ. of North Carolina (1968).

9. A. H. Sparrow, L. A. Schairer, R. C. Sparrow, W. F. Campbell, *Radiat. Bot.* **3**, 169 (1963); F. G. Taylor, Jr., *ibid.* **6**, 307 (1965).

10. E. Gorham and A. G. Gordon, *Can. J. Bot.* **38**, 307 (1960); *ibid.*, p. 477; *ibid.*, **41**, 371 (1963).

11. F. H. Tschirley, *Science* **163**, 779 (1969).

12. R. H. Whittaker and G. M. Woodwell, *Amer. J. Bot.* **54**, 931 (1967).

13. F. H. Bormann, G. E. Likens, D. W. Fisher, R. S. Pierce, *Science* **159**, 882 (1968).

14. These relationships have been summarized in detail by J. Phillipson [*Ecological Energetics* (St. Martin's Press, New York, 1966)]. See also L. B. Slobodkin, *Growth and Regulation of Animal Populations* (Holt, Rinehart and Winston, New York, 1961) and J. H. Ryther, *Science* **166**, 72 (1969).

15. C. F. Wurster, *Science* **159**, 1474 (1968).

16. G. M. Woodwell, *ibid.* **138**, 572 (1962).

17. Research carried out at Brookhaven National Laboratory under the auspices of the U.S. Atomic Energy Commission. Paper delivered at 11th International Botanical Congress, Seattle, Wash., on 26 August 1969 in the symposium "Ecological and Evolutionary Implications of Environmental Pollution."

4
Population, Natural Resources, and Biological Effects of Pollution of Estuaries and Coastal Waters

Bostwick H. Ketchum

Introduction

Estuaries and coastal waters are of great importance to the human race, and man uses these waters in many ways, some of which are compatible and others are in conflict. From the earliest days of history, seaports have been essential for transportation, and many of the great cities of the world have developed because of ready access to the sea. In the United States, more than half of the population lives in the coastal states, including those bordering the Great Lakes. With our increasing affluence, the demands of the population for recreation in estuaries and coastal waters are steadily increasing. A major share of the world's marine fisheries is obtained from the coastal waters, and estuaries are essential as breeding grounds for many species of coastal fishes as well as serving as the home throughout the life cycle of many of our seafood delicacies, such as the oyster and the crab. These waters are also used for the disposal of the waste products of our civilization and technology, a use which is in conflict with the fisheries and recreational demands for these waters. Many of our estuaries are so heavily polluted that species of fish which used to abound are now eliminated, and the taking of shellfish in many areas is prohibited because of hazards to human health.

The World Population

The pollution problems of the world have been exacerbated by the rapid growth of the human population and even more rapid growth of our technology. More people use more resources and produce more pollution. Manlike species have lived on earth for about a million years, and *Homo sapiens* evolved some 100,000 years ago. By 1830 a population of 1 billion people had developed. In another century the population had reached 2 billion people, and in the next thirty years, in 1960, the population was 3 billion people. Today our population is 3.5 billion people and still increasing.

Prepared for SCEP.

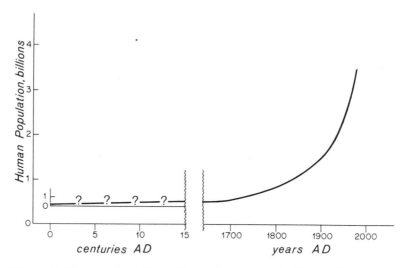

Figure 4.1 The growth of the world population since the birth of Christ
The scales on the left of the figure, both for population and for time, are one-fifth of the scale on the right.

A curve of the growth of the human population from the start of the Christian Era to the present is shown in Figure 4.1. The scales on the left-hand part of this figure, both for population and for time, are one-fifth of the scale shown on the right. The data prior to 1630 are hypothetical since no adequate information on global populations for this period exists. For these sixteen centuries the human population is presumed to have doubled, on the average, about every two hundred years. This is in sharp contrast to the doubling of our population in the last century and the more rapid rate of increase in recent decades.

During the early stages of man's life on earth, the population was kept in check by pestilence, famine, war, and the rigors of the environment. With the exception of war, we have learned to control these external factors, at least in the more civilized parts of the world, and no other controlling influence has appeared to limit population growth. Studies of other species, both plants and animals, show that the population size will level off at a maximum value which is determined by the carrying capacity of the environment in which the organism is grown. We do not know the carrying capacity of the planet for the human population, but even if we did, it is probable that the *maximum* human population which could be maintained on a subsistence level on earth is probably far greater than the *optimum* population size. Hulett (1970) has estimated that the present world production of food would provide the diet enjoyed in the United States for only 1.2 billion people, about one-third of the present world population. Even fewer people could be supplied with nonrenewable resources, such as fossil fuels, steel, and aluminum, if we attempted to extend U.S. levels of affluence to all of the people in the world. We now recognize that the resources of the earth are not infinite and that the length of time we can continue to use the nonrenewable resources of earth is short in comparison to the length of time man has lived on earth. Even, to maintain the human population at its present level indefinitely, we must learn to recycle those materials that are now considered wastes and reuse them over and over again.

Studies of population dynamics have also shown that if a species is allowed to grow to the maximum population size that the carrying capacity of the environment will allow, wild fluctuations or a crash of the population to extinction inevitably follows. It is also known, however, that a population can be maintained indefinitely at a lower level, provided the resources of the environment are renewed. The population will remain constant when, and only when, the rate of reproduction is equal to the rate of mortality. For some commercial fish crops or for laboratory cultures, the population level is commonly maintained by adjusting the harvest rate so that the sum of harvest and natural mortality equals the rate of reproduction. For ethical reasons, it is obvious that one cannot suggest increasing the mortality rate of the human population, but the constant population size can be achieved by a deliberate limitation of the birth rate so that it comes into balance with natural mortality. If birth rate limitations were done perfectly and immediately, however, it would take several decades before the population would reach a constant level. In the meantime, we are inevitably faced with increasing pollution problems because of the increases of our population and development of our technology.

Food from the Sea
This growing population must also be fed. Our agricultural technology has increased the productivity from the land at a rate in excess of the rate of population growth. In the United States we

can now feed an average human a vegetable diet which can be produced on about a quarter of an acre of good farmland and provide his demand for animal protein with an additional acre. Even with our massive production of foodstuffs, however, malnutrition still exists in this country, and there is reason for concern that mass starvation may occur in many parts of the world before our population growth is brought under control.

While the sea is not a good source for total calorie requirements of the world population, it is a good source for animal protein. Deficiency of animal protein in the diet is a serious cause of malnutrition. While it is true that a well-balanced mixed vegetable diet can provide all of the ten essential amino acids for human nutrition, in many parts of the world a single agricultural crop is the major element of the diet. Under these conditions protein deficiency may be severe, and if protein deficiency persists for more than a year or two of a childs life, recovery is never complete. Addition of comparatively small amounts, about 10 grams daily, of animal protein to the diet can prevent the development of protein deficiency.

The sea now provides half or more of the animal protein consumed by about 2 billion people in the world. Many countries, Japan and Portugal for example, do not have enough arable land within their boundaries to permit replacing the protein they now harvest from the sea with terrestrial animal protein. Various estimates have been made evaluating the potential harvest from the oceans. Over the last decade, the harvest has increased at a cumulative rate of about 8 percent per annum. We now harvest about 60 million tons of fisheries products annually. Schaeffer (1965) estimated that harvesting all of the fish in presently fished areas might double this commercial catch and that exploitation of new areas might double the catch again. He concluded that a conservative estimate of the potential harvest from the sea was about 200 million tons of fresh fish. Ryther (1969) evaluated the geographical distribution of the production of organic material by the plants in the ocean and concluded that 100 million tons would be a more realistic estimate of the potential.

Neither of these estimates of the potential harvest from the sea include any consideration of the possible effects of pollution in decreasing the harvest. In freshwater lakes and streams, the Federal

Water Pollution Control Administration (1968a) reported that over 11 million fish were killed by pollution in 1967. These fish kills affected over a thousand miles of streams and nearly two thousand acres of lakes and reservoirs. Comparable statistics are not readily available for the marine environment, but it is well known that massive mortalities of the biota occurred in connection with the oil spills from the *Torrey Canyon,* the Santa Barbara Channel, and other spills throughout the world. Harvesting of mackerel off the coast of California has recently been prohibited because of excessive accumulations of pesticides, and it is well known that throughout the country many areas are closed for shellfishing because of excessive pollution of the waters. Unless man learns to control his pollution of the environment, the estimates of the potential harvest from the sea may be unduly optimistic.

Biological Effects of Pollution

In addition to causing mortalities of natural populations, pollution may have indirect effects that, in the long run, may be more important than the obvious fish kills. One indirect effect is a change of the diversity of the natural populations. This results because only the more resistant species can survive in polluted areas; the other species are eliminated or excluded by the pollution. In extreme, anoxic areas the living population may be limited to the anaerobic bacteria. A decrease in diversity is associated with a parallel decrease in the stability of the natural population. Any additional stress may have a major and catastrophic effect on the population.

Another indirect effect of pollution may be a modification of the breeding capability of the population, even though the adults can survive in the area. This may be an important effect of thermal pollution, which will be discussed later. It is known that temperature is an important factor in controlling the timing of breeding of many marine species. The soft-shell clam, for example, can be induced to breed at any time of the year by appropriate manipulation of the environmental temperature. If organisms breed prematurely in nature, the appropriate food may be unavailable for larval development. Other pollutants may cause an inhibition of breeding or decrease the survival of larvae, which are generally more sensitive to environmental stress than are the adults.

Pollutants may also have a major effect on the behavior of organisms in the marine environment. Communication among organisms in the aquatic environment is frequently mediated by chemotaxis. The anadromous fish, those which migrate to fresh water in order to breed, such as the alewife, the shad, the salmon, the striped bass, probably depend to a large extent upon chemical stimuli in moving to their breeding grounds. Pollution may obscure these stimuli or may actually exclude the migrating fish from the polluted section of the estuary or river.

There may also be chronic sublethal effects of pollution which have no obvious immediate consequence. For example, an organism may gradually accumulate a toxin throughout its life-span. The low initial concentration may be unimportant to the organism, but it may accumulate to a level where the behavior or physiology is drastically altered. This may be augmented by the biological multiplication in the food web, a process by which members of each trophic level may accumulate more of a particular toxin that is found in the food that they eat. This appears to be the case with DDT, which may accumulate along the food chain and have drastic effects on predatory birds, which are at the top of the pyramid. In this case, the accumulation affects the calcium metabolism so that eggs with soft shells are laid that do not develop, and, as a result, many species of these birds are declining in many parts of the world.

Another problem directly associated with domestic pollution is the process of eutrophication. This is a natural process in the aging of lakes, but it is greatly accelerated by the nutrients which are added in domestic pollution. Increased fertility of the water and the change in the balance of essential nutrients frequently leads to suppression of the natural population and the development of species of phytoplankton that are either malodorous or are unsuitable as food for the animal members of the community. An example of this chain of events associated with pollution by ducks was described by Ryther (1954) in Great South Bay on Long Island. The duck feces fertilized the water excessively, eliminated the natural marine species of phytoplankton, and led to the development of very dense growths of small green algae. Great South Bay was once famous for its oysters, but these small algae were not suitable as food for the oyster, and the oyster harvest declined precipitously.

Some people may prefer roast duckling to oysters, but in an ideal world both should be available.

Examples of Pollution Problems

Excessive domestic pollution is probably the most ubiquitous pollution problem facing the world today. The problem has become acute because, until recently, the engineering approach was based upon the premise that "dilution is the solution to pollution." Actually it is not dilution per se that can solve the problems, but the ability of natural processes to decompose organic material and to recycle the nutrients released to the water. Dilution is important to keep the concentration of the pollutant at or below a level which permits this natural recovery. In many confined bodies of waters such as lakes, rivers, and estuaries we have exceeded the dilution and recovery capacity of the environment.

The consequences of aquatic disposal of domestic sewage are presented in Table 4.1. The immediate benefits, the hidden costs, and some alternate methods of operation are listed. Aquatic disposal of domestic sewage has been commonly used because it is the cheapest method of disposal. In many places it was satisfactory as long as the population was small and the loads were not excessive. The immediate benefits are easily defined and evaluated in economic terms. Pollution, however, develops insidiously by the accumulation of small changes, any one of which alone may have an insignificant effect. Without adequate monitoring the additive effects may cause disaster suddenly, and the river, lake, or estuary is dead.

The hidden costs are those that are imposed upon the population as a whole by an irresponsible polluter. The economist calls these hidden costs the externalities of the system and has not developed methods for placing an economic value upon them. As long as a municipality or industry can assess these hidden costs on the population as a whole without increasing operating costs, it is human nature to do so. Experience has shown, however, that cleaning up a polluted aquatic environment is much more expensive than it would have been to keep the environment clean from the beginning. Industries and municipalities are beginning to realize that they must share in repaying the debt to the environment that has accumulated over long periods of time.

Table 4.1 Consequences of the Discharge of Excessive Amounts of Domestic Sewage into Natural Waters

Benefits:	For untreated wastes:
	lowest cost
	repurification by nature
	dispersion and transport away from the source
	For treated wastes:
	reduction of:
	solids
	bacterial counts
	biological oxygen demand
	obnoxious odors
	sludge banks
	easier dispersion and transport
Hidden costs:	Increases downstream water purification costs
	Decreases oxygen content of the water
	Water unsuitable for bathing and recreation
	Sludge banks may eliminate bottom populations
	Large areas closed for shellfishing
	Land values decreased
	Eutrophication results in:
	elimination of desirable species
	growth of obnoxious algae
	development of anoxic conditions
Alternatives:	Improved treatment, including removal of minerals
	Recycle and reuse liquid wastes for:
	irrigation
	agriculture
	silvaculture
	aquaculture
	Dry sludge solids for:
	soil conditioners
	fertilizers

Some of the problems of the disposal of domestic sewage have been ameliorated by the development of sewage treatment plants. The treatment is designed to remove solids and organic matter by processes similar to those provided in nature and to do this faster and under better supervision and control than can be achieved in nature. Primary treatment removes some of the solids; secondary treatment removes most of the solids and decomposes most of the organic matter; tertiary treatment is required if fertilizing elements are also to be removed. Sewage treatment generally reduces the biological oxygen demand (the oxygen required to decompose the added organic matter) considerably so that the undesirable effects of complete removal of oxygen from the aquatic environment are avoided. It is not an ultimate solution, however, unless the fertilizing elements in the sewage are also removed. The natural plant populations of the aquatic environment can use these elements, primarily phosphates and nitrates, to synthesize about as much organic matter as was removed at considerable cost in the treatment plant. Since some time is required for the development of this organic material, however, the problem is moved downstream and the hidden costs may be imposed upon a population that was not consulted in advance about the procedure. Providing each municipality and town with a sewage treatment plant will not solve the domestic pollution problems of our heavily contaminated waters unless the nutrients are also removed from the effluent.

The only ultimate solution to the problems of domestic pollution, and to most other pollution problems as well, is the recycling and reuse of the materials. The very elements which produce malodorous and obnoxious conditions in the aquatic environment from pollution are the elements that are required as fertilizers. Wolman (1956) states that disposal of sewage wastes on the land in an effort to encourage plant growth was the first approach used in England, and it proved to be a dismal failure. In part this was because of the wet character of the climate and in part because of fine soils that could not absorb the liquids and solids without clogging. He laments that this effort probably delayed the "successful artificial procedures for treating sewage by at least a quarter of a century." Now that we do have adequate secondary treatment processes that remove most of the solids, disposal on land for recycling and reuse of fertilizing elements deserves reconsideration.

Use of a clear, sterilized effluent for aquaculture, silvaculture, or irrigation and agriculture would solve the problems of eutrophication of natural waters caused by pollution, would produce a product of value to man, and would not create a public health hazard. The engineer and the economist will reply that it is too expensive to do this, and this conclusion is probably true if only the immediate benefits and costs are considered. Considering the hidden costs or the costs associated with cleaning up a damaged environment may well reverse this conclusion. In any case, we must follow this course of action if we are to leave an environment to posterity that is suitable for human habitation.

We can define the limits of fertilizing elements that may be added to estuaries or coastal waters without producing the undesirable effects of overfertilization. This is possible because the amount of oxgyen that can be dissolved in the water is determined by its temperature and salinity. If the water is fertilized to an extent that will produce more organic material than this amount of oxygen can decompose, the obnoxious effects will inevitably be produced.

Figure 4.2 shows the amount of chlorophyll produced in various marine waters as a function of the amount of phosphate available in the water. Phosphate is used as an indicator, assuming that the other essential elements are present in the pollutant that was added to some of these waters, and chlorophyll is used as an indication of the density of the plant population. Excessive fertilization clearly leads to large plant populations and polluted conditions. To avoid these conditions the phosphorus content must be kept below a concentration of $2.5\mu g$ atom/liter (2.5×10^{-6} moles per liter). This is about the concentration found in deep-ocean water, showing how delicately the ocean system is balanced.

Observations made a few years ago in the Hudson River show that the pollution entering this estuary exceeds the permissible limits by a factor of about 5. Figure 4.3 shows the stations at which the observations were made, and the amount of phosphorous found in both surface and deep waters at these stations is shown in Figure 4.4. The distribution of any pollutant in an estuary is related to the distribution of fresh and salt water in the estuary

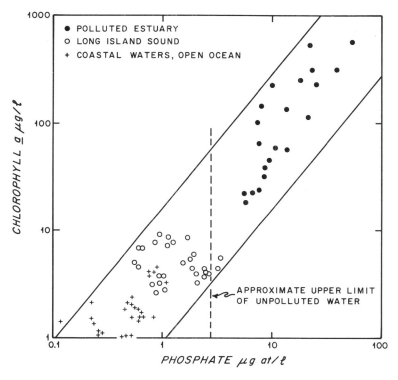

Figure 4.2 The relationship between the phytoplankton population (as indicated by chlorophyll *a*) and the phosphate content of marine waters
Phosphate is expressed as microgram atoms (micromoles) per liter; chlorophyll as micrograms per liter.

(Ketchum, 1955) since the pollutant is dispersed and mixed by the same mechanisms that disperse and mix fresh and salt water. The fraction of salt water in each sample is given by the ratio of salinity, S_x/S_o, in which S_o is the salinity in the water at the source of pollution and S_x is the salinity at any other location. The freshwater fraction (F_x) is given by $(1 - S_x/S_o)$. The phosphorus data in Figure 4.4 are plotted as a function of the salt content upstream of the source of pollution, Station C, at the lower end of Manhattan. Downstream the data are plotted as a function of the freshwater content. The relationship shows clearly that the distribution of pollution is closely related to the distribution of fresh and salt waters in the system. At the source, the concentration of phosphorus added to the Hudson is about five times greater than the

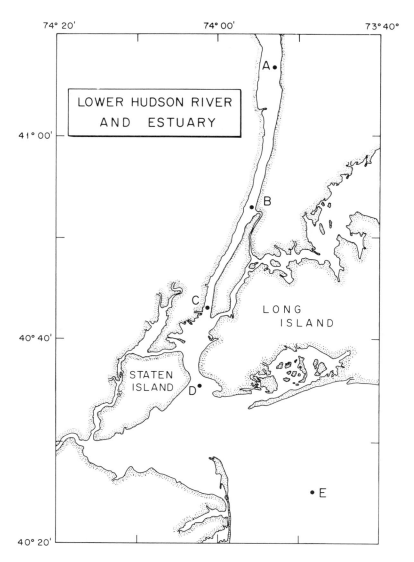

Figure 4.3 The Hudson River Estuary showing locations where the observations of Figure 4.4 were made

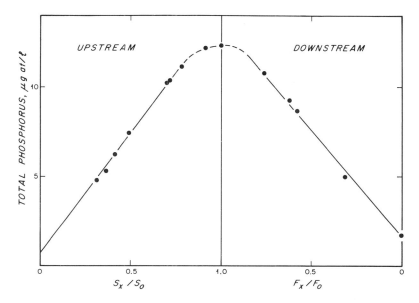

Figure 4.4 The distribution of total phosphorus in the Hudson River Estuarine waters
The maximum phosphorus pollution was found in the surface waters at station C (Figure 4.3) off the tip of Manhattan Island. The fresh and salt water fractions in other samples are calculated relative to conditions at this location.

permissible concentration. This critical concentration is exceeded for a distance of about 40 miles of the estuarine system.

Another example of pollution effects, summarized in Table 4.2, is the disposal of excess heat from power plants to the marine environment. Again, immediate benefits, hidden costs, and alternate solutions are given. Today heat pollution is not a problem of the same magnitude as domestic pollution, but it may become increasingly serious in the future. In the United States the demand for electric power is increasing more rapidly than the growth of the population, with the demand doubling about every decade. We are becoming more and more dependent upon nuclear power plants, which produce more excess heat for each unit of electrical power, and are thus less efficient than fossil-fuel plants. An increasing number of these power plants will be placed on estuaries and coastlines because of the availability of large volumes of water for cooling purposes. In this situation we must anticipate the problems and develop solutions and adequate methods of operation

Table 4.2 The Consequences of Disposing Excess Heat to the Aquatic Environment

Benefits:	Helps meet electrical power demands which double each ten years
	Simplest disposal method
	Least expensive
	Gradual release to atmosphere
Hidden costs:	Marine life killed at lethal temperature
	Reduces oxygen capacity of water; thus may reduce self-purification
	Causes premature spawning (clams, oysters, fish)
	Modifies fish behavior
	Excludes anadromous fish
	Increases metabolism rate
	Synergistic increase in toxicity of other pollutants
Alternatives:	Cooling towers, but air pollution, fog production, more expensive
	Use for beneficial purposes, for example:
	warm cold beaches for humans
	control temperature of lagoons for aquaculture
	induce upwelling of nutrient rich deep water
	provide optimum temperature for sewage digestion
	develop processes to use heat for industrial purposes
	Recycle cooling water in large artificial ponds

beforehand. Warming natural waters can have many biological effects, some of which are detrimental and others beneficial. Our objectives in disposing of this excess heat should obviously be to maximize the beneficial effects while minimizing the detrimental ones.

The hot water released from a power plant may be as much as 25°F warmer than the natural ambient temperature. The water quality criteria issued by FWPCA (1968b) specify that, beyond a reasonable mixing zone, the temperature should not be increased by more than 4°F in fall, winter, and spring and 1.5°F in summer. Neglecting heat exchange with the atmosphere, the heated effluent at 25°F would have to mix with about eight times its own volume in the colder months and seventeen times its own volume in summer before this standard would be met. The rate of mixing will depend both upon the design of the outfall and upon the local circulation. Before the site for a new power plant is selected, studies of the local environment should be made to evaluate and predict the extent of the heated water and the ecological consequences of the operation.

Temperature of the environment is of great importance to all biological processes. As mentioned previously, it may control the breeding cycle of many species, and it is certainly of great importance in determining the geographical distribution of plants and animals. The species present in any given environment are obviously adapted to the normal seasonal variations in temperature. Even so, excessively hot or cold spells can kill off large numbers of individuals. Excess heat at the warmest times of year could be detrimental and cause fish kills, but it is equally possible that warming the waters at the coldest time of year might be beneficial to many organisms. In temperate latitudes it is quite conceivable that one could extend the growing season of many edible species by warming the water during the coldest months of the year. The tolerance of organisms to an increase of heat, however, is reduced as the natural temperature of the environment increases. Thus, the discharge of hot water into a lagoon where desirable species are grown may be beneficial during the wintertime and detrimental in summer. Different methods of heat disposal may be required at these two times of year.

Heated effluents from coastal power plants are being used in "fish farming" in the British Isles. Nash (1969) reports that the effluent, which is about 10° to 23°C, provides plaice and Dover sole with a more "amenable habitat than the sea (3° to 16°C)." The growing season has been extended throughout the year and has produced fish of marketable size at least twelve months before the most advanced individuals in nature. Nash further points out that the fish farming plans must be considered in the very earliest stages of power plant design if ideal conditions are to be provided.

A variety of other ways can be conceived to use this excess heat beneficially. Many coastal waters in the temperate zone are too cold for enjoyable swimming even in the summertime, and the use of this excess heat might improve the recreational desirability of these colder waters. It is customary, because of greater efficiency of operation, to use the colder deep water as a supply and to discharge the heated water at the surface. The less dense, warmer water spreads out as the surface lens and decreases the rate of mixing because of the increased temperature gradient. An artificial upwelling could be produced if the warmer waters were released at depth instead of at the surface. This would produce vigorous mixing and bring some of the nutrient rich deeper waters to the surface where the fertility would be increased. It is worth pointing out that the upwelling of natural waters does not present the hazards of overfertilization because the concentration of nutrients in these waters is within the limits that the natural populations can utilize advantageously.

Another beneficial use of excess heat, suggested by Mihursky (1967), would be to provide optimum temperature for sewage digestion in treatment plants. If this should prove to be feasible from an engineering viewpoint, it might help to solve two problems simultaneously. Conceivably some industrial processes could use the excess heat, just as some now use the effluent from a sewage treatment plant (Wolman, 1956).

There are obviously some unresolved problems before the excess heat produced by power plants can be used wisely. In the tropics or subtropics, the waters may be close to lethal temperatures for many species during parts of the year. The problems of the disposal of excess heat in these waters are quite different from the problems in temperate waters. We have time to resolve some of these difficulties, and if we do it wisely we can insure that thermal pollution never becomes as serious a problem as some of our other kinds of pollution.

These two examples of pollution have been used to show the benefits, hidden costs, and alternative solutions. It is obvious that many other types of pollutants could have been treated in a similar way. Pesticides, industrial wastes, solid wastes, and detergents are all placing cumulative stresses on the environment. If we are to survive and to maintain an environment of suitable quality for human use and enjoyment, our methods of control of these materials require reevaluation.

Multiple Uses of an Estuary
Each environment is subject to multiple demands by the human population. The effects of various uses on the quality of the environment of an estuary are shown in Figure 4.5. Uses are ranged across the top of the figure from industrial uses at the left to the more personal uses on the right. The effects of each use on the quality of the water and of the habitats in the estuary, upon natural resources and upon human uses are shown by different degrees of shading. The industrial uses are generally detrimental to environmental quality, or at least must be pursued with caution in order to avoid damage. They are, of course, economically valuable to the population living in the immediate area. In contrast to this, the more personal, recreational uses of the estuary can generally be pursued with little or no detrimental effect upon the environment.

It is clear that some uses are compatible with each other while others are not. The compatibility of these various uses of the environment is shown in Figure 4.6. Industrial uses may be compatible with each other even though they are damaging to the environment. Clearly the personal and recreational uses of the environment are compatible, but they are not compatible with the industrial development of the estuarine region.

What Can Be Done?
For generations man has considered growth of the population, of industrialization, and of the gross national product to be the prime desiderata of civilization. We must now recognize that uncontrolled growth carries within it the potential for disaster. More and more people are recognizing that we are placing stresses upon all parts of our environment from which nature cannot recover without our help. At the same time, just because of these growth factors which have been so important to us as a civilization, more and more people have the time and the finances which permit enjoyment of the environment both for recreational purposes and for the natural aesthetic values. It is not too late to preserve what we have left undamaged and to recover the quality of some environments which we have destroyed.

76 Bostwick H. Ketchum

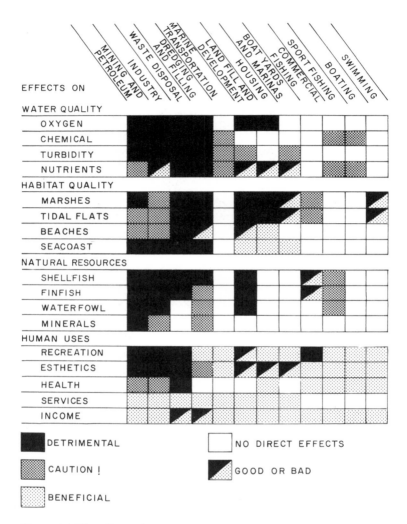

Figure 4.5 The effects of various human uses of an estuary on the quality of the estuarine environment

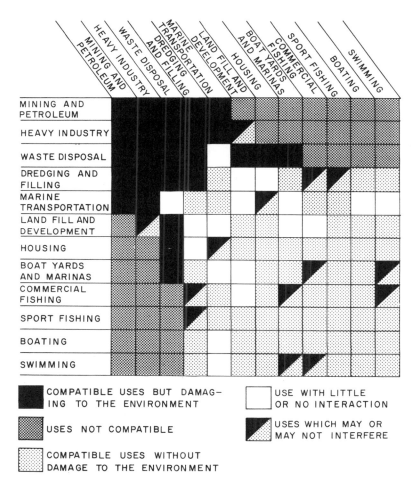

Figure 4.6 The compatibility of various human uses of an estuary, and the effects on other possible uses

The first thing that must be done is to reorient our thinking concerning the waste products of our industry and our civilization. The resources of our planet are *not* unlimited. We have not yet run out of minerals, of fossil fuels, or of food, but the known reserves of many of these will, at the rate of consumption in the United States, maintain the supply for time spans ranging from a few decades to a few centuries. If we attempt to provide the populations of developing nations with the standard of living enjoyed in the United States, 2 billion people would need about fifty times as much natural resources as they now use. This would reduce the known reserves that much more rapidly. If the human race looks forward to survival on this planet for another million years, it is clear that we must learn to reuse and recycle those things which are essential for our civilization. We must convert our thinking from considering wastes as an objectionable material to be gotten rid of at the minimum cost to the concept that a waste is a natural resource out of place in the environment. Recycling and reuse of materials is the only ultimate solution to the problems of pollution; all other concepts of waste treatment are mere palliatives that will postpone the problems for a brief period of time.

The economist must learn to place a realistic evaluation on the hidden costs or the externalities of given operations. As we slowly but surely change the earth from a planet of abundant natural resources to one of limited resources, these hidden costs will become progressively more and more important. A glass of water, for example, is of little economic value if one can turn on the faucet and have an abundant supply on hand. To a man dying of thirst in the desert, however, a glass of water is of greater value than everything else that he possesses.

In spite of his magnificent achievements, man has not yet learned to live in harmony with his environment. He has mastered the environment for the "benefit" of mankind, and by doing so he has eliminated the natural controls of unbridled population growth. This increasing population, the increasing needs of the population for the essentials of life, the increasing demand of people for material things, have all combined to place man's footsteps in a path which can lead only to disaster. We must learn to control the population so that we do not exceed the carrying capacity of this planet. We must learn to solve some of the problems of our urban ghettos which are increasing drastically as the crowding increases. We must learn to control our own activities so that we recycle and reuse those materials that are essential to our civilization, and we must learn to maintain an environment that is suitable for man's livelihood. It is not a question of whether or not we can do it or can afford it; we must do it or run the danger of following the dinosaurs into extinction.

References

Federal Water Pollution Control Administration (FWPCA), 1968a. Pollution caused fish kills—1967, *Eighth Annual Report* (Washington, D.C.: U.S. Government Printing Office).

Federal Water Pollution Control Administration (FWPCA), 1968b. *Report of the Committee on Water Quality Criteria* (Washington, D.C.: U.S. Government Printing Office).

Hulett, H. R., 1970. Optimum world population, *Bioscience, 20* (3): 160–161.

Ketchum, Bostwick H., 1955. Distribution of coliform bacteria and other pollutants in tidal estuaries, *Sewage and Industrial Wastes, 27* (11): 1288–1296.

Mihursky, Joseph, 1967. On possible constructive uses of thermal additions to estuaries, *Bioscience, 17:* 698–702.

Nash, C. E., 1969. Thermal aquaculture, *Sea Frontiers, 5* (5): 268–276.

Ryther, John H., 1954. The ecology of plankton blooms in Moriches Bay and Great South Bay, Long Island, New York, *Biological Bulletin, 106* (2): 198–209.

Schaeffer, Milner B., 1965. The potential harvest of the sea, *Transactions of the American Fisheries Society, 94:* 123–128.

Wolman, A., 1956. Disposal of man's wastes, *Man's Role in Changing the Face of the Earth,* edited by W. L. Thomas, Jr. (Chicago: The University of Chicago Press).

5
Health Effects Research Program of the National Air Pollution Control Administration

Vaun A. Newill, Gory J. Love, F. Gordon Hueter, and Robert J. M. Horton

The primary basis for urban air pollution control in the United States has been the adverse effects of air pollutants on human health. One of the purposes of the Clean Air Act that has not been altered by the several amendments to the original legislation reads as follows: ". . . to protect and enhance the quality of the nation's air resources so as to promote the public health and welfare and the productive capacity of the population. . . ." The amended Act of 1967 was more specific (U.S. Government, 1963). It provided for the production of air quality criteria documents to summarize what is known about pollutants or combinations of pollutants and established that the publication of these documents would trigger action programs for setting ambient air quality standards as goals and the developing of implementation plans to serve as the blueprint for reaching these goals. Such an approach utilizes existing knowledge concerning health effects and reasonable extrapolations therefrom for the standard-setting process, the process by which the health and welfare of the population is to be protected.

More recently, the National Environmental Policy Act of 1969 was enacted into law (U.S. Government, 1970). This Act extends the breadth of the considerations used to formulate our environmental protection policy. However, it does not alter the responsibility to protect the health of the general public as a minimum requirement.

The Division of Health Effects Research (DHER), Bureau of Criteria and Standards, National Air Pollution Control Administration (NAPCA), has as its stated mission the supplying of health-related intelligence for inclusion in air quality criteria documents. The activities of the division, in keeping with its mission, are restricted to a limited portion of the ecosystem of which man is a part, that portion describing the deleterious effects of pollutants on the health of man. It is recognized that man may not be the most sensitive receptor in the ecosystem. Thus, it does not follow logically that protection of the health of man will assure

Prepared for SCEP.

the protection of the ecosystem. Nevertheless, the protection of the health of man is the noncompromisable minimum requirement for our urban environmental pollution control programs. To assure that ambient air levels for specific pollutants do, in fact, afford protection for the health of our population is a difficult task and demands strong research and surveillance programs. When more detailed information has been gathered, it may well be shown that there are other receptors in the ecosystem more sensitive to exposure to environmental pollutants than is man himself. The protection of these more sensitive receptors will require even greater control of ambient air pollution than is necessary to protect man's health.

The program of the Division of Health Effects Research consists of three major tasks, namely: (1) supplementation of existing intelligence concerning the health effects of single or combinations of pollutants for air quality criteria documents; (2) community health effects surveillance to assure that ambient air pollution levels, after the implementation of approved standards, do in fact protect human health; and (3) provision of consultation or technical assistance services where both a knowledge of health effects of pollutants and medical or biological expertise is required. During the time period 1971–1976 twenty-four additional air quality criteria documents are planned for issuance. Nine of the public documents will be considered for revision if need be (see Table 5.1). The research plans and priorities of the DHER are determined completely by the set of deadlines established by this schedule for production of documents.

Parenthetically, you may be interested to learn that a preliminary report for each of these air quality criteria documents will be prepared by the National Academy of Sciences. A study of lead was begun in early 1971.

The health effects information needs for any pollutant, the needs that determine the DHER program, require qualitative and quantitative identification of the prime effect. Usually this effect is identified via an acute exposure, that is, high-level short-term exposure, and must be verified subsequently to result from long-term low-level exposure; that is, dose-time effects research. At the same time, the investigator must be cognizant of and watch for other possible effects from the long-term low-level exposure. Real-

Table 5.1 Tentative Order of Publication of Air Quality Criteria Documents*

Year	Pollutants
1969†	Particulate matter and sulfur oxides
1970‡	Carbon monoxide, hydrocarbons, and photochemical oxidants
1971	Fluorides, lead, nitrogen oxides, and polycyclic organics
1972	Asbestos, beryllium, chlorine gas, hydrogen chloride, and odorous compounds
1973	Arsenic, cadmium, copper, manganese, nickel, vanadium, and zinc
1974	Barium, boron, chromium, mercury, selenium, particulate matter,‡ and sulfur oxides‡
1975	Pesticides, radioactive substances, carbon monoxide,‡ hydrocarbons,‡ nitrogen oxides,‡ and photochemical oxidants‡
1976	Aeroallergens, fluorides,‡ lead,‡ and polycyclic organic matter‡

* In some cases, it may be desirable to publish a state-of-the-art review for a particular pollutant in lieu of a criteria document.
† Criteria documents are available.
‡ Reevaluate.

istically, no pollutant in the ambient atmosphere ever exists as a single entity and, therefore, the identified effects must be studied in relation to the interaction of other components of the total milieu. Such information provides a firm base for the initial establishment of meaningful air quality standards. Subsequent modification of such standards may then be required on the basis of additional research that identifies the potentiation or attenuation of known effects by other environmental parameters, for example, temperature or humidity. Research related to the identification of susceptible populations as determined by age, existing disease, stress, or special sensitivities may dictate the need for modifying established standards.

Meaningful health effects parameters require input from the integration of three scientific approaches. These are (1) laboratory animal toxicological studies, (2) laboratory clinical investigations of human subjects, and (3) epidemiological studies under naturally occurring conditions. Animal toxicology is needed to identify prime effects via controlled experimental exposures that block out confounding environmental variables normally existing under field conditions. These studies also can utilize exposure concentrations unsuitable for human studies due to ethical as well as medical-legal limitations. Animal toxicology also allows for in vitro and various manipulative approaches that can develop techniques for subsequent application to clinical and epidemiological investigations. Laboratory clinical investigations on human subjects are utilized to test techniques resulting from animal toxicology studies and to verify, in a cause-and-effect manner under controlled conditions, correlative epidemiological results. The final test of any indicated effect is its expression as measured by sophisticated epidemiological techniques in human populations exposed to ambient pollution. Such field studies are the final proving ground for all laboratory research. In addition, epidemiological results provide an important input into the designating of the direction of laboratory animal and human clinical experimentation. This triphasic integrated research effort is deemed essential for the most expeditious acquisition of meaningful health effects information.

The responsibilities of the division assigned by the Clean Air Act as amended in 1967 necessitated the development of new program attitudes and direction. These attitudes have been built more specifically into the program plan for the fiscal year 1971 and have culminated in the establishment of Pollution Project Groups, each of which is charged with the responsibility for gathering particular information relative to an assigned group of pollutants (see Table 5.2). These responsibilities include the collection, review, and evaluation of data on the following points:

1. What is the lowest standard for each pollutant that can now be supported and defended confidently?
2. What data need to be collected to support the establishment?
3. What studies are needed to improve, quantitatively and qualitatively, measurements of individual exposure to pollutants?

Table 5.2 Department of Health Effects Research Pollutant Project Groups

Group number	Pollutants included
1	Cd, V, Mn, Zn
2	As, Ni, Cr, Cu
3	B, Se, Ba, Hg
4	Polycyclic organics, Be, asbestos
5	Odors (H_2S), HCl, Cl_2
6	SO_2, particulates
7	CO
8	O_x, NO_x, reactive hydrocarbons
9	Pb, F

4. What studies are needed to improve the detection and quantitation of initial effects produced by air pollution?
5. What studies are needed to determine short-term exposure limits?

During the fiscal year 1971 it is planned that a significant portion of professional staff time within the division will be devoted to these efforts. It is anticipated that during the fiscal year 1971 each of the Pollution Project Groups established will

1. Develop a state-of-the-art review of the health effects literature with regard to each of the pollutants assigned.
2. Identify the additional information needed to assure that the health of the population will be protected or to provide solid support for the air quality standards established.

After the fiscal year 1971 each Pollution Project Group will update the review and the list of needed studies on an annual basis.

The advantages that are visualized as coming from this method of planning are most significant. First, it emphasizes divisional programs relating air pollution to health effects rather than animal studies, laboratory studies, or field studies. Second, it relates the division's studies more directly to the administration's program in that our investigations now relate to pollutants as do criteria documents or emissions from either stationary or mobile sources. Third, it provides for an integrated divisional program in which the personnel work jointly on specific problems rather than separately with particular kinds of studies.

For the fiscal year 1972 and subsequent years, it is planned that the division's program will be tied even more closely to the output of the Pollution Project Groups. This output is to include a list of studies needed to indicate the standards that should be established or should be more stringent. The list will be placed in priority order. Within the capabilities of our staff, facilities, and funding, the studies at the top of the list will be undertaken as the in-house program. Once the in-house studies have been decided upon and are removed from the list of needed studies, there remains, still in priority order, the list of studies to obtain under contract insofar as funding will allow.

The assignment of responsibility for additional pollutants has engendered a recognition of the fact that the philosophy for ambient air quality standards when applied to a trace substance, for example, lead or cadmium, must be different from what it is for a substance that enters the body only by inhalation, for example, ozone. When studying the effects of trace material, we are interested in an air transport system that permits the pollutants in question to enter the body via inhalation and via pathways in the food and water chain (see final section). The effects in this situation are dependent upon the total amount of material absorbed from all routes of entrance into the body and are manifested at particular sites within the body either directly or indirectly through other metabolites. This need for a change in philosophy for the setting of air quality standards also alters the information requirements for the Air Quality Criteria Documents.

It is sometimes difficult to perceive the relationship between research requirements and public health action programs. In this regard the DHER fully accepts the philosophy expressed by Sir Austin Bradford Hill (1962): "All scientific work is incomplete—whether it be observational or experimental. All scientific work is liable to be upset or modified by advancing knowledge. That does not confer upon us a freedom to ignore the knowledge we already have, or to postpone the action that it appears to demand at a given time." It is, however, equally true that establishment of an active public action program with regard to air pollution control does not alleviate or ameliorate the need for a strong effects research program. The lead times necessary for researchers to supply a more firm basis for the public action program are long, possibly eight to ten years, and lack of timely investment in support of them assures that the information will not be available when required.

The investigative procedures of the division are demonstrated most readily by discussion of a few of the materials found suspended in the form of particles in our ambient atmosphere. There is no intention to suggest that they are the most important pollutants, but they are among those in which there is great interest at the present time.

Certain metal ions, like sodium, potassium, calcium, and magnesium, are present in high concentrations in all living organisms, while others, the so-called trace elements, occur in minute quantities. Some of these trace elements are indispensable to the living cell. Iron and copper are essential to various respiratory pigments (hemoglobin, myoglobin, hemocyanins, and so on) and in oxidative-enzymes which play a central role in metabolism (cytochrome, catalases, peroxidases, various metallo-flavoproteins, and so on). Cobalt occurs in vitamin B_{12}, and zinc is an integral part of enzymes such as carbonic anhydrase and several dehydrogenases. The requirement for certain metal ions may vary with the type of organism; for example, only some of the higher plants have been shown to require aluminum, boron, or vanadium, while molybdenum appears to be essential only to those organisms that derive nitrogen from inorganic sources. Other elements occur in small quantities, but no definite function has been established (cesium, chromium, nickel, rubidium, strontium, tin); still others not only are nonnutritive but produce toxic effects (antimony, arsenic, barium, beryllium, bismuth, cadmium, mercury, lead, selenium, silver, tellurium, and thorium) (Nilsson, 1970).

The procedure for deciding upon safe levels for such trace metals entails (1) determining what level of intake, if any, is required for normal body function; (2) determining what levels show the earliest signs of toxicity; (3) attempting to locate a safe zone with a safety margin so that neither deficiency or toxicity will occur; (4) establishing a public action program to keep the public in the safe zone. Three specific examples of this procedure are discussed now.

Lead

Lead is a ubiquitous metal not known to be necessary for normal life. It is extensively used by industry and is the major additive in gasoline.

Lead is absorbed into the body via both the gastrointestinal tract and the respiratory tract and excreted from the body in the feces, urine, and skin (shedding of tissue, nails, hair, and through skin secretions). With low and constant levels of exposure, intake and excretion are approximately equal. Some lead does accumulate in the body, and this is designated the body burden. Blood levels are used as an estimate of the body burden. Persons with no unusual exposure normally will have lead levels of 10 to 30 μg/100 g blood.

In adults, signs and symptoms of lead toxicity have not been observed at blood lead levels less than 80 μg/100 g blood, although they have been observed at somewhat lower levels in children. Symptoms probably are a function of the rate of accumulation of lead since in instances in which the accumulation has occurred slowly, levels have reached 200 μg/100 g blood without clinical toxic manifestations.

Symptoms and signs of lead toxicity are abdominal cramps, headache, constipation, loss of appetite, fatiguability, anemia, motor nerve paralysis, and encephalitis. The last sign is seen in children particularly and may leave permanent damage.

We have less understanding of subclinical illness. The formation of hemoglobin and red blood cells is affected, and there is urinary excretion of delta amino levulinic acid. Subclinical effects in the nervous system, liver, or kidney have scarcely been investigated. There have been very few follow-up studies of exposed groups, and there is no information concerning susceptibility either temporary or permanent.

Given knowledge of the content of lead in food and water and in the ambient atmosphere, it is possible to estimate the amount absorbed. But absorption is a complex process. Only about 10 percent of the available lead is absorbed in the gastrointestinal tract, while the proportion absorbed from the lung is larger. Several factors are important in determining the amount of absorption from the lung.

For a particle to be absorbed, it must reach the proper part of the lung and must be deposited there. Both the quantity of material deposited and location in which it is deposited are particle-size dependent. If the particles are greater than 5 microns in diameter, they are deposited in the nose and throat. Later they may be swallowed. Particles in the range of 0.5 to 5 microns are most likely to be deposited in the trachea and bronchioles. About one-half are deposited and half exhaled (Task Group on Lung Dynamics, 1966). The particles are cleared from these surfaces by ciliary action and swallowed (Albert et al., 1967). Exposure to particles in this size range gives no increase in blood lead levels, but there is detectable increase of lead in the feces. If the particle size is less than 0.5 micron, about one-half is deposited on the walls of the alveolar ducts or alveoli (Kehoe, 1961; Nozaki, 1966). Here clearance from the lung is less rapid, less complete, and absorption can occur. The absorption is thought to be a function of solubility. More recently gathered information seems to indicate that there is a slow phase clearance mechanism so that all of the lead retained probably is not absorbed. Lead particles in the ambient air or urban areas have a mean particle size of 0.25 micron (Robinson and Ludwig, 1967). Therefore, much of the material is likely to be retained and to be absorbed.

Recent experiments of prolonged exposure of man to low levels (10 $\mu g/m^3$) of small lead particles have shown no evidence of increased absorption as measured by blood lead levels and lead excretion (Kehoe, 1966). Similar experiments using 20 $\mu g/m^3$ are under way (Lutmer, Busch, and Miller, 1967) but have not been reported.

The position of the Bureau of Criteria and Standards of the NAPCA with regard to lead is summarized in the following quotes from a memorandum written recently (Barth, 1970).

We would agree that there is no proven deleterious human health effect from present ambient lead air concentrations. However, for reasons discussed below we still feel that steps should now be taken which will ultimately lead to the complete removal of lead from gasoline. Important facts bearing on this problem are:
a. Lead is not known to be required in normal human nutrition and metabolism whereas cases of fatal lead poisoning have been reported for centuries.
b. Symptoms of acute lead poisoning in humans have been reported with blood lead levels as low as 60 $\mu g/100$ grams of blood.
c. Normal human blood lead levels now lie in the range 10–30 $\mu g/100$ grams of blood.
d. It is likely that chronic or subclinical acute deleterious human health effects on the blood forming system or the nervous system result at blood lead levels less than 60 $\mu g/100$ grams of blood.
e. Air uptake of lead by humans is quite significant even though it is probably less than uptake from food and beverages combined.
f. Control of human lead intake via the air is much more feasible than control of lead intake from food and beverages.
g. The major source of lead in the atmosphere is the exhaust from gasoline-fueled vehicles.
h. The total amount of lead being used in motor vehicle gasoline is steadily increasing each year.
i. The cost of removing lead from gasoline is not prohibitive.
From the above we conclude that the existing safety margin

between normal human blood levels and those of humans suffering from clinical lead poisoning may not be adequate. This is particularly true in view of possible chronic or subclinical acute deleterious health effects of lead as well as the yearly increasing usage of leaded gasolines. The phased removal of lead from gasoline, which will not be prohibitively expensive, is a prudent protective action which can and should be undertaken as soon as practicable. We must not wait for proven deleterious human health effects before removing probable harmful pollutants from the human air environment. This is especially true when that pollutant can be removed to a major extent without any significant negative effect on the quality of life of our average citizen.

Cadmium

Cadmium is a metal much used industrially. It is found wherever zinc is found. Cadmium can cause lethal acute poisoning and may cause insidious and slowly developing chronic disease. Its deleterious nature was recognized in 1956 by the adding of chronic cadmium poisoning to the list of diseases covered under the National Insurance (Industrial Injuries) Act of the United Kingdom (Bonneli, 1965). Cadmium coating of cooking utensils is an old practice and one that is prohibited by law in several countries.

Cadmium has an affinity for sulfhydryl groups but also affects hydroxyl groups and ligands containing nitrogen. It is a potent inhibitor of several enzyme systems in intermediary metabolism (Hewitt and Nicholas, 1963) (alpha-ketoglutaric acid dehydrogenase, alpha-dihydrolipoic acid dehydrogenase, and transhydrogenase). Further, on certain substrates the metal is a potent inhibitor of L-aminoacid decarboxylase (Nilsson, 1970). Cadmium inhibits the biosynthesis of chlorophyll.

Schroeder fed 5 ppm of Cd in the drinking water of rats (Schroeder and Vinton, 1962; Schroeder, 1964). Hypertension appeared in some of the animals after one year, and the hypertension increased with age. Once hypertension was present, it remained until death. When he gave zinc to hypertensive animals to replace the Cd, after one injection that also contained a chelating agent with a strong affinity for cadmium, the animals were normotensive in one week. This was verified with intraperitoneal injection of zinc.

The absorbed cadmium is eventually transported to the kidney and stored in the form of a metallo-protein complex, and this eventually leads to an abnormal renal function. Such exposure in man causes excretion of low molecular weight proteins.

Because of Schroeder's work and the human evidence of kidney involvement, Dr. Robert Carroll (1962), then in our division, examined the correlation between cadmium content of suspended particulate air pollution and death rates from hypertension and arteriosclerotic heart disease. Data were available for twenty-eight cities and demonstrated a high correlation. There were no correlations with other indexes of air pollution so this is not thought to be a general air pollution effect.

Many other effects of cadmium toxicity have been demonstrated in animals, but only two others have been demonstrated in man (Tsuchiya, unpublished), decalcification and "Itai-Itai" found in Japan. Dr. Tsuchiya (1967) from Japan feels that the anemia is one of the earliest manifestations of Cd toxicity among exposed industrial workers.

As with lead there appears to be no useful purpose for cadmium in man, and the potential for harm is great. The public action program against cadmium has not yet been started, but in all probability it is not far away.

Polychlorinated Biphenyl Compounds

Even less is known about these specific particles (Lichtenstein et al., 1969; Koeman, Ten Noever de Brauw, and de Vos, 1969; Risebrough et al., 1968; Holmes, Simmons, and Tatton, 1967; and Street et al., 1969). In 1966 a group of compounds were first found in gas chromatograms of tissue of fish and wildlife. In Sweden it was determined that these materials corresponded to some of the polychlorinated biphenyls. In 1967 the same compounds were reported in fish and wildlife in Great Britain and the Netherlands.

Since the period after the Second World War, there have been declining populations of fish-eating birds in the United Kingdom and North America. These populations have produced thin-shelled eggs and apparently as a consequence have been unsuccessful in reproducing. A widespread change in the chemical environment that affected the calcium physiology of these species evidently occurred at that time. The chlorinated hydrocarbons, which came into use in the 1940s, may now be the most abundant synthetic pollutants present in the global environment. Thin eggshells have been found only in species that accumulate high concentrations of these compounds: relatively uncontaminated populations of these species continue to produce normal eggs.

In both Great Britain and North America it was the decline of the peregrine falcon which initiated concern about the extent of the harmful effects of environmental contamination with these compounds. In the United States the eastern population of falcons was extinct before competent observers were aware of a general, widespread decline. Breeding peregrines persist in apparently normal numbers in British Columbia and in the Arctic.

It was then found that several of these compounds increase the toxicity of dieldrin and DDT. Since these potentiating compounds are found in the environment, their potential effects on biological systems, especially in combination with other synthetic chemicals, cannot be disregarded.

Polychlorinated biphenyl compounds are widely used in industry as lubricants, heat transfer media, insulators, and plasticizers. The compounds are being widely disseminated in the atmosphere because many of the materials into which they are incorporated are discarded and incinerated. In the incineration process the substances volatilize in the initial heating and later condense into a fine particle that can be dispersed from the stack if there is no afterburning process.

As a means of looking into this problem, we have one study underway in DHER, and others are anticipated. One thousand human sera collected in Charleston, South Carolina, are being analyzed for polychlorinated biphenyls. Thus far, one hundred of the analyses have been completed and about one-third are showing these compounds in the 1 to 2 ppb range. This is a minimum estimate because we are using blood, which is bound to give lower yields than fat tissue, and because only two or three of the major peaks on the chromatograms are included in the tests. Soon we will start to analyze high-volume filter samples from various parts of the United States. Also, human serums are available from a study on military recruits in Chicago, and eventually they will be run for these compounds. Initial findings indicate that we must launch a more specific program to determine if specific health effects due to these compounds can be identified in man.

Community Air Pollution Health Effects Surveillance
Another aspect of the division's program for which high hopes are held is the Community Air Pollution Health Effects Surveillance (Shy, 1970). This concept has been developed because the effects of air pollution and socioeconomic levels are too frequently confounded in community studies. Industrialized, polluted areas of a city usually are inhabited by residents of lower economic levels, while the middle- and upper-middle-class populations tend to reside in the relatively clean suburban fringe. Attempts to compare populations of clean and polluted sectors of the same city too often result in a contrast between lower-class, polluted and upper-class, clean sectors.

A second problem in community studies of air pollution is the multiplicity of individual pollutants in the air of the more polluted area when clean versus polluted comparisons are made. If a health effect is more prevalent in the polluted area, it is often impossible to specify which of the many components of atmospheric pollution is responsible for the effect and should, therefore, be controlled.

The Community Air Pollution Health Effects Surveillance Program is designed to obtain comparable socioeconomic groups and to isolate the effects of individual pollutants insofar as this is possible in community studies.

Various health indicators that are effects of atmospheric pollution are being compared in trios of cities selected to represent a concentration gradient for an individual pollutant. In each trio, comparable areas with respect to socioeconomic characteristics and climatology are selected for study. One trio will be selected for each major pollutant, that is, particulate matter, sulfur dioxide, photochemical oxidants, and nitrogen dioxide. For methodologic reasons, community studies of carbon monoxide have been excluded from this list.

From the three largest American conurbations, New York, Chicago, and Los Angeles, trios of communities will be considered for study within each metropolis. These trios will not, in the case of New York City and Chicago, allow isolation of the effects of individual pollutants but will provide a framework for studying effects of pollutant combinations, when contrasted with the effects of the individual pollutants studied in the rest of the surveillance program.

The specific objectives of the program are:
1. To determine dose-response relationships for major pollutants.
2. To provide a mechanism for detecting and evaluating the effects of air pollution episodes.
3. To evaluate the effect on human health of changes in ambient air quality resulting from the introduction of air pollution control measures.

The kind of health indicators under evaluation or consideration for evaluation are:
1. Pulmonary function in schoolchildren.
2. Frequency of acute lower respiratory tract disease in infants and primary grade children.
3. Incidence of acute upper respiratory tract disease in household panels.
4. School absence rates and proportions of absences due to respiratory illness.
5. Prevalence of chronic respiratory disease symptoms in parents of high school age children.
6. Frequency of disease epidsodes in panels of subjects with asthma, chronic bronchitis, or emphysema.
7. Disease incidence and serum antibody titer in panels after exposure to major influenza epidemics.

Several of these evaluations are about to be completed and will be reported at scientific meetings at a later date.

In summary, the program of the Division of Health Effects Research is developed for the purpose of assuring that every citizen of this country can breathe the air in which he lives without fear or concern for the consequences. The administrative, technical, sociological, and political problems are complex and difficult to solve, but the dedicated staff of professional scientists working on the problems are determined, and we are certain with adequate financial support the battle will be won.

Research Needs to Delineate Impact of Air Pollution on Man and His Environment *

The fundamental air pollution effects research problem may be described as the need to quantitate exposure-dose-effect relationships in the field for air pollutants from a variety of field sources acting singly or in combination upon a variety of receptors under a variety of meteorological conditions (this section is a summary of Barth 1970b). In the general case the solution of this problem requires quantitation of the transfer functions associated with all steps of the transport of various air pollutants from their sources to some ultimate site at which they exert an effect on a receptor.

Some of the complexity of this situation is outlined in Figure 5.1. In general, air pollutants proceed from their sources via air transport to the total air environment. They then proceed via various routes as indicated to cause some ultimate total exposure, for example, to man, who is generally the primary receptor of concern. This exposure then leads to some dose to a critical portion of man's body which may be a molecule, a cell, a tissue, an organ, an organ system, or the whole body. In addition, social and economic impacts in the broadest sense must be taken into consideration to assess the total effect on man. As a minimum such consideration for social impact must include an assessment of man's perception of air pollution and the resulting effects upon his social habits, mores, attitudes, and institutions. For economic impact one must consider effects on vegetation, materials, and domestic and wild animals as well as soiling and general aesthetic effects resulting in economic cost to man.

As can be seen in Figure 5.1, recycling occurs at various places. Also in the general case the air pollutants can undergo complex time- and space-dependent physical and chemical transformations throughout the entire process. In addition various concentrating mechanisms can occur.

To fully establish air pollution control needs that are based on effects on man and are scientifically defensible, it is necessary to quantitate each transfer function associated with the various steps outlined in the schematic diagram. Obviously, an extensive long-range systems approach is needed to insure that all required

---

* Summary of a presentation given by Dr. D. S. Barth, Director (1970b), BCS, NAPCA at the 1970 NAPCA Program Planning Conference, June 2–4, 1970.

95 Health Effects Research Program of NAPCA

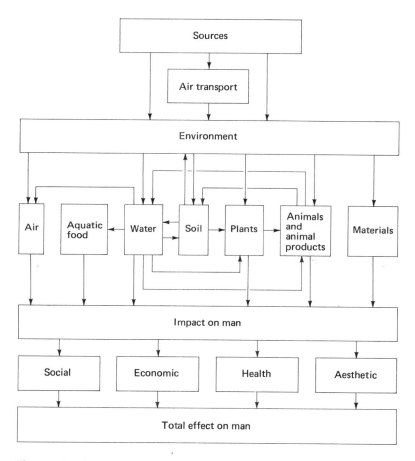

Figure 5.1 Delineation of pathways by which air pollution impacts on man and his environment

research will be accomplished in an orderly and timely fashion. It should be emphasized that it is not always necessary to understand fully the details of each and every mechanism taking place within each representative box: In many cases an input-output quantitative analysis may suffice.

The foregoing discussion has given no specific regard to total ecological impact of air pollutants, a problem orders of magnitude more difficult but one that must ultimately be addressed and solved in order to assess the total environmental impact of air pollutants.

References

Albert, R. E., Lippmann, M., Spiegelman, J., Liuzzi, A., and Nelson, N., 1967. The deposition and clearance of radioactive particles in the human lung, *Archives of Environmental Health, 14:* 10.

Barth, D. S., 1970a. Suggested response to question concerning removal of lead from gasoline, dated March 27, 1970, memorandum from D. S. Barth, Director, Bureau of Criteria and Standards, National Air Pollution Control Administration, to Leighton A. Price, Director, Office of Education and Information, National Air Pollution Control Administration.

Barth, D. S., 1970b. Presentation at the 1970 NAPCA Program Planning Conference, June 2–4, 1970.

Bonneli, J. A., 1965. Cadmium poisoning, *The Annals of Occupational Hygiene, 8:* 45.

Carroll, R. E., 1966. The relationship of cadmium in the air to cardiovascular disease death rates, *Journal of the American Medical Association, 198:* 177.

Hewitt, E. J., and Nicholas, D. J. D., 1963. Cations and Anions: inhibitions and interactions in metabolism and in enzyme activity, *Metabolic Inhibitors,* vol. 2, edited by R. M. Hochster and J. H. Quasel (New York: Academic Press), pp. 311–436.

Hill, A. B., 1962. The statistician in medicine, *Journal of the Institute of Actuaries, 88:* 178.

Holmes, D. C., Simmons, J. H., and Tatton, J. O'G., 1967. Chlorinated hydrocarbons in British wildlife, *Nature, 216:* 227.

Kehoe, R. A., 1961a. The Harbom lectures, 1960. The metabolism of lead in man in health and disease (lectures 1, 2b, and 3), *Journal of the Royal Institute of Public Health and Hygiene, 24:* 81–97, 129–143, 177–203.

Kehoe, R. A., 1961b. The Harbom lectures, 1960. The metabolism of lead in man in health and disease (lecture 2a), *Journal of the Royal Institute of Public Health and Hygiene, 24:* 101.

Kehoe, R. A., 1966. *Proceedings of the 15th International Congress on Occupational Health,* vol. 3 (Vienna), p. 83.

Koeman, J. H., Ten Noever de Brauw, M. C., and de Vos, R. H., 1969. Chlorinated biphenyls in fish mussels and birds from the river Rhine and the Netherlands coastal area, *Nature, 221:* 1126.

Lichtenstein, E. P., Schulz, K. R., Fuhremann, T. W., and Liang, T. T., 1969. Biological interaction between plasticizers and insecticides, *Journal of Economic Entomology, 62:* 761.

Lutmer, R. F., Busch, K. A., and Miller, R. G., 1967. Lead from automobile exhaust: effect on mouse bone lead concentration, *Atmospheric Environment, 1:* 585.

Nilsson, R., 1970. Aspects on the toxicity of cadmium and its compounds, *Ecological Research Committee, Bulletin No. 7* (Stockholm: Swedish Natural Science Research Council).

Nozaki, K., 1966. Method for studies on inhaled particles in human respiratory system and retention of lead fumes, *Industrial Health, 4:* 118.

Risebrough, R. W., Rieche, P., Peakall, D. B., and Herman, S. G., 1968. Polychlorinated biphenyls in the global ecosystem, *Nature, 220:* 1098.

Robinson, E., and Ludwig, F. L., 1967. Particle size distributions of urban lead aerosols, *Journal of the Air Pollution Control Association, 17:* 664.

Schroeder, H. A., 1964. Cadmium hypertension in rats, *American Journal of Physiology, 207:* 52.

Schroeder, H. A., and Vinton, W. H., 1962. Hypertension induced in rats by small doses of cadmium, *American Journal of Physiology, 202:* 515.

Shy, Carl M., 1970. Health effects surveillance network, In-house report to the Division of Health Effects Research, National Air Pollution Control Administration, February 1970.

Street, J. C., Urry, F. M., Wagstaff, D. J., and Blau, A. D., 1969. Paper presented at the 158th Meeting of the American Chemical Society (Pesticide Chemistry Division), September 8, 1969.

Task Group on Lung Dynamics, 1966. Deposition and retention models for internal dosimetry of the human respiratory tract, *Health Physics, 12:* 173.

Tsuchiya, K., 1967. Protein uria of workers exposed to cadmium fume, *Archives of Environmental Health, 14:* 875.

Tsuchiya, K. Unpublished preliminary report.

Working Group on Lead Contamination, 1965. Public Health Service Publication No. 999-AP-12 (Washington, D.C.: U.S. Government Printing Office).

U.S. Government, 1963. *The Clean Air Act.* Public Law 88-206, 88th Congress H.R. 6518, December 17, 1963. Revised as *The Clean Air Act of 1967,* Public Law 90-148. 90th Congress, S-780, November 21, 1967 (Washington, D.C.: U.S. Government Printing Office).

U.S. Government, 1970. *National Environmental Policy Act of 1969,* Public Law 91-190, 91st Congress, S-1075, January 1, 1970 (Washington, D.C.: U.S. Government Printing Office).

Part II   Pollution and
                                            Terrestrial Ecosystems

Pollution can have both direct and indirect effects on terrestrial ecosystems and their components. In this section, some of the direct effects of air pollutants and pesticides will be treated, and the indirect effects caused by climatic change will be discussed in Part III. Although the literature contains many references to acute damage to vegetation resulting from pollution, relatively little is known about the less obvious chronic effects. The present state of knowledge is reviewed in the following four papers—three of which were prepared especially for SCEP by scientists in the U.S. Department of Agriculture.

The implications of chronic injury to crop plants are particularly serious in terms of agricultural losses and the expenses involved in breeding resistant plants. In the first paper of this series, Dr. Heggestad summarizes the literature dealing with the effects of air pollution on cultivated plants, evaluates the state of the art, and identifies areas where further research is required.

Trees are dominant components in the global mass of living matter, and they greatly influence the climate and the hydrologic cycle on land. Dr. Hepting's paper discusses the effects of air pollution on trees at all levels of analysis—from individual leaf damage to regional changes in climate. In addition to noting how trees serve as ameliorators of pollution, the paper presents the intriguing prospects for using sensitive genetic strains as air pollution monitors and for breeding resistant strains for planting in urban environments.

In her paper Dr. Dixie Smith has pulled together scattered information on a wide variety of pollutants to produce this first survey of the effects of pollution on rangelands. Her treatment of the role of rangelands as sources of pollution complements the papers by Drs. Harrold and Lull in Part IV of this volume. It will be noted that although the information on herbicides, sediments, and to some degree radionuclides is reasonably good, much more work is required for these and other pollutants.

The ecological effects of pesticides, good and bad, have been the subject of many papers. Too often the bias of the writer influences the selection of references, resulting in a less than ideally objective analysis. The paper by Dr. Moore, reprinted here in part, is as close as any to being an impartial review while at the

100 Pollution and Terrestrial Ecosystems

same time treating the subject in a thorough fashion. It is regrettable that the length of the original, published paper required the deletion of several sections, since all of them are informative and each contributes to a balanced treatment. The sections selected for inclusion in this volume are closest in subject matter to the more general, global concern expressed in the report of the SCEP Work Group on Ecological Effects.

6
Air Pollution and Plants   H. E. Heggestad

This paper is primarily a review of air pollution problems caused by gaseous and particulate pollutants on cultivated plants. The effects of pesticides or radioactive materials are not included. Some thought was given to the possibility of global air pollution problems on vegetation, but evidence is lacking that global transfer of pollutants is sufficient to injure vegetation. There is some concern about effects in the future on vegetation due to changes in the distribution of rainfall and alteration of temperature that may result from global increase in levels of carbon dioxide and particles.

Movement of pollutants across state and national boundaries has caused serious vegetation problems. Industrial developments and the demands of people everywhere for the same standard of living will result in more and similar air pollution problems on vegetation in all countries. For example, sulfur dioxide and fluorides injure vegetation in many countries. Now there is an awareness that injury to vegetation by photochemical smog may be a worldwide problem. The severity of this latter type of air pollution in any given area will be related, primarily, to density of motor vehicles and existence of meteorological parameters favorable to the accumulation of pollutants and to photochemical reactions.

Only in the past decade in the United States did we recognize that photochemical oxidant injury to vegetation was a problem except in California. There is some evidence of greater damage to agriculture and ornamental plantings in the eastern United States than in California (Middleton, 1961). Agricultural losses due to air pollution are an estimated $500 million annually and are increasing (Heggestad, 1968). Most of the loss is due to growth suppression or to chronic injury as indicated by relative growth of plants in ambient air containing pollutants and carbon-filtered air with most oxidants and some other pollutants removed.

Photochemical Oxidants
Fuel combustion, especially the burning of gasoline in motor vehicles, results in emission of many compounds, including hydro-

Prepared for SCEP.

carbons and nitrogen oxides. The action of sunlight on these compounds causes the production of very toxic pollutants, such as ozone and peroxyacetyl nitrate (PAN).

In 1944 Middleton, Kendrick, and Schwalm (1950) recognized vegetation injury from photochemical smog in the Los Angeles basin. The effects of this type of pollution on plants were reviewed by Middleton (1961), Darley et al. (1963), and Taylor (1968). Research by Haagen-Smit et al. (1952) established the role of automobile exhaust in the air pollution problem. Considering effects on vegetation, the major toxicants in the oxidant complex are ozone, peroxyacetyl nitrate (PAN), nitrogen dioxide, and various aldehydes. Each causes somewhat different plant responses. When total oxidants in ambient air are high, ozone may approach 99 percent or more of the total (Cherniak and Bryan, 1965; Renzetti and Romanovsky, 1956). As ozone levels increase, the PAN levels also increase (Tingey and Hill, 1967). Some unidentified toxicants may be present in the oxidant complex. Although their concentration may be too low to cause injury as a single toxicant, possible additive, synergistic, or antagonistic effects need to be considered.

White-flowered petunias cv. (White Cascade), tobacco variety Bel-W3, and pinto beans are very sensitive to the pollutants in photochemical smog and are useful bioindicators (Heggestad and Darley, 1969). Undersurface glazing of leaves is a typical symptom of PAN. The tobacco variety Bel-W3 is, perhaps, the most sensitive biologic indicator of ozone (Taylor, 1968).

A wide range exists in symptomatology, depending upon the host plant and mixtures of pollutants present. Ozone was primarily responsible for an unusual disorder of lilacs in New York City, characterized primarily by leaf roll and bronzing (Hibben and Walker, 1966). Clones from lilacs with the leaf roll symptoms produced healthy foliage when removed from New York City, but parent lilacs in Brooklyn continued to develop typical symptoms. Use of carbon filters reduced the injury to a light, abaxial bronzing.

Activated carbon filters are excellent for removing photochemical oxidants from enclosed areas. Such air filtration usually prevents visual injury to plants and permits more rapid growth than comparable plants grown in unfiltered air (Darley and Middleton, 1961; Hull and Went, 1952; Menser, Heggestad, and Grosso, 1966; Taylor, 1958). At Beltsville, Maryland (Heggestad, 1970), leaf injury, attributed to oxidants, developed to a varying degree on eight potato (*Solanum tuberosum* L.) varieties grown in ambient air but not in carbon-filtered air. In addition, tuber yields of the varieties Haig, Irish Cobbler, and Norland were about 50 percent lower in the ambient air. These varieties also were the most sensitive to ozone in controlled fumigations. Yields of Avon, Superior, and Peconic in ambient air were reduced 11, 20, and 24 percent, respectively. Kennebec and Alamo were the most tolerant to the air pollution and produced about the same yield in both environments. About 50 percent reduction in yields have been reported from California in studies with citrus and grapes (Thompson and Taylor, 1969; *California Air Environment*, 1970).

Ozone

The first report of ozone damage to a specific crop was on grapes in 1957 in California (Richards, Middleton, and Hewitt, 1958). In 1959 it was reported that the air pollutant ozone caused weather fleck, a leaf spot problem, on cigar-wrapper tobacco in the eastern United States (Heggestad and Middleton, 1959). In 1959 losses on tobacco in the Connecticut Valley were an estimated $5 million, and most losses occurred over a single weekend. About twenty-four hours are required for development of the stipple, or fleck, lesion after an air pollution episode. For some unknown reason, the palisade cells near the upper surface of leaves are first to be injured although ozone enters primarily through the stomata on the lower surface. The partially expanded youngest leaves are resistant. In the laboratory the tobacco variety Bel-W3 (Heggestad and Menser, 1962) is injured after exposure for four hours at 0.05 part per million or two hours at 0.10 ppm. Alfalfa, spinach, pinto beans, potatoes, tobacco, lilacs, and cereal crops are among the species most sensitive to ozone. Varietal differences within species in sensitivity to ozone are a common characteristic; consequently, the selection and planting of varieties with most resistance to ozone injury is an effective method to reduce losses. Without knowledge of air pollutants, plant breeders, for example, of cigar-wrapper tobacco, have increased the tolerance of plants to pollutants by selecting for highest productivity and least leaf injury.

Few studies have been conducted to determine effects of ozone on fungi and development of plant diseases. Hibben (1966) found that 0.1 ppm ozone is toxic to moist fungus spores of some species. Exposures to 0.5 and 1 ppm reduced or prevented germination on all species tested. Comparable exposure to 10 ppm sulfur dioxide had no effect, and spores of some species were viable after 60- and 100-ppm doses.

Recent results reveal that ozone reduces development of an obligate parasite, such as crown rust of oats, but it increases development of a weak parasite, such as *Botrytis* on potatoes and geranium. Heagle (1970) exposed different varieties of oats to 0.1 ppm ozone for ten days after infection with crown rust (*Puccinia coronata*). Pustules containing urediospores were much reduced by exposure to this low level of ozone. It is not known whether the effect was directly on the fungus or resulted indirectly from changes in the host plant. When Manning, Feder, and Perkins (1970) exposed geranium leaves to ozone before inoculating with *Botrytis cinera,* the amount of infection increased. Apparently, the lesions caused by ozone served as infection courts for the fungus.

Peroxyacetyl Nitrate (PAN)

The identification in 1960 (Stephens et al., 1961) of PAN as the cause of a type of plant injury previously attributed to ozonated hydrocarbons was a significant development. PAN, as well as ozone, is formed by action of sunlight on nitrogen oxides and simple olefins.

The distinguishing characteristics of PAN injury are glazing and bronzing of the lower leaf surfaces, attack of young leaves, and a tendency to produce transverse banding on individual leaves (Stephens et al., 1961). Chlorosis, growth retardation, and leaf abscission are caused by PAN as well as ozone.

Field levels of PAN between 0.02 and 0.05 ppm for a few hours injure sensitive plants (Tingey and Hill, 1967). Although PAN seems to be the principal toxicant in a homologous series, some closely related compounds in the series are four to eight times more toxic (Taylor, 1968). Romaine lettuce, Swiss chard, pinto beans, petunias, and African violets are sensitive species.

Nitrogen Dioxide

Interest in nitrogen dioxide as a phototoxicant is of recent origin.

In 1966 Taylor and Eaton (Taylor and Eaton, 1966) found a marked depression in growth when pinto beans and tomatoes were exposed to 0.5 ppm nitrogen dioxide for ten to twelve days. The stunted plants were a darker green, indicating the possibility of increased accumulation of nitrogen. At higher concentrations, greater than 3 ppm, the injury on these species resembled necrotic lesions caused by sulfur dioxide and high levels of ozone (Taylor, 1968; Taylor and Eaton, 1966).

Ethylene
Orchids are reported to be injured by ethylene concentrations as low as .04 ppm for eight hours (Davidson, 1949) and carnations by 0.1 ppm for six hours (Darley, Nichols, and Middleton, 1966). Ethylene is a plant hormone; however, when present in air, at indicated or higher levels, symptoms on plant species may include epinasty, chlorosis, necrosis, leaf and bud abscission, and failure of flower buds to open. Injury to cotton in the field, similar to that on roses and other plants in greenhouses, resulted when ethylene was released in the manufacture of certain plastics. Heck, Pires, and Hall (1962) determined the response of eighty-nine plant species to ethylene and reported that cowpeas, privets, peas, blackberries, and roses were among the most sensitive.

The variation in injury patterns on plants exposed to irradiated automobile exhaust suggests presence of additional phytotoxicants in photochemical smog. Typical PAN and ozone injury was identified, but, in addition, Hindawi, Dunning, and Brandt (1965) noted bronzing on the upper leaf surface of young leaves and meristematic activity between the lower epidermis and mesophyll layer of leaves. The cause of the latter symptoms is not known.

Sulfur Dioxide
Sulfur dioxide is emitted as a consequence of burning sulfur-containing fuels, such as coal, and by smelting of ores. Damage to vegetation is common but not as severe as in the early part of this century. Several thousand acres near Copper Hill, Tennessee, are still a barren wasteland due to sulfur dioxide pollution that occurred more than fifty years ago. When concentrations of sulfur dioxide in leaf tissue are low, it is oxidized to the relatively non-

toxic sulfate form; however, when concentrations are high, the more toxic sulfite ion accumulates. Effects of sulfur dioxide on vegetation were reviewed recently by Daines (1968), Wood (1968a), and Barrett (1968).

The plants most sensitive to sulfur dioxide tend to have succulent leaves and a rapid growth rate. Alfalfa, barley, and cotton are among the most sensitive species. Injury may develop on sensitive species following exposure to 0.48 ppm for four hours or 0.28 ppm for twenty-four hours. In a recent study on white pine, current needles between four and five weeks old were acutely injured by SO_2 at dosages as low as 0.05 ± 0.015 ppm for one hour (Castonis, 1970). This is lower than established for any other vegetation.

Fluoride
Fluoride acts as a cumulative poison on some plant species. Other species accumulate fluoride without apparent injury. Fluoride is adsorbed and rapidly translocated to leaf tips and margins (Jacobson et al., 1966). The toxicant remains in a soluble form and seems to retain the chemical properties of free inorganic fluoride. Levels of 0.5 ppb fluoride in air for prolonged periods will cause injury to several species of plants. Under some conditions, a sensitive variety of gladiolus may develop necrosis at the leaf tip and margins as soon as 20 ppm fluoride accumulates. A tolerant plant species, such as cotton, may show no injury even though leaves contain 4,000 ppm of the toxicant (Jacobson et al., 1966). Depending on dosage, injury may appear in a few hours, but injury develops most commonly after several weeks of exposure. Analysis of leaf tissue for fluoride content is essential to diagnosis of a fluoride problem.

Among the plant species most sensitive to fluoride injury are certain varieties of gladioli, apricots, prunes, corn, and grapes (Hill, 1969). Celery, alfalfa, tomatoes, tobacco, and some other species are resistant to fluoride but susceptible to sulfur dioxide (Zimmerman and Hitchcock, 1956).

Other Pollutants
Accidental spills of chemicals, such as chlorine, from manufacturing operations have resulted in vegetation damage. Apparently

the relative toxicity of the minor pollutants, chlorine, ammonia, hydrogen cyanide, and hydrogen sulfide, are in the order listed (Thornton and Satterstrom, 1940). Radishes and alfalfa were the most sensitive to chlorine of twenty-six species tested by Brennan, Leone, and Daines (1965). Injury from chlorine develops after exposure of sensitive species to 0.10 ppm chlorine for two hours. Carbon monoxide has very little effect on plant foliage except at about 1,000 ppm (Zimmerman, Crocker, and Hitchcock, 1933). Nitrogen fixation by red clover, however, is retarded by 100 ppm carbon monoxide and almost stopped by 500 ppm (Lind and Wilson, 1941). Mercury vapors at concentrations of 8 ppb have been known to be injurious to roses (Hitchcock and Zimmerman, 1957).

Combined Effect of Two or More Gases
Although rarely is just a single phytotoxicant present in polluted air, there has been relatively little research to determine the combined effects on plants of two or more pollutants. Evidence of the interference of sulfur dioxide with other air pollutants was reported in 1952 by Haagen-Smit et al. (1952). No plant injury developed after exposure to mixed air containing 0.1 ppm sulfur dioxide and various toxic levels of the reaction products of ozone and gasoline. Middleton, Darley, and Brewer (1958) found no sulfur dioxide interference with injury induced by ozone. The sulfur dioxide and ozone levels were 1.5 and 0.3 ppm, respectively. However, when the sulfur dioxide–ozone ratio in these experiments was 4:1 or less, they reported that ozone seemed to interfere with the development of sulfur dioxide injury. Symptoms of both gases developed on the same leaves if the sulfur dioxide–ozone ratio was increased to 6:1 or more.

Menser and Heggestad (1966) in 1966 obtained evidence of synergistic action of sulfur dioxide and ozone in studies with tobacco. The level of ozone (0.03 ppm) was about one-tenth of levels in the California studies (Middleton, Darley, and Brewer, 1958). No visible injury developed after exposing tobacco to the individual gases at the concentrations present in mixed air. Synergistic action of sulfur dioxide and ozone helped provide an explanation for tobacco injury in spring and fall when ozone levels were too low.

Significance of Environmental Factors

The more important environmental factors affecting plant response are light intensity, light quality, photoperiod, temperature, humidity, wind, soil moisture, soil texture, and nutrient levels (Leone, Brennan, and Daines, 1965; Heck, Dunning, and Hindawi, 1965; Heck, 1968). Plants grown under eight-hour photoperiods are more sensitive to ozone than those grown under longer photoperiods. Injury tends to increase to a maximum, with increased humidity and intensity of light during exposure.

Biochemical and physiological effects of air pollutants are the subject of much research. Considering photochemical oxidants, there are many significant reports, such as studies by Mudd (1963) and by Hill and Littlefield (1969). Dugger, Koukol, and Palmer (1966) reviewed the information in 1966.

Although more effort will be made to control air pollution at the source, we can anticipate continuing and serious problems. Sources of plant pathogenic pollutants and future trends to the year 2000 are discussed by Wood (1968). We may ask what can be done to minimize losses to food, fiber, and ornamental plants. Air pollution injury to vegetation would be more apparent and serious now, except for the continuing selection and propagation of plants that show least injury. Injured plants and those that are least productive are automatically eliminated from culture by breeders, horticulturists, homeowners, and so on, even though the causes of injury and reduced growth are unknown. Air pollution losses can be reduced and production increased by greater application of plant breeding procedures to crop and ornamental species known to be damaged by pollutants.

In addition to the effects of pollutants on crop plants, there is increased concern about effects on natural plant communities. This information was reviewed by Treshow (1968).

Summary and Discussion of Needs for Research*

The problems caused by air pollution on vegetation are variable in severity and nature in different regions of the country because

---

* Developed by the staff, Plant Air Pollution Laboratory, Crops Research Division, as a part of a joint planning effort in cooperation with the Agricultural Branch, National Air Pollution Control Administration. I wish to acknowledge the interest and encouragement of W. C. Shaw, Crops Research Division, in our identifying these research needs.

of the variation in (1) the amount and kinds of pollutants present; (2) the meteorology that influences photochemical reactions and either favors the accumulation or dispersal of pollutants; and (3) the kinds, distribution, and relative sensitivity of plants in the area. Furthermore, the sensitivity of plant species to pollution injury is controlled by genetic factors, but this sensitivity may be altered by cultural conditions and environmental factors. Southern California has the highest concentration of photochemical oxidants in the country. However, the eastern United States may suffer greater losses because a greater land area is involved. Also, in this latter area moisture conditions tend to be more favorable for plant growth and for increasing the sensitivity of plants to pollutants. Some locations in the country have acute problems on vegetation because crops or plants are grown in close proximity to industrial operations that emit toxic pollutants, such as sulfur dioxide and fluorides.

In order of significance to vegetation, the ten most important pollutants or groups of pollutants are

1. Ozone
2. Sulfur dioxide
3. Peroxyacyl nitrates (PANs)
4. Ethylene and other hydrocarbons
5. Fluoride
6. Pesticides
7. Chlorine and hydrogen chloride
8. Nitrogen dioxide
9. Heavy metals
10. Particles

Other pollutants of concern are acid aerosols, ammonia, hydrogen sulfide, aldehydes, carbon monoxide, large air ions, and radioactive materials.

The rating of the relative importance of these pollutants to vegetation injury is based on very little information. Even less is known about the importance of various mixtures of the gaseous pollutants, such as commonly exist in community atmospheres.

The pollutants can be grouped into primary and secondary pollutants, depending on whether they originate from a point source or are formed in the atmosphere by photochemical reactions that involve nitrogen oxides and certain reactive hydrocarbons. Ozone, a secondary pollutant, is a powerful oxidant and extremely toxic to vegetation. Concentrations sufficiently high to injure sensitive plants seem to be present in most states, at least for a few days during summer. We rated ozone as the single most important pollutant affecting vegetation because of increased evidence of growth suppression, as well as acute injury to the plant.

Some air pollutants are of concern because they enter the food chain. This may occur with or without visible injury to vegetation. These include fluorides, various heavy metals such as lead, arsenic, beryllium, cadmium, chromium, nickel, and vanadium, and pesticides and radioactive materials. There is need for more research to determine possible reduction in quality of foods that results from the altered metabolism in plans following pollutant uptake.

This review has focused attention on these pollutants that affect plants. There are air pollutants, such as pollens and spores, originating from plants. These cause serious allergic reactions in a portion of the population.

On the plus side, we recognize that plants generate life-sustaining oxygen and also remove undesirable pollutants from the atmosphere. If the concentrations are not too high, it seems that plants may utilize for growth such pollutants as sulfur dioxide and nitrogen dioxide. Carbon dioxide is increasing in the atmosphere because of fuel combustion. Plants may benefit from the increased amounts of carbon dioxide. Under experimental conditions, young seedlings of herbaceous plants respond remarkably well to additions of CO_2 up to at least 2,000 ppm. Increased height of plants, growth of lateral buds, and flowering are associated with CO_2 enrichment (Krizek, Bailey, and Kleuter, 1968).

Areas of research needing most attention are as follows:
1. Determine with controlled experiments and careful observations the economic impact of air pollution on crop and ornamental plants.
2. Field studies: (a) Detailed studies of key field situations are needed for both cultivated and natural occurring plant communities, including studies on monitoring of pollutants. In conjunction with these investigations, data should be obtained on (i) significant meteorological parameters, such as prevailing wind

direction and speed during high pollution episodes; (ii) the determination of yield and quality of plants; and (iii) changes in plant populations which are related to air pollutants. Special attention should be given to the response of plants to air pollutants during different stages of their growth.

(b) The monitoring of total oxidants or, if possible, ozone specifically, and in some areas sulfur dioxide, should be sufficiently extensive to identify problem areas in rural America and serve as background information on which to base the extent that air quality may change (improve or deteriorate) in the future even in regions most remote from urban and industrial situations. Information on pollution concentrations may also be a basis for selecting the kind of plants to grow in an area. For example, the most sensitive truck crops may be grown more profitably to the south and west of urban centers because pollution levels tend to be greater on the north and east of the urban centers. The prevailing wind is from the south or southwest during pollution episodes that involve photochemical oxidants in eastern United States.

3. Determine the relative injury to major crop, ornamental, and shade tree species caused by pollutants, singly and in various combinations. Information is needed on response of more species and varieties within species, with emphasis on response to pollutant concentrations in the upper range in areas where the species may be grown.

4. Determine effects of nutrition and cultural practices on altering plant response to pollutants. Deficiency or excess of certain mineral elements alters plant physiology and, to some extent, changes the response of plants to pollutants. More attention needs to be given to the water relations of plants because both air and soil moisture affect plant responses. Also, pollution injury may alter the uptake and utilization of water.

5. Determine pathogen-pollutant interactions with broadly representative pathogens and plant species. Little is known about the extent air pollution may increase or decrease injury to plants from diseases, or how diseases may alter plant response to pollutants.

6. Special studies are needed to determine the biochemical and physiological mechanisms of injury to vegetation by different

pollutants. Results of such studies may expedite the development of chemical and genetic controls. Studies on the nature of resistance of plants to pollutants may be very productive.

7. Develop varieties in major plant species that will be more tolerant of air pollution. This tolerance, or resistance, to damage by pollution should be coupled with resistance to disease and to the incorporation of desirable horticultural or agronomic qualities into the species.

8. Evaluate the extent of cleansing action found in plant species to particulate and gaseous pollutants, taking into account environmental factors, such as the effect wind speed has on absorption of pollutants by leaves.

9. Determine effects of particulate pollutants on crop production, giving attention to possible effects of reduced light on growth and subsequent yield, as well as on the altered market value due to presence of various deposits. Some attention should also be given to the interacting effects of particulate and gaseous pollutants, such as ozone. The presence of the particles on leaves might reduce injury by ozone.

10. There is need for more information on ecological changes related to air pollutants. The response of representative wild plants to various pollutants should be known.

11. Global monitoring of certain gases would be desirable to determine fate of pollutants and to detect significant environmental changes. In addition to carbon dioxide and carbon monoxide, we need more information on global dispersal of ozone, sulfur dioxide, acid aerosols, and fluorides.

The eleven research areas are not listed in a rigid order of priority, and they are only an estimation of the research needs on vegetation. In some areas of the country specialized research needs exist.

References

Barrett, Thomas W., 1968. Air quality standards for sulfur dioxide in agriculture and forestry, paper presented at Sixty-first Annual Meeting of the Air Pollution Control Association, Saint Paul, Minnesota.

Brennan, E. G., Leone, I. A., and Daines, R. H., 1965. Chlorine as a phytotoxic air pollutant, *International Journal of Air and Water Pollution, 9:* 791–797.

California Air Environment, 1970. Photochemical smog reduces grape yields 60%, p. 3.

Castonis, A. C., 1970. Acute foliar injury of eastern white pine induced by sulfur dioxide and ozone, *Phytopathology, 60:* 994–999.

Cherniak, I., and Bryan, R. J., 1965. A comparison of various types of ozone and oxidant detectors which are used for atmospheric air sampling, *Journal of the Air Pollution Control Association, 15:* 351–354.

Daines, R. H., 1968. Sulfur dioxide and plant response, *Journal of Occupational Medicine, 10:* 84–92.

Darley, E. F., and Middleton, J. T., 1961. Carbon filter protects plants from damage by air pollution, *Florists Review, 127:* 15–16, 43, 45.

Darley, E. F., Dugger, W. M., Jr., Mudd, J. B., Ordin, L., Taylor, O. C., and Stephens, E. R., 1963. Plant damage by pollution derived from automobiles, *Archives of Environmental Health, 6:* 761–770.

Darley, E. F., Nichols, C. W., and Middleton, J. T., 1966. Identification of air pollution damage to agricultural crops, *California Department of Agriculture Bulletin, 55:* 11–19.

Davidson, O. W., 1949. Effects of ethylene on orchid flowers, *Proceedings of the American Society for Horticultural Science, 53:* 440–466.

Dugger, W. M., Jr., Koukol, Jane, and Palmer, R. L., 1966. Physiological and biochemical effects of atmospheric oxidants on plants, *Journal of the Air Pollution Control Association, 16:* 467–471.

Haagen-Smit, A. J., Darley, E. F., Zaitlin, M., Hull, H., and Noble, W., 1952. Investigation on injury to plants from air pollution in the Los Angeles area, *Plant Physiology, 27:* 18–34.

Heagle, A. S., 1970. Effect of low-level ozone fumigations on crown rust of oats, *Phytopathology, 60:* 252–254.

Heck, W. W., 1968. Factors influencing expression of oxidant damage to plants, *Annual Review of Phytopathology, 6:* 165–188.

Heck, W. W., Pires, E. G., and Hall, W. C., 1962. *Effect of Ethylene on Horticultural and Agronomic Plants*, MP-613 (College Station: Texas Agricultural Experiment Station), pp. 1–12.

Heck, W. W., Dunning, J. A., and Hindawi, I. J., 1965. Interactions of environmental factors on the sensitivity of plants to air pollution, *Journal of the Air Pollution Control Association, 15:* 511–515.

Heggestad, H. E., 1968. Diseases of crops and ornamental plants incited by air pollutants, *Phytopathology, 53:* 1089–1097.

Heggestad, H. E., 1970. Variation in response of potato cultivars to air pollution (Abstract), *Phytopathology, 60:* 1015.

Heggestad, H. E., and Middleton, J. T., 1959. Ozone in high concentrations as a cause of tobacco leaf injury, *Science, 129:* 208–210.

Heggestad, H. E., and Menser, H. A., 1962. Leaf spot sensitive tobacco strain Bel-W3, a biological indicator of the air pollutant ozone (Abstract), *Phytopathology, 52:* 735.

Heggestad, H. E., and Darley, E. F., 1969. Plants as indicators of the air pollutants ozone and PAN, *Proceedings of First European Congress on the Influence of Air Pollution on Plants and Animals* (Wageningen, The Netherlands: Centre for Agricultural Publishing and Documentation).

Hibben, C. R., 1966. Sensitivity of fungal spores to sulfur dioxide and ozone (Abstract), *Phytopathology, 56:* 880.

Hibben, C. R., and Walker, J. T., 1966. A leaf roll-necrosis complex of lilacs in an urban environment, *Proceedings of the American Society for Horticultural Science, 89:* 636–642.

Hill, A. C., 1969. Air quality standards for fluoride vegetation effects, *Journal of the Air Pollution Control Association, 19:* 331–336.

Hill, A. C., and Littlefield, N., 1969. Ozone: effect on apparent photosynthesis, rate of transpiration, and stomatal closure in plants, *Environmental Science and Technology, 3:* 52–56.

Hindawi, I. J., Dunning, J. A., and Brandt, C. S., 1965. Morphological and microscopical changes in tobacco, bean, and petunia leaves exposed to irradiated auto exhaust, *Phytopathology, 55:* 27–30.

Hitchcock, A. E., and Zimmerman, P. W., 1957. Toxic effects of vapors of mercury and of compounds of mercury on plants, *Annals of New York Academy of Sciences, 65:* 474–497.

Hull, H. M., and Went, F. W., 1952. Life processes of plants as affected by air pollution, *Proceedings of the Second National Air Pollution Symposium* (Los Angeles: Stanford Research Institute), pp. 122–123.

Jacobson, J. S., Weinstein, L. H., McCune, D. C., and Hitchcock, A. E., 1966. The accumulation of fluorine by plants, *Journal of the Air Pollution Control Association, 16:* 412–417.

Krizek, D. T., Bailey, W. A., and Kleuter, H. H., 1968. Accelerated growth of seedlings and precocious flowering under controlled environments at elevated CO_2, light, and temperature, *Plant Physiology, 43* (Supplement): S32.

Leone, I. A., Brennan, E. G., and Daines, R. H., 1965. Factors influencing SO_2 phytotoxicity in New Jersey, *Plant Disease Reporter, 49:* 911–915.

Lind, C. J., and Wilson, P. W., 1941. Mechanism of biological nitrogen fixation. VIII. Carbon monoxide as an inhibitor for nitrogen fixation by red clover, *Journal of the American Chemical Society, 63:* 3511–3514.

Manning, W. J., Feder, W. A., and Perkins, J., 1970. Ozone injury increases infection of geranium leaves by *Botrytis cinera*, *Phytopathology, 60:* 669–670.

Menser, H. A., and Heggestad, H. E., 1966. Ozone and sulfur dioxide synergism injury to tobacco plants, *Science, 153:* 424–425.

Menser, H. A., Heggestad, H. E., and Grosso, J. J., 1966. Carbon filter prevents ozone fleck and premature senescence of tobacco leaves, *Phytopathology, 56:* 466–467.

Middleton, J. T., 1961. Photochemical air pollution damage to plants. *Annual Review of Plant Physiology, 12:* 431–448.

Middleton, J. T., Kendrick, J. B., and Schwalm, H. W., 1950. Injury to herbaceous plants by smog or air pollution, *Plant Disease Reporter, 34:* 245–252.

Middleton, J. T., Darley, E. F., and Brewer, R. F., 1958. Damage to vegetation from polluted atmospheres, *Journal of the Air Pollution Control Association, 8:* 9–15.

Mudd, J. B., 1963. Enzyme inactivation by peroxyacetyl nitrate, *Archives of Biochemistry and Biophysics, 102:* 59–65.

Renzetti, N. A., and Romanovsky, J. C., 1956. A comparative study of oxidants and ozone, *Archives of Industrial Health, 14:* 458–467.

Richards, B. L., Middleton, J. T., and Hewitt, W. B., 1958. Air pollution with relation to agronomic crops. V. oxidant stipple of grape, *Agronomy Journal, 50:* 559–561.

Stephens, E. P., Darley, E. F., Taylor, O. C., and Scott, W. E., 1961. Photochemical reaction products in air pollution, *International Journal of Air and Water Pollution, 4:* 79–100.

Taylor, O. C., 1958. Air pollution with relation to agronomic crops. IV. Plant growth suppressed by exposure to airborne oxidants (smog), *Agronomy Journal, 50:* 556–558.

Taylor, O. C., 1968. Effects of oxidant air pollutants, *Journal of Occupational Medicine, 10:* 485–499. (Includes discussion by H. E. Heggestad and W. W. Heck.)

Taylor, O. C., and Eaton, F. M., 1966. Suppression of plant growth by nitrogen dioxide, *Plant Physiology, 41:* 132–136.

Thompson, C. R., and Taylor, O. C., 1969. Effects of air pollutants on growth, leaf drop, fruit drop, and yield of citrus trees, *Environmental Science and Technology, 10:* 934–940.

Thornton, N. C., and Satterstrom, C., 1940. Toxicity of ammonia, chlorine, hydrogen cyanide, hydrogen sulphide, and sulphur dioxide gases. III. Green plants, *Contributions from Boyce Thompson Institute, 11:* 343–356.

Tingey, D., and Hill, A. C., 1967. The occurrence of photochemical phytotoxicants in the Salt Lake Valley, *Proceedings of the Utah Academy of Sciences, Arts, and Letters, 44* (Part 1): 387–395

Treshow, M., 1968. The impact of air pollutants on plant populations, *Phytopathology, 48:* 1108–1113.

Wood, F. A., 1968a. Discussion of paper by Dr. Daines. Sulfur dioxide and plant response, *Journal of Occupational Medicine, 10:* 92–102.

Wood, F. A., 1968b. Sources of plant-pathogenic air pollutants, *Phytopathology, 58:* 1075–1084.

Zimmerman, P. W., Crocker, W., and Hitchcock, A. E., 1933. The effect of carbon monoxide on plants, *Contributions from Boyce Thompson Institute, 5:* 195–211.

Zimmerman, P. W., and Hitchcock, A. E., 1956. Susceptibility of plants to hydrofluoric acid and sulphur dioxide gases, *Contributions from Boyce Thompson Institute, 18:* 263–279.

7
Air Pollution and Trees   George H. Hepting

Abstract

Coniferous trees, bearing the same foliage year-round and often for several years, are notable victims of air pollution. Not only is there much variation among tree species in sensitivity to toxic airborne gases, but there is great variation from tree to tree within the same species. Thus, some sensitive white pine trees are being propagated to provide a clonal line useful in monitoring air pollution, while other neighboring resistant white pines have enabled us to produce clonal lines of trees that can withstand field levels of certain toxic gases. The main air-pollutant gases toxic to trees are sulfur dioxide, fluorides, ozone, peroxyacyl nitrate, and oxides of nitrogen. Heavy damage to forests has, in the past, resulted mainly from sulfur oxides produced in ore reduction. Today the main damage to forest and shade trees is resulting from stack gas from coal-burning power plants and from the photochemical toxicants in urban smog. Where trees were killed out over thousands of acres around a smelter, the climate of the area was definitely altered adversely. Trees help screen out polluting industries and serve as fixed monitors of dangerous pollution levels. Trees have also been considered as sometimes contributing to atmospheric haze through photochemical combinations involving their aromatic hydrocarbons such as terpenes. Trees wisely used can help in many ways in the struggle against air pollution.

The Problem

Trees present some special problems as victims of air pollution damage and provide some special opportunities in combating air pollution. On the damage side, the deciduous tree in leaf reacts, in general, like any other plant. Some tree species are more susceptible to damage from certain gases than some annuals, while others are more resistant. However, the tree-to-tree variation within a tree species, for example, in resistance of eastern white pine to sulfur dioxide, is often much greater than the difference between species of pine.

The evergreen, whether conifer or broadleaf, behaves quite differently from the deciduous tree. The leaves of the evergreen

Prepared for SCEP.

are exposed year-round, not just for a few months, including the winter when strong air inversions on clear days can keep toxic emissions near ground level for many hours each day. Then, too, many conifers keep the same foliage for several years, thus prolonging their time span of gas exchange, and thus their likelihood of damage. This explains why larch, a deciduous conifer, is sometimes listed as pollution resistant, when considered on a year-round basis, and sometimes as sensitive, when considered only during that part of the year that it is in leaf.

Differences in length of time of leaf retention also explain why an evergreen and a deciduous tree in leaf, which react the same in a chamber, following an experimental fumigation, may react very differently on a year-round basis in the field.

There are other reasons why the approach to air pollution damage to trees differs from that of so many other plants. For example, the slight necrosis to orchid sepals caused by ethylene is ruinous to the orchid grower, but such injury would be inconsequential with a forest tree. Similar slight marginal burning of leafy vegetables from SO_2 or fluoride which often causes much loss in quality and value would hardly affect wood production in a tree.

The Pollutants

To appreciate the problem of air pollution damage to trees, it is probably best to consider first the gases that are known to injure trees. The sulfur oxides have, in the past, been the worst offenders. Sulfur dioxide and, to a lesser extent volumewise, sulfur trioxide are major constituents of the stack gases from ore-reduction. The trioxide combines with moisture, upon emission, to form the very damaging sulfuric acid aerosol. These sulfur oxides are also emitted in great quantity but in greater dilution from the stacks of soft-coal-burning power plants and from the consumption of high-sulfur fuel oils. Sulfur oxides from space heating and hydrogen sulfide from vegetation decomposition and industrial processing are less important as phytotoxicants, quantitatively.

Fluorine has caused important damage to trees in many areas. It is highly toxic in parts per billion to some plants yet is less toxic than SO_2 to others. Ponderosa pine was killed out by fluoride fumes over an area of fifty square miles around an aluminum ore reduction plant at Spokane. Southern pines and citrus have been injured around the phosphate mines and fertilizer plants in Florida; and fluorides causing injury to plants have also emanated from other aluminum and phosphate reduction, from the smelting of other ores, and from ceramics industries. Treshow, Anderson, and Harner (1967) describe the killing of Douglas fir in Idaho over a 200-acre area near a phosphate-reduction plant. They were also able to show a significant reduction in tree growth at greater distances from the extraction plant, where there was no necrosis of needle tissue, but an elevation in fluoride content of the foliage. Important reduction in the growth of eastern white pine within thirty miles of smelters in the Sudbury region of Ontario has been documented by Linzon (1958).

Silicon tetrafluoride, a ceramics industry emission, is known to be toxic to some plants in concentrations as low as 0.1 part per billion.

It is worth noting that fumigations with hydrogen fluoride, conducted both in daylight and darkness, showed larch foliage to be as sensitive as gladiolus to fluoride, the latter being the standard test plant for bioassay of ambient fluoride. The sensitivity to fluoride, of three pine species, three *Prunus* species, arborvitae, elm, mulberry, willow, and maple have also been determined by Adams, Hendrix, and Applegate (1957).

The term smog was first used to connote a mixture of smoke (suspended particles) and fog. However, smog is now used in referring to any mixture of gases and particles that pollute city air. The London-type smog is mainly a mixture of smoke and sulfur oxides and is chemically a reducing agent. The soot particles also serve as nuclei upon which air moisture condenses, and thus there is a synergistic accentuation of the opacity of the air resulting in smoke plus fog.

The Los Angeles-type smog is oxidizing in nature, and plant damage results largely from the photochemical products ozone and peroxyacetyl nitrate (PAN). These reactive gases are produced mainly from ultraviolet irradiation of automobile engine exhaust, with nitrogen oxides and unconsumed hydrocarbons involved in the chemistry of their formation. Photochemical smog from Los Angeles has played a major role in the decline of commercial citrus production south and east of the city, making citrus growing unprofitable in much of California's main citrus belt. It also is now known to be responsible for heavy mortality of ponderosa pine, and to a lesser extent other forest species, in the San Bernardino National Forest, fifty miles east of the city, and at elevations over 5,000 feet above sea level (Miller, Parameter, and Taylor, 1963).

In the San Bernardino case, many of the smog-injured ponderosa pines are attracting and being killed by bark beetles, and with respect to the Arrowhead Lake–Crestline area involved, Cobb and Stark (1970) have recently reported that they "have concluded that should air pollution continue unabated ponderosa pine will be virtually eliminated from forest stands in this area."

In the East, we know that white pine suffers from (1) SO_2 injury resulting from exposure to industrial and power plant stack gases; (2) from a pollution problem called chlorotic dwarf (Dochinger, Seliskar, and Bender, 1965); and (3) from ozone damage. The last, ozone injury called emergence tipburn, has occurred during periods when ambient ozone, monitored by Mast recorder, reaches about 7 pphm for four hours (Berry and Ripperton, 1963; Costonis and Sinclair, 1969), and during periods when tobacco sensitive to ozone fleck showed its characteristic markings on the upper leaf surface.

I have already indicated the major known phytotoxicants in the air. There are others including nitrogen oxides, chlorine, mercury, hydrogen cyanide, ammonia, and hydrogen sulfide (Thomas, 1961).

Damage—the Past
Some of the most devastating and dramatic cases of air pollution damage to living things have involved trees. Early in this century, tens of thousands of acres of timber were destroyed, mainly by sulfur dioxide from smelters both in the East and the West. Had people been killed, rather than trees, these cases would by far have eclipsed the mortality associated with episodes of the Meuse Valley in France in 1930; Donora, Pennsylvania in 1948; London in 1952; and the chronic smog problem in Los Angeles.

The smelters at Copper Hill, Tennessee, over a period of several years just before and after 1900, resulted in the complete denuding of 17,000 acres of forest land (7,000 acres bare and 10,-

000 acres in eroding broom-sedge type), and the injuring of an additional 30,000 acres. Most of the denuded area is still bare of vegetation, with the exposed red hills deeply gullied and the remaining soil being carried away in great washes of mud after every rain.

Even the climate of the area has been measurably changed as a result (Hursh, 1948). The bare zone, as compared with the contiguous forest, averages 2° to 4°F higher in air temperature during the summer; 0.6 to 2°F lower in winter; soil temperature is 22°F higher in summer; wind movement five to fifteen times higher; and even rainfall is consistently lower in the bare area.

The smelter at Trail, British Columbia, by destroying timber for a distance of forty miles down the Columbia River Valley in the United States during the 1920s, created a major international incident (Scheffer and Hedgcock, 1955).

Similar notorious damage to large acreages of trees from smelter fumes have occurred at Anaconda, Montana; in parts of Pennsylvania; at Sudbury, Wawa, and other localities in Ontario, Canada; and in many other areas in this country. Pollution abatement practices, often resulting in valuable by-product recovery, have greatly reduced but have not eliminated damage to the forest from ore smelting.

Damage—the Present
The few examples I have given of devastation around smelters are well known to most concerned with air pollution. Within the past twenty years, however, as smelter damage was being reduced, new air pollution impacts to trees began to make themselves felt. I refer mainly to (1) photochemical smog, with the formation of ozone and PAN, resulting from the enormous increase in automobile engine exhausts; (2) the rapid increase in number and capacity of soft-coal and, to lesser extent, oil-burning power plants; (3) to new industrial processing, including polluting incineration of wastes, for example, of plastics, coatings, insulation, and other materials that produce toxic products on burning; and (4) prescribed fire in agriculture and forestry.

I hardly need enlarge upon the situation in California where, particularly in Los Angeles, the combination of an enormous concentration of automobiles plus sunshine has brought tears, discomfort, and illness to people and millions of dollars in losses to agriculture and forestry mainly through the formation of PAN and ozone. These gases, particularly ozone, are now known to also occur widely in the East, and to cause damage to many leafy vegetable crops, to tobacco, and to white pines, in particular.

The rapid increase in thermal power production is one of our current concerns in the buildup of air pollution. Up to 1954 the United States had no power plant with a boiler capacity of 2 million pounds of steam per hour. By 1962 over 60 percent of our power was generated in such huge new plants, and the percentage continues to rise (Frankenberg, 1963).

Investigations and experiments have shown a serious foliage blighting and killing of white pines that occurred within a twenty-mile radius of the largest of the steam plants charted by Frankenberg (1963). These generating units can use a low grade of coal, high in sulfur, with today's effective pulverizing of fuel and efficient combustion.

A plant burning fuel oil with, for example, 2 percent sulfur, will emit about the same amount of sulfur oxides as one using coal with 2 percent sulfur. While there is less documentation of SO_2 damage from oil-burning power plants than coal plants, Peirson (1964) recently cited the dying of pine around an oil-burning power plant in Maine, and Linzon (1965) reports damaging concentrations of SO_2 that have severely injured trees growing in the vicinity of oil refineries in Ontario and Manitoba, Canada.

In the case of chronic injury to trees from air pollution, in virtually every case there has been very obvious tree-to-tree variation in susceptibility, within a given species, whether the offending gas be SO_2 injuring spruce in Westphalia, Germany; SO_2 killing white pine in Tennessee; fluoride blasting ponderosa pine near Spokane, Washington; ozone mottling pine foliage in our East or West; or city smog blighting horse chestnut in Paris, France.

The British pine arboretum at Kew, outside London, had to be moved to Kent because of London smog damage, and decline of the stone pines of Rome, Italy, has been the subject of an intensive scientific inquiry that concluded that the cause was mainly photochemical smog (Balsi and Montolla, 1963). New York, among its other air quality problems, also has its bus stop disease, a pollution induced decline of urban trees associated with photochemical action on unsaturated hydrocarbons from large numbers of stopping and starting vehicles.

Pennsylvania has just completed its first statewide survey of air pollution damage to vegetation and conservatively estimates an annual loss of $3.5 million, attributable mainly to oxidants, sulfur oxides, hydrogen chloride, particles, herbicides, and ethylene, depending upon the crop and the locality (Hepting, 1964; Weidensaul and Lacasse, 1969).

Many additional examples of pollution damage to trees could be cited. Scurfield (1960) cites 258 references to the subject. As far as Texas is concerned, I have heard accounts of considerable tree damage associated with oil refining and power production, but forests and the oil industry do not occur together in much of Texas. McKee (1964) gives an account of tree damage around industrial areas in the Houston area and provided lists of trees sensitive, moderately so, and resistant to fume injury in that area. He considers SO_2 from petroleum refineries, chemical manufacturing, and metallurgical industries to be the main cause of the damage.

The Future for Trees in the Struggle for Clean Air

At the rate that air pollution abatement was progressing up to about 1965, the outlook was anything but bright for the environment. In spite of the increase in nuclear-powered steam plants, the number and capacity of soft-coal-burning plants have been increasing far more rapidly, with oil also playing an increasing role, as this country's power demands exceeded its clean air demands. The lack of what has been termed "financially feasible" means of removing a high percentage of the sulfur oxides from power-plant stack gas and smelter emissions has been a major stumbling block.

I have no intention of getting into the engineering involved in the many processes of sulfur oxide removal now under trial around the world, not to mention the predesulfurization of fuel, and the attempts to use only low-sulfur fuel. The fact remains that any means now known will greatly increase the cost of power or commodity production. While engineers and chemists have labored over the technology involved in removing a high percentage of the sulfur oxides, economists and businessmen have been wondering what they would do with the enormous extra tonnage of sulfuric acid that would be produced if upwards of 90 percent of the total sulfur oxides were to be removed from all the power plants, ore-reduction plants and others that emitted these gases in quantity. Many states are now considering legislation that would require such high percentages of SO_2 removal from stack gases, regardless of the cost.

Let us now come back to trees and take a look at where they fit into the air quality picture in the years to come. We know, for example, that they can be a source of air pollution. Smoke from the burning of logging slash, particularly in the Northwest, has been a source of concern for a long time. It interferes with transportation, creates a dirty nuisance, and is considered by some to interfere with the ripening of fruit in the great valleys of Oregon and Washington. Prescribed fire, used as a silvicultural tool on several million acres of southern pine land, may not be tolerated many years hence. Nassau County on Long Island, New York, has already gone to burying the approximately 3,000 elms that die every year from Dutch elm disease, to avoid the smoke nuisance accompanying burning (Cusumano and Wasser, 1965).

Went (1963) has postulated that the oxides of nitrogen in the air, when acted upon by sunlight in the presence of aromatic hydrocarbons emitted mainly from woody plants (for example, the terpenes from conifers), produce the haze that develops over the Great Smoky Mountains of North Carolina and Tennessee, over sagebrush areas of the West, and over the eucalyptus forest of Australia. He notes that such hazes never form over deserts, large bodies of water, or areas lacking suitable vegetation. Went's theories have been widely publicized, but there are some unexplained anomalies that have prevented their general acceptance as the full explanation of natural hazes.

Belts of trees have long been used to screen out particles such as dust, soot, and fly ash around industrial sites (Weisser, 1961). Trees resistant to a given type of pollution can be used to mask factories and to provide protective soil cover while also producing a harvest of wood products. This use of trees involves recognizing not only how tree species differ from each other in resistance to specific airborne toxicants but also how individual trees, within a species, vary. German foresters in some highly industrialized areas are now changing the species composition of nearby forests to lessen damage from air pollution. Thus, they mention replacing, for example, the sensitive spruces and firs with the more resistant Austrian pine and with hardwoods (Wentzel, 1963).

In our own work on power-plant stack gas injury to white pine (Berry and Hepting, 1963), we discovered trees that were very sensitive, moderately so, or very resistant, and sometimes all of those types had their roots and branches intertwined. Grafts (ramets) made from such trees have the resistance or susceptibility characteristics of the parent tree (ortet). We have produced ramets from resistant trees, in quantity, both for planting in polluted areas and also in establishing a seed orchard that we expect will produce seed yielding progeny much more resistant to the stack gas in question than nursery-run white pine seedlings.

We have also discovered that individual white pine trees react differently to different gases. Ramets from these trees have produced clonal lines of white pine each of which is sensitive to only one of the following three pollutants: sulfur dioxide, fluoride, or Washington's photochemical oxidative smog. Thus, evergreen trees can be used to monitor for these gases, providing cheap, year-round highly sensitive detectors that show, by their foliage markings, what a recording instrument would show on its chart. This principle of bioassay for air pollution has been used by exposing gladiolus to detect airborne fluoride, and the method can also be applied by exposing Bel W-3 tobacco to detect ozone (Heck and Heagle, 1970; Heggestad and Menser, 1962).

Noble and Wright (1958) engaged in a comprehensive oxidant monitoring program in Los Angeles, using annual bluegrass and petunia, and Middleton and Paulus (1956) monitored in the same area using the pinto bean.

One of the major groups of lichens embraces the most sensitive plants known to air pollution, and sometimes lichens may be widely used as monitors (Jones, 1952). In addition to reports in the literature, many of us have noticed or have had reported to us the disappearance of lichens from trees in the vicinity of major sulfur oxide emissions.

I believe that we should closely examine claims of air purification by trees. We hope that research will prove this to be a significant factor in pollution abatement, but I know of no experimental evidence that the foliage of wooded areas around major emission sources can pick up enough sulfur or fluorine to affect the atmosphere materially. We know even less about the absorption of photochemical smog. So far as oxygenation of the air by trees is concerned—a subject that keeps cropping up in the literature—the proponents of this means of purification seem not to take into account that while a foliated tree in daylight uses CO_2 and gives off oxygen through photosynthesis, this same tree when it dies uses oxygen and gives off CO_2 in the oxidation involved in the decomposition of its wood. It also uses oxygen each year in decomposing the foliage, flowers, fruits, and twigs that fall, in addition to the oxygen used in normal respiration.

I have indicated forest damage from air pollution in the past and the situation as it is today with respect to the major airborne plant toxicants. We now have many new "entries" in the villain field. I have also indicated how, on the one hand, trees add to the pollution load in some areas, and, on the other hand, how they can be used as monitors in pollution detection and damage abatement through species and clonal selection.

Looking ahead, I see trees playing an increasingly more important role in the fight against air pollution. I have mentioned some of the amenities that trees provide, in addition to their use as monitors. Polluting industries tend to avoid the cities and "take to the woods" where their toxic emissions are recorded on the vegetation around them.

Finally, I would like to remind biologists of the valuable characteristics of most woody vegetation that make it so suitable for studies in pollution damage. Genetic variation from plant to plant is so striking in air pollution effects on most trees that we can easily select a resistant or susceptible individual from a population exposed to injury, run off a set of grafts from it (ramets) constituting a clonal line, and have every replicate react to a gas virtually exactly as its neighbor. This minimizes experimental error and reduces the number of replicates needed in experimental work.

We can also graft a susceptible shoot on a resistant tree or vice versa, and the shoot and the tree it is grafted to will each continue to show its original resistance or sensitivity. Thus we can find out often within a year whether a given problem has its origin in the air or in the soil, and then, if in the air, through the use of activated carbon filters find out whether or not the culprit is or is not an oxidant. Grafting techniques with trees and the use of the homogeneity of the clonal line have given us the tools that have established such widespread tree declines as emergence tipburn and chlorotic dwarf of white pine, as caused by air pollution, rather than by either fungi, insects, virus, bacteria, nematodes, or soil or climatic influences.

Ecosystem Changes through Forest Damage from Air Pollution
With the background of information already provided, we can see how air pollution damage to forest vegetation is altering the environment in many areas. I have pointed out Hursh's study (1948) in the Copper Hill area of Tennessee, and how the climate over an area of about 50,000 acres has been unfavorably changed as a result of the removal of forest vegetation by smelter fumes.

In California, photochemical air pollution from the Los Angeles area has not only made citrus growing and the culture of other crops unprofitable or impossible in a large part of the Los Angeles air shed that formerly produced these crops, but the oxidants originating from this area are being carried up the mountains fifty or more miles away and are leading to the decimation of much of the forest, especially ponderosoa pine in the San Bernardino mountains. This vegetation is vital to the protection of soil and water values, and if its destruction continues, drastic environmental changes can be expected.

To compound the problem of tree losses from air pollution in both eastern and western pine types, we now have clear evidence that pines weakened by air pollution often become prey to bark beetles, as these destructive insects are resulting in heavy subsidiary kills of conifers already suffering fume damage.

Since smelters are usually located in remote and wooded areas, damage to forests is common around ore-reduction plants. Large coal- or oil-burning power plants also are commonly located in forest areas to avoid cities and to be near the required fuel, water, and other resources. Ecologically, one of the results of this situation is, if not destruction of forests, at least a change in their composition, involving the gradual loss of the more sensitive species, especially evergreens, accompanied by an increase in deciduous species, many of which are less effective, for example, in intercepting snowfall and controlling snowmelt than are the conifers. The foliage cover of conifers is a factor in protecting snow cover from too rapid buildup or dissipation, the latter leading to flood conditions.

The heavy damage to forests from stack gas air pollution in the Ruhr Valley and other heavily industrialized parts of Europe is leading to deliberate attempts by foresters to change species composition to those kinds of trees least susceptible to injury from sulfur oxides. In these lower-elevation areas, such changes need not necessarily result in any deterioration of the environment, such as would a shift from conifers to hardwoods where control of the snowpack, a delicate process, is important.

Finally, we must recognize that the spread of metropolitan conditions, leading to a megalopolis such as we are witnessing develop from New York to Washington, D.C., with its attendant increase in automobile exhaust volume and industrial fumes, will doubtless lead to more damage from oxidants and other pollutants. One authority recently reported that now, in New Jersey, he considers ozone damage to commercial crop plants to occur in every county in that state.

With the increase in soft-coal-burning power plants throughout the southern Appalachian region, where large volumes of cheap, low-grade, high-sulfur coal and the necessary water are available, we can expect much more forest damage from sulfur oxides, unless measures are taken soon to cope with this problem.

In general then, we are certain to increase the output of both stationary and mobile sources of pollutants capable of deteriorating the environment. Therefore, we must counter this trend with effective means of rendering these sources innocuous either by changes in fuels, alteration in means of combustion, stack gas treatments to remove toxicants, or the substitution of alternative means of power production or chemical processing if we are not going to see our forest resources suffer additional depletion.

References

Adams, D. F., Hendrix, J. W., and Applegate, H. G., 1957. Relationship among exposure periods, foliar burn, and fluorine content of plants exposed to hydrogen fluoride, *Agriculture and Food Chemistry, 5:* 108–116.

Balsi, Giannetto, and Monttola, Paolo, 1963. The desiccation of the pines of Rome (translated title), *L'Italia Agricola, 100:* 3–7.

Berry, C. R., and Ripperton, L. A., 1963. Ozone, a possible cause of white pine emergence tipburn, *Phytopathology,* 53: 552–557.

Berry, C. R., and Hepting, G. H., 1964. Injury to white pine by unidentified atmospheric constituents, *Forest Science, 10:* 2–13.

Cobb, Fields W., Jr., and Stark, R. W., 1970. Decline and mortality of smog-injured ponderosa pine, *Journal of Forestry, 68:* 147–149.

Costonis, A. C., and Sinclair, W. A., 1969. Relationships of atmospheric ozone to needle blight of eastern white pine, *Phytopathology, 59:* 1566–1574.

Cusumano, Robert D., and Wasser, George L., 1965. A survey of practices in regard to the air pollution aspects of Dutch elm disease eradication, *Journal of the Air Pollution Control Association, 15:* 230–234.

Dochinger, L. S., Seliskar, C. E., and Bender, F. W., 1965. Etiology of chlorotic dwarf of eastern white pine (Abstract), *Phytopathology, 55:* 1055.

Frankenberg, T. T., 1963. Air pollution from power plants and its control, *Combustion, 34:* 28–33.

Heck, Walter W., and Heagle, Allen S., 1970. Measurement of photochemical air pollution with a sensitive monitoring plant, *Journal of the Air Pollution Control Association, 29:* 97–99.

Heggestad, H. E., and Menser, H. A., 1962. Leaf spot sensitive tobacco strain Bel W-3, a biological indicator of air pollutant ozone (Abstract), *Phytopathology, 52:* 735.

Hepting, G. H., 1964. Damage to forests from air pollution, *Journal of Forestry, 62:* 630–634.

Hepting, G. H., 1968. Diseases of forest and tree crops caused by air pollutants, *Phytopathology, 58:* 1098–1101.

Hursh, C. R., 1948. Local climate in the Copper Basin of Tennessee as modified by the removal of vegetation, *U.S. Department of Agriculture Circular, 744.*

Jones, Eustace W., 1952. Some observations on the lichen flora of tree boles, with special reference to the effect of smoke, *Revue de Bryologique et Lichenologique, 1951–1952* (Paris: Musée National d'Histoire Naturelle, Laboratoire de Cryptogamie), Section 2, pp. 96–115.

Linzon, S. N., 1958. The influence of smelter fumes on the growth of white pine in the Sudbury region, *Department of Agriculture of Canada, Forest Biology Division,* Contribution 439, 45 p.

Linzon, S. N., 1965. Sulphur dioxide injury to trees in the vicinity of petroleum refineries, *Forestry Chronicle, 41:* 245–250.

McKee, Herbert C., 1964. Air pollution and its effect on trees, *Fortieth International Shade Tree Conference Proceedings, 1964:* 149–163.

Middleton, J. T., Darley, E. F., and Brewer, R. F., 1958. Damage to vegetation from polluted atmospheres, *Journal of the Air Pollution Control Association, 8:* 9–15.

Middleton, J. T., and Paulus, A. O., 1956. The identification and distribution of air pollutants through plant response, *Archives of Industrial Health, 14:* 526–532.

Miller, P. R., Parameter, J. R., Jr., and Taylor, O. C., 1963. Ozone injury to the foliage of *Pinus ponderosa, Phytopathology, 53:* 1072–1076.

Noble, W. M., and Wright, L. A., 1958. A bio-assay approach to the study of air pollution, *Agronomy Journal, 50:* 551–553.

Peirson, H. B., 1964. Oil fumes as a possible cause of dying pines, *Scientific Tree Topics, 2*(10): 4.

Scheffer, T. C., and Hedgcock, G. G., 1955. Injury to northwestern forest trees by sulfur dioxide from smelters, *U.S. Department of Agriculture Technical Bulletin 1117.*

Scurfield, G., 1960. Air pollution and tree growth, *Forestry Abstracts, 21:* 339–347, 517–528.

Thomas, M. D., 1961. Effects of air pollution on plants, *Air Pollution,* World Health Organization Monograph Series 46: 233–278.

Treshow, Michael, Anderson, Franklin K., and Harner, Frances, 1967. Responses of Douglas-fir to elevated atmospheric fluorides, *Forest Science, 13:* 114–120.

Weidensaul, Craig T., and Lacasse, Norman L., 1969. Statewide survey of air pollution damage to vegetation, *Pennsylvania State University Center for Air Environment Studies Publication 148–70,* 52 p.

Weisser, D., 1961. Woods as a natural barrier against air pollution in areas of concentrated industrialization and urbanization, *Air Pollution Control Association Abstracts, 7*(4): 6.

Went, F. W., 1963. Blue hazes in the atmosphere, *Nature, 187:* 641–643.

Wentzel, K. F., 1963. Forest protection against gas emissions, *Allgemeine Forstzeitschrift* (February): 101–106.

8
Pollution and Range Ecosystems

Dixie R. Smith

The range ecosystem contains a characteristic biotic community that functions together with the nonliving environment as a complex system. Grasses and shrubs dominate the vegetation; herbivores are a major influent. The abiotic environment varies widely. Representative subdivisions of the range, their locations, and general environmental conditions are listed in Table 8.1.

The range is an open system, not a closed one. Energy and matter continually enter and leave the system. The flux of energy and matter connects the range with other ecosystems. Dice (1952) deduced four general principles that apply to interecosystem relatonships: (1) the influence of an ecosystem increases with its area; (2) the influence of an ecosystem decreases with distance; (3) ecosystems compete with one another for space; and (4) ecologically, the world is a unit. From this perspective, the range is a very large ecosystem, but much of it is remote from the urban systems and their critical pollution problems.

In the context of pollution, the range is a source and a recipient of pollutants. A fundamental distinction should be drawn between two types of pollutants. The natural pollutants are present in natural ecosystems and in fact become pollutants only when the system is overloaded because the rate of introduction is increased. Artificial pollutants are not naturally present in the ecosystem; they include such examples as insecticides and herbicides.

Major Theories on Pollution

Despite the multitude of reports on the effects of various pollutants on various components of ecosystems, only Woodwell (1970) has attempted to generalize the effects of pollution on ecosystems. Woodwell noted that the changes in ecosystems as a result of pollution are complex, but regardless of the pollutant the changes follow in aggregate patterns that are similar in different ecosystems and are, therefore, broadly predictable. The changes, Woodwell noted, are similar to the familiar ones along natural gradients of increasingly severe conditions. The parameters most changed

Prepared for SCEP.

by pollution are diversity, structure, and the ratio of photosynthesis to respiration.

Diversity is the first parameter affected by pollution. Sensitive species are eliminated from the flora and fauna, and a lower diversity means shorter food chains, fewer cases of symbiosis (Margalef, 1968), and less stability (Watt, 1968). The species first affected are those with a low reproductive rate, those linked by multiple bonds to other elements of the ecosystem, and those with little ecological amplitude.

Structure of the ecosystem is next affected by pollution. The pattern is toward plants of lesser stature. Trees give way to shrubs; shrubs to herbs; herbs to still smaller herbs and frequently to mosses and lichens. Woodwell (1970) suggests that one important mechanism involved in this structural change is the ratio of gross production to respiration in different strata of the community. Chronic disturbance frequently damages the capability for photosynthesis without reducing appreciably the total amount of respiration. Large plants with their large respiration requirement are more vulnerable than small plants.

Pollution that affects the structure of an ecosystem results in a corresponding reduction of nutrients in the total system. Decay of the larger plants releases a pool of nutrients that the soil cannot retain, and they are lost to aquatic ecosystem through a detrital chain.

Woodwell (1970) summarized the effects of loss of structure, which are aggravated by almost any pollutant, in the following way:

The loss of structure involves a shift away from complex arrangements of specialized species toward the generalists; away from forest toward hardy shrubs and herbs; away from those phytoplankton of the open ocean . . . , toward those algae of the sewage plants that are unaffected by almost everything . . . ; away from diversity in birds, plants, and fish toward monotony; away from tight nutrient cycles toward very loose ones with terrestrial systems becoming depleted, and with aquatic systems becoming overloaded; away from stability toward instability . . . ; away from a world that ruins itself through a self-augmentative, slowly moving evolution, to one that requires constant tinkering to patch it up, a tinkering that is malignant in that each act of repair generates a need for further repairs to avert problems generated at compound interest.

Table 8.1 Locations and General Environmental Conditions for Certain Subdivisions of the Range Ecosystem

Climax Ecosystem Type	Principal Locations	Precipitation Range (inches/year)	Temperature Range (°F) (daily maximum and minimum)	Soils
Tropical savanna	Central America (Pacific coast) Orinoco basin Brazil, south of Amazon basin Northern Central Africa East Africa Southern Central Africa Madagascar India Southeast Asia Northern Australia	10–75 Warm season thunderstorms Almost no rain in cool season Long dry period during low sun	Considerable annual variation; no really cold period Rainy season (high sun) Maximum 75–90 Minimum 65–80 Dry season (low sun) Maximum 70–90 Minimum 55–65 Dry season (higher sun) Maximum 85–105 Minimum 70–80	Some laterites considerable variety
Broad-sclerophyll vegetation	Mediterranean region California Cape of Good Hope region Central Chile Southwestern Australia	10–35 Almost all rainfall in cool season Summer very dry	Winter Maximum 50–75 Minimum 35–50 Summer Maximum 65–105 Minimum 55–80	Terra rossa, noncalcic red soils; considerable variation
Temperate grasslands	Central North America Eastern Europe Central and Western Asia Argentina New Zealand	12–80 Evenly distributed through the year or with a peak in summer Snow in winter	Winter Maximum 0–65 Minimum −50–50 Summer Maximum 70–120 Minimum 30–60	Black prairie soils Chernozems Chestnut and brown soils Almost all have a lime layer

133 Pollution and Range Ecosystems

Warm Deserts	Southwestern North America Peru and Northern Chile North Africa Arabia Southwest Asia East Africa Southwest Africa Central Australia	0–10 Great irregularity Long dry season, up to several years in most severe deserts	Great diurnal variation Maximum 80–135 Minimum 35–75 Frosts rare	Reddish desert soils often sandy or rocky Some saline soils
Cold Deserts	Intermountain Western North America Patagonia Transcaspian Asia Central Asia	2–8 Great irregularity Long dry season Most precipitation in winter; some snow	Great diurnal variation Winter Maximum 20–60 Minimum −40–25 Frosts common ½–¾ of year Summer Maximum 75–110 Minimum 40–70	Gray desert soils, often sandy or rocky Some saline soils
Alpine Tundras	Western North America Northern Appalachian North America European mountains Asian mountains Andes African volcanoes New Zealand	30–80 Much winter snow; long-persisting snowbanks	Winter Maximum −35–30 Minimum −60–10 Summer Maximum 40–70 Minimum 15–35	Usually rocky Some turf and bog soils Polygons and stone nets Some permafrost
Arctic Tundra	Northern North America Greenland Northern Eurasia	4–20 Shallow snowdrifts, but many bare and dry areas in "High Arctic"	Winter Maximum −40–20 Minimum −70–0 Summer Maximum 35–60 Minimum 30–45	Rocky or boggy Much patterned ground Permafrost

Source: Condensed from *Plants, Man and the Ecosystem*, second edition, by W. D. Billings. © 1964, 1970 by Wadsworth Publishing Company, Inc., Belmont, California, 94002. Reprinted by permission of the publisher.

Specific Pollutants and the Range

Carbon Dioxide

The range ecosystem is linked to the atmosphere through photosynthesis and respiration:

$$CO_2 + H_2O \;\underset{\text{Respiration}}{\overset{\text{Photosynthesis}}{\rightleftarrows}}\; \text{Carbon compounds of plants} + O_2$$

Each year the range ecosystem (grassland and desert) transforms photosynthetically some 1.2×10^9 tons of carbon per year from CO_2 to plant material. This transformation represents less than 0.2 percent of the CO_2 content of the atmosphere. To give further perspective to the range ecosystem, cultivated lands transform about 4.0×10^9 tons of carbon each year; forest 10×10^9 tons (Lieth, 1963); and oceans 135×10^9 tons (Bonner and Galston, 1952).

For the range ecosystem a state of near equilibrium probably exists; the uptake of CO_2 in photosynthesis being offset by CO_2 liberation in the process of respiration. Organic matter is being buried in sediment today, but worldwide the rate is only 1×10^9 tons per year (Peterson, 1969).

The atmosphere now contains about 325 ppm, or 0.032 percent CO_2. However, man's activities—especially the burning of fossil fuels—is reducing the O_2 content and increasing the CO_2 content of the air. From 1958 to 1962 atmospheric CO_2 content increased by 1.13 percent, or 3.7 ppm (Keeling and Bolin, 1963). The President's Science Advisory Committee (PSAC) (1965) projected a 25 percent increase in CO_2 content of the atmosphere during the period 1950–2000.

Obviously the role of range in reducing atmospheric CO_2 is a very limited one. And the possibilities for increasing that role are even more limited. Increases in population and urbanization suggest a decrease in range area is more likely than an increase (U.S. Department of Agriculture, Economic Research Service, 1968). But more importantly, the nature of range is such that standing crop does not increase in annual increments as it does in forests. Each year the aboveground portions of grasses and forbs are consumed by animals or lost to litter and humus. In either case, animal or soil respiration tends to return them to the atmosphere as CO_2. Shrubs, of course, do add annual increments to their aboveground standing crop. However, shrublands are now being converted to grasslands to increase their forage or water yields (Pechanec et al., 1965; Pase and Fogel, 1967).

Where CO_2 concentration is limting, an increase in CO_2 will generally increase, at least temporarily, the rate of photosynthesis. However, in most range ecosystems, carbon dioxide is probably not the limiting factor in plant growth. Lack of sufficient water is the predominant limitation (Thomas and Ronningen, 1965).

The range ecosystem is potentially a contributor of large amounts of CO_2 to the atmosphere. Large quantities of carbon are stored in the surface soils of the temperate grasslands and alpine and arctic tundras. Certain management practices are capable of accelerating the decomposition of this soil organic matter and thereby releasing large quantities of CO_2.

Pesticides

In the United States, herbicides were applied to about 71 million acres in 1962. Only about 4 percent of the total acreage represented forest or range ecosystems (Shepard, Mahan, and Fowler, 1966). Nevertheless, large acreages of the range ecosystem are sprayed occasionally with herbicides. In 1966, 10.5 million pounds of herbicides were applied to ranges and pastures; 2,4-D (2,4-D-chlorophenoxyacetic acid) and 2,4,5-T (2,4,5-Trichlorophenoxyacetic acid) composed over 90 percent of the total (U.S. Department of Agriculture, Economic Research Service, 1970). Use of 2,4-D was about 9.1 million pounds; 2,4,5-T, 0.4 million pounds.

The two herbicides can be used in the range ecosystem without seriously polluting adjacent waters. Some herbicides will appear in nearly all streams associated with the treated area. But, nearly all of the pollution results from the direct application of spray to the water surface. Contamination can be minimized by simply avoiding the situations that lead to direct application to streams or lakes. Norris (1967) monitored the use of herbicides in western Oregon and concluded that their use constituted little or no threat to fish populations or downstream water users.

Herbicides have probably not accumulated in soils of the range ecosystem. No published account is known. And, unlike the buildup of some insecticides, accumulation of most herbicides is evident to the observant individual because of phytotoxicity (Sheets, 1966).

Heat

Range is not a source of thermal pollution, but may be affected greatly by it. Man produces energy, primarily through burning of fossil fuels, at a rate of about 5×10^{19} ergs/sec, or 25/1000 of 1 percent of the total radiated by the earth. The direct effect of this energy output on the current average temperature of the earth's surface is totally insignificant. But, with the annual growth rate of 7 percent, man's production of energy could increase the mean surface temperature by 1°C in 91 years or 3°C in 108 years (Cole, 1969, 1970). Even the 1° increase could cause real changes in the boundaries between plant communities (Waggoner, 1966). A 15°C increase, possible in 130 years, would make the earth uninhabitable (Cole, 1969).

N-P-K

Artificial pollution arising from N-P-K fertilizers on ranges is not a problem now, nor is it likely to become a major problem in the future. Results from range fertilization in the United States have been highly variable, but even where substantial increases in forage production have been obtained, the cost of the added forage is frequently prohibitive (U.S. Department of Agriculture, Agricultural Research Service-Soil Conservation Service, 1961).

Both 2,4-D and 2,4,5-T are biodegradable. The former is relatively short-lived in the soil and is inactivated by microorganisms; the latter is more resistant (Sheets, 1966). Even 2,4,5-T does not usually carry over from one growing season to the next (Montgomery and Norris, 1970); where conditions are favorable, it may be detoxified in one month (Warren, 1954).

Herbicides may be broken down more slowly in plants than in the microbial environment of the soil. Morton, Robinson, and Meyer (1967) found 90 to 300 ppm 2,4,5-T on grasses immediately following application of 2,4,5-T at the rate of 2 lb/acre. Residues diminished to one-half of the maximum level in 1.6 to 2.6 weeks. Precipitation reduced the residues markedly—indicating that the residue was primarily a surface deposit.

Residues of 2,4-D as acid in forage dropped from a maximum

of 58 to 5 ppm in seven days after spraying at the rate of 2 pounds and equivalent per acre (Klingman et al., 1966).

There is no accumulation of 2,4-D or 2,4,5-T residue in animal tissues as a result of their feeding on contaminated plants. The major route of elimination of 2,4,5-T from pigs, calves, and rats dosed with 100 mg/kg was in the urine. Repeated doses did not result in the accumulation of herbicide (Erne, 1966). Milk from cattle grazing on 2,4-D treated forage contained a maximum of 0.03 ppm during the two days after spraying and decreased to less than 0.01 ppm after four days (Klingman et al., 1966). Mitchell and coworkers (1946) have reported no detectable residues in milk from cows fed 2,4-D.

In 1966, about 342,000 pounds of insecticides were applied to pastures and ranges in the United States. Methoxychlor was the insecticide most commonly used; it was applied to 193,000 acres of hay and pasture crops (including ranges) in 1966. A total of 563,000 acres was treated with insecticides (U.S. Department of Agriculture, Economic Research Service, 1970).

Radioactive Compounds

Radioisotopes occur naturally in the range ecosystem. Most of the radioactivity is associated with potassium 40 (^{40}K) in the soils. The average soil contains about 20,000 millicuries per square mile to a depth of 1 foot (Alexander, Hardy, and Hollister, 1960). Since potassium is a common component of forages, ^{40}K is the principal source of internal irradiation of grazing animals that originates in the soil.

Prior to the limited test ban treaty of 1963, a wide variety of radioactive elements was deposited in range ecosystems. Many of the short-lived isotopes have decayed and are no longer present in significant quantities. Cesium-137 and strontium-90 are now the most abundant fission products in the environment. The average cesium-137 and strontium-90 soil levels in the United States are about 240 and 150 millicuries per square mile (Reitemeier, Hollister, and Alexander, 1967).

Strontium-90 is the radionuclide of most concern to man, especially in ecosystems having high rainfall, low calcium and mineral concentrations, and heavy grazing (Odum, 1963). Because it behaves chemically like calcium, strontium-90 is taken up by plants more readily than most fission products and becomes concentrated in the bones of animals at the higher trophic levels.

Cesium-137 is similar to potassium in chemical properties. It also enters the food chains of range ecosystems. But, unlike strontium-90, it accumulates largely in the muscle tissues of animals and has a relatively short residence time (Reitemeier, Hollister, and Alexander, 1967). Cesium-137 is fixed in the mineral soil horizons in a nonexchangeable form (Sawhney, 1966). This widespread fixation results in a low uptake by plants.

Prior to the test ban treaty, cesium-137 entered the food chain by direct deposition on plants. On arctic ranges, lichens and mosses contained more radioactivity than their associated plants (Rickard et al., 1965). Accumulation in lichen and mosses appears to be related to the concentration of detritus and fallout by shrinking snowbanks with a subsequent filtering action by these lower plants (Osburn, 1963). Concentration through the food chain, lichen to caribou to man, resulted in high accumulations of cesium-137 in the Eskimos of northern Alaska (Palmer et al., 1964).

Strontium-90 was also concentrated in the food chain of arctic ranges (Watson, Hanson, and Davis, 1964), and because of its mobility it remains a pollutant of major concern even though testing in the atmosphere has ended.

Sediment

Sediment, the solid material and organic materials, is the greatest pollutant of surface waters. Each year some 4 billion tons of sediment are washed from terrestrial ecosystems into tributary streams in the United States. About 1 billion tons are transported to the sea (Wadleigh, 1968).

Sediment is the most abundant pollutant arising from the range ecosystem. Chaparral ranges in southern California produce sediment at an annual rate of 2,000 to 5,000 tons per square mile. Wildfires may increase the rate to over 100,000 tons per square mile (Glymph and Storey, 1966). Depleted ranges in New Mexico have an annual sediment yield of 1,500 tons to 5,600 tons per square mile (Dortignac, 1960).

Sediment, the product of erosion, is damaging in many ways. First, there is the depletion of soil in the range ecosystem. Second, there is the decrease in water quality. And third, there is the physical damage of sediment deposition. In addition to its physical impact on the environment, sediment is a primary vehicle for transporting chemicals from the range. Phosphorus, pesticides, and radioactive materials have a marked affinity for the small soil particles that are easily eroded.

Particulate Matter

In the United States alone, the internal-combustion engine used in transportation adds 73 million tons of mixed pollutants to the atmosphere each year; the generation of electricity adds another 16 million tons, largely sulfur dioxides. Industry adds 23 million tons of mixed pollutants. Miscellaneous burning, including forest fires, contributes 10 to 12 million tons. Thirty million tons of natural dusts also enter the atmosphere (Wadleigh, 1968).

From research on radioactivity we have learned: (1) particulate matter introduced into the lower atmosphere enters air currents that move around the world in periods of fifteen to twenty-five days in the middle latitudes; (2) the half-time residence of particulate matter carried in these currents ranges between a few days and a month; (3) particulate matter tends to be removed from air by precipitation; and (4) the patterns apply to any particulate matter entering the air currents of the troposphere (Woodwell, 1969).

Air pollutants have direct and indirect effects upon the range ecosystem. The direct effect upon range ecosystems is not documented. The studies to date have focused on high-value crops such as forest trees (Scheffer and Hedgcock, 1955; Hepting, 1964; Scurfield, 1960), and vegetables, fruit, and horticultural crops (Daines, Leone, and Brennan, 1967). The absence of documentation on the effects of particulate matter in range ecosystems probably reflects only our failure to study them. The basic plant processes and mechanisms affected by air pollution run throughout the plant kingdom—trees, shrubs, and herbs alike.

Air pollution may have indirect effects upon the range ecosystem through modification of the weather and precipitation. Schaefer (1969) believed that the "misty rain" and "dusty snow" of the Northeast was due to the overseeding of clouds—a process that inhibited precipitation by stabilizing clouds. Increased precipitation and thunderstorm frequency has been noted around La Porte, Indiana, downwind from the Chicago urban complex

(Changnon, 1968). Urban-produced thunderstorms occur at St. Louis and Chicago (Changnon, 1969), and urban-produced increases in precipitation have been noted at Tulsa, Oklahoma (Landsberg, 1956), and various European cities (Kratzer, Vieweg, and Braunshweig, 1956). More recently, Hobbs and Radke (1970) reported a 30 percent increase in annual precipitation downwind of industrial complexes in the state of Washington.

In Perspective

The range ecosystem is an open system connected to other systems by a flux of energy and matter. In the case of oceans, the flux is largely an undirectional transfer in the form of detritus.

Exploitation by man usually accelerates the flow of energy and matter to other systems. At the current rate of exploitation, artificial pollutants rarely stem from the range ecosystem. Natural pollutants, primarily sediment and the associated nutrients, do stem from the system and pollute other environments.

Pollutants flow to the range ecosystem from other systems. Their direct effects are largely, but not entirely, undocumented. Examples include the lichen-caribou-Eskimo food chain that concentrates radionuclides and the pesticide-eggshell thinning relationship that has endangered the peregrine falcon and threatens the prairie falcon.

Indirect effect of pollution, the modification of climate, cannot be accurately predicted. Many of the pollutants have the capability of changing climate but tend to have counterbalancing effects.

References

Alexander, L. T., Hardy, E. P., Jr., and Hollister, H. L., 1960. Radioisotopes in soils; particularly with reference to strontium-90, *Radioisotopes in the Biosphere*, edited by R. S. Caldecott and L. A. Snyder (Minneapolis: University of Minnesota Center for Continuation Study), pp. 3–22.

Billings, W. D., 1964. *Plants, Man and the Ecosystem* (Belmont, California: Wadsworth Publishing Co., Inc.).

Bonner, J., and Galston, A. W., 1952. *Principles of Plant Physiology* (San Francisco: Freeman and Company).

Changnon, S. A., Jr., 1968. The La Porte weather anomaly—fact or fiction, *Bulletin of the American Meteorological Society, 49:* 4–11.

Changnon, S. A., Jr., 1969. Recent studies of urban effects on precipitation in the United States, *Bulletin of the American Meteorological Society, 50:* 411–421.

Cole, L. C., 1969. Thermal pollution, *BioScience, 19:* 989–992.

Cole, L. C., 1970. Cole's reply, *BioScience, 20:* 72.

Daines, R. H., Leone, S. A., and Brennan, E., 1967. Air pollution and plant response in the northeastern United States, *Agriculture and the Quality of Our Environment,* edited by N. C. Brady, American Association for the Advancement of Science, Publication 85 (New York: American Book Company).

Dice, L. R., 1952. *Natural Communities* (Ann Arbor: University of Michigan Press).

Dortignac, E. J., 1960. The Rio Puerco—past, present, and future. Paper presented at 5th Annual New Mexico Water Conference, New Mexico University, November 1-2.

Erne, K., 1966. Distribution and elimination of chlorinated phenoxyacetic acid in animals, *Acta Veterinaria Scandinavica, 7:* 240.

Glymph, L. M., and Storey, H. C., 1966. Sediment—its consequences and control, *Agriculture and the Quality of Our Environment,* edited by N. C. Brady, American Association for the Advancement of Science, Publication 85 (New York: American Book Company).

Hennigan, R. D., 1969. Water pollution. *BioScience, 19:* 976–978.

Hepting, G. H., 1964. Damage to forests from air pollution, *Journal of Forestry, 62:* 630–634.

Hobbs, P. V., and Radke, L. F., 1970. Cloud condensation nuclei from industrial sources and their apparent influence on precipitation in Washington State, *Journal of Atmospheric Sciences, 27:* 81–89.

Keeling, D. C., and Bolin, B., 1963. Large scale atmospheric mixing as deduced from the seasonal and meridianal variations of carbon dioxide, *Journal of Geophysical Research, 68.*

Klingman, D. L., Gordon, C. H., Yip, G., and Burchfield, H. P., 1966. Residues in the forage and in the milk from cows grazing forage treated with esters of 2,4-D, *Weeds, 14:* 164–167.

Kratzer, P. A., Vieweg, F., and Braunshweig, S., 1956. *Das Stadtklima* (translated from German by the American Meteorological Society [AMS]) (Boston: AMS).

Landsberg, H. E., 1956. The climate of towns, *Man's Role in Changing the Face of the Earth* (Chicago: University of Chicago Press).

Lieth, H., 1963. The role of vegetation in the carbon dioxide content of the atmosphere, *Journal of Geophysical Research, 68:* 3887-3898.

Margalef, R., 1968. *Perspectives in Ecological Theory* (Chicago: University of Chicago Press).

Mitchell, J. W., Hodgson, R. E., and Gaetjens, C. F., 1946. Tolerance of farm animals to feed containing 2,4,-dichlorophenoxy acetic acid, *Journal of Animal Science, 5:* 226.

Montgomery, M. L., and Norris, L. A., 1970. A preliminary evaluation of the hazards of 2,4,5-T in the forest environment, *U.S. Department of Agriculture Forest Service Research Note PNW-116.*

Morton, H. L., Robinson, E. D., and Meyer, R. E., 1967. Persistence of 2,4-D, 2,4,5-T and dicamba in range forage grasses, *Weeds, 15:* 268–271.

Norris, L. A., 1967. Chemical brush control and herbicides in the forest environment, *Herbicides and Vegetation Management in Forests, Ranges, and Non-crop Lands* (Corvallis, Oregon: Oregon State University, School of Forestry and Oregon State System of Higher Education, Division of Continuing Education), pp. 103–123.

Odum, E. P., 1963. *Ecology* (New York: Holt, Rinehart and Winston, Inc.).

Osburn, W. S., Jr., 1963. The dynamics of fallout distribution in a Colorado alpine tundra snow accumulation ecosystem, *Radioecology*, edited by V. Schultz and A. W. Klement, Jr. (New York: Reinhold Publishing Corp.; Washington, D.C.: American Institute of Biological Sciences).

Palmer, H. E., Hanson, W. C., Griffin, B. I., and Fleming, D. M., 1964. Radioactivity measurements in Alaskan Eskimos in 1963, *Science, 144:* 859–860.

Pase, C. P., and Fogel, M. M., 1967. Increasing water yield from forest chaparral and desert shrub in Arizona, *International Conference on Water for Peace, 2:* 753–764.

Pechanec, J. F., Plummer, A. P., Robertson, J. H., and Hall, A. C., Jr., 1965. *Sagebrush Control on Rangelands*, U.S. Department of Agriculture Handbook No. 277.

Peterson, E. K., 1969. Carbon dioxide affects global ecology, *Environmental Science and Technology, 3:* 1162–1169.

President's Science Advisory Committee, 1965. *Report of the Environmental Pollution Panel* (Washington, D.C.: U.S. Government Printing Office).

Reitemeier, R. F., Hollister, H., and Alexander, L. T., 1967. The extent and significance of soil contamination with radionuclides, *Agriculture and the Quality of Our Environment*, edited by N. C. Brady, American Association for the Advancement of Science, Publication 85 (New York: American Book Company), pp. 269–282.

Rickard, W. H., Davis, J. J., Hanson, W. C., and Watson, D. G., 1965. Gamma-emitting radionuclides in Alaskan tundra vegetation 1959, 1960, 1961, *Ecology, 46:* 352–356.

Sawhney, B. L., 1966. Kinetics of cesium sorption by clay minerals, *Soil Science Society of America Proceedings, 30:* 565–569.

Schaefer, V. J., 1969. Some effects of air pollution on our environment, *BioScience, 19:* 896–897.

Scheffer, T. C., and Hedgcock, G. C., 1955. Injury to northwestern forest trees by sulfur dioxide from smelters, *U.S. Department of Agriculture Technical Bulletin, 1117*.

Scurfield, G., 1960. Air pollution and tree growth, *Forestry Abstracts, 21:* 1–20.

Sheets, T. J., 1966. The extent and seriousness of pesticide buildup in soils, *Agriculture and the Quality of Our Environment*, edited by N. C. Brady, American Association for the Advancement of Science, Publication 85 (New York: American Book Company), pp. 311–330.

Shepard, H. H., Mahan, J. J., and Fowler, D. L., 1966. *The Pesticide Review—1966*, U.S. Agriculture Stabilization and Conservation Service.

Thomas, G. W., and Ronningen, T. S., 1965. Rangelands—our billion acre resource, *Agricultural Science Review, 3:* 11–17.

U.S. Department of Agriculture, Agriculture Research Service—Soil Conservation Service, 1961. Range fertilization workshop, Denver, Colorado, August 22–24.

U.S. Department of Agriculture, Economic Research Service, 1968. *Major Uses of Land and Water in the United States with Special Reference to Agriculture, Summary for 1964,* Agriculture Economic Report No. 149.

U.S. Department of Agriculture, Economic Research Service, 1970. *Quantities of Pesticides Used by Farmers in 1966,* Agriculture Economic Report No. 179.

Wadleigh, C. H., 1968. *Wastes in Relation to Agriculture and Forestry,* U.S. Department of Agriculture Miscellaneous Publication No. 1065.

Waggoner, P. E., 1966. Weather modification and the living environment, *Future Environments of North America,* edited by F. F. Darling and J. F. Milton (Garden City, New York: Natural History Press).

Warren, G. F., 1954. Rate of leaching and breakdown of several herbicides in different soils, *North Central Weed Control Conference Proceedings, 11:* 5.

Watson, D. G., Hanson, W. C., and Davis, J. J., 1964. Strontium-90 in plants and animals of arctic Alaska, 1959–61, *Science, 144:* 1005–1009.

Watt, K. E. F., 1968. *Ecology and Resource Management* (New York: McGraw-Hill Book Co.).

Woodwell, G. M., 1970. Effects of pollution on the structure and physiology of ecosystems, *Science, 168:* 429–433.

Woodwell, G. M., 1969. Radioactivity and fallout: the model pollution, *BioScience, 19:* 884–887.

9
A Synopsis of the Pesticide Problem (Abridged) N. W. Moore

Introduction

Definitions

This review is about pesticides and ecology. Both terms need defining. The term pesticide, as employed here, means any chemical agent used to kill any living organism which is free-living or has a free-living stage in its life history.

The term ecology raises important questions of methodology. One branch of ecology deals with the relationship between individual organisms and their physical environment. Population ecology on the other hand is the study of groups of individuals rather than of separate organisms. I take the view that there is a fundamental difference between these two types of ecology. The first is an adjunct of other disciplines, e.g. of physiology, but population ecology is a scientific discipline in its own right. Unfortunately the terms autecology and synecology do not coincide completely with the two divisions of the subject; for population ecology includes the population aspects of autecology, as well as the synecology subjects of population ecology of groups of species, of community ecology and ecosystem ecology. Accordingly the term population ecology is used throughout the paper when reference is made to the study of groups of organisms, and the term ecology is retained as a general inclusive term, covering all relationships between organisms and their environment.

I should state briefly my personal views on population control since they underlie some of the arguments in this paper but are not discussed in it. I believe a useful distinction can be made between density-independent and density-dependent factors (Howard and Fiske, 1911, etc.). Experimental evidence is lacking to show which type of factor usually controls animal populations. Density-independent factors operate continuously and one or more of them may keep a population at a relatively low level for a long period of time, but in the long run it seems unlikely that they could produce the stability in populations which is observed in nature. Therefore I favour the thesis that ultimately populations are controlled by those factors in the total complex which are density-dependent. I believe that behaviour mechanisms control population density more often than is usually supposed.

The Study of Pesticides as an Ecological Problem

The widescale use of chemicals to control plants and animals is a recent development: within the space of about twenty years a new type of ecological factor has come into prominence, which is likely to affect all the ecosystems of the world. The situation is potentially of extraordinary interest to biologists. Yet, while great interest has been shown in the use and abuse of pesticides, the fundamental and theoretical aspects of the problem have been given less attention. It is possible to guess some of the reasons. The chemical revolution of agriculture has developed gradually from small beginnings, and there has been no particular moment when it could be said that a qualitative difference in the situation had arisen. The new ecological factor may be ignored because it is technological. The medieval notion that man exists "outside" nature still survives; its unfortunate corollary is that man-made factors like pesticides are considered to be unsuitable subjects for fundamental study. The pesticide problem is essentially about population ecology, and the principal difficulties of pest control and the control of pesticides result from the lack of fundamental ecological knowledge. It is, therefore, particularly unfortunate that it has received relatively little notice from the ecologists.

On the other hand the general public has not been slow to grasp the significance of pesticides; it has taken an almost obsessional interest in the subject. Few technical problems have received so much attention in the press and elsewhere. One of the reasons for this state of affairs is simple; there is the fear of being poisoned. Others are more complex and deserve objective study. There is a widespread fear of too rapid technological advance and change in the environment. It appears that the killing of wildlife by pesticides has become symbolic of the modern predicament and as a result pesticides appear to have become a whipping boy for all modern changes. Public interest in the subject may have had the effect of frightening away some scientists in the same way that popular interest in certain taxa, notably birds and Lepidoptera, at one time reduced the amount of scientific work done on these groups which otherwise are particularly suitable for investigation.

Despite the lack of theoretical interest a great deal has been published on pesticides, mainly by those concerned with crop protection and by those concerned with the harmful side effects to man, domestic animals and wildlife. Chemical firms are concerned primarily with making profits from the sale of pesticides, therefore most of their scientific resources go towards the discovery of new pesticides and towards testing them for short-term efficiency and safety. The emphasis in biological work is on field trials to discover the effects of new pesticides on pest species and on toxicological studies on laboratory animals to discover hazards to man and domestic animals. Very little of the work of chemical firms producing pesticides is ecological. Government agricultural research reinforces the work of the firms and extends it to cover some ecological aspects, but again the emphasis is on the immediate effects of chemicals on the pests, and so there is relatively little work on population ecology. In recent years concern with the side effects of pesticides on wildlife species has led to numerous studies on this subject by both Government and private organizations, mainly in the U.S.A. and Great Britain.

We have to conclude that most work on the effects of pesticides on living organisms does not progress further than toxicology or the field trial. The ecological aspects of the problem have been studied mainly by those working for agricultural advisory organizations and for conservation organizations. The former have been concerned largely with arthropods, the latter with vertebrates. Therefore we have a situation in which there are links between scientific discipline, taxonomic groups and organization:
Toxicological effects on vertebrates—chemical firms;
Effects on arthropods in the field—chemical firms and agricultural advisory organizations;
Effects on vertebrates in the field—conservation organizations.
Both specialization and group interest has helped to prevent a unified approach to the subject of pesticide effects.

Pesticide research involves chemistry, biochemistry, physiology, ecology and related disciplines. Few workers are trained in more than two of these disciplines, and therefore it is not surprising that the divisions of science have divided the study of pesticides. For example, those trained in chemistry do not easily see the pesticide problem as one of population ecology. It is more surprising that the subdivisions of biology erect barriers to understanding. For example, work on the effects of pesticides on avian predators rarely takes into account the extensive literature on the effects of pesticides on invertebrate prey/predator relationships in crops. Since the pesticide problem involves several scientific disciplines and is very complicated, it is one in which it is extremely easy to miss seeing the wood for the trees. The aim of this review is to describe the nature of the wood. The picture given is a crude one; but I hope that the great need for a logical and total view of the pesticide problem will be accepted as an excuse for the detailed errors of commission and omission which it inevitably contains. In particular I hope that this review will help to clarify the problem to pesticide specialists who are not ecologists, and at the same time provide a rational introduction to the subject for ecologists who know little about pesticides. Accordingly the effects of pesticides will be classified and described in ecological terms.

Mr. R. Crompton, who acted as Information Officer of the Toxic Chemical and Wildlife Division of the Nature Conservancy from 1961–65, estimated that about 300 papers relevant to the study of pesticides in the environment were published every month. I must emphasize that I have only read a small fraction of the vast literature on this subject and am in no position to attempt to assess it. Therefore this paper is not a review of the subject but of its nature. My only qualification for attempting the task is that over the last six years my work has given me an unusual opportunity to see it as a whole. In this paper references to some important works are given, but many others are omitted. I have not attempted to review the literature other than to mention some key reviews, books and papers, but have merely used examples of the literature to illustrate the themes. Wherever possible I have referred to the most recent work known to me when this contains references to earlier publications.

The reasons that have hindered a unified approach to pesticide problems have doubtless discouraged workers from writing general reviews of the subject. The nearest approach to an ecological review is R. L. Rudd's valuable book *Pesticides and the Living Landscape,* 1964. There are many excellent works on the various subdivisions of the subject. Of the text books the following are particularly valuable: Brown (1951), Metcalf (1955) and Martin (1964). From the ecologist's point of view Martin's book is especially valuable for its summaries of the chemistry and mode of action and methods of application of pesticides, Metcalf's for the summaries of the mammalian and insect toxicology of insecticides, Brown's for information on the older insecticides and on the physical aspects including aerial spraying. The volume on biological control edited by DeBach (1964) provides an excellent introduction to the ecology of crop protection. More has been written about DDT than about any other pesticide: the volumes edited by Müller (1955, 1959) on DDT summarize the extensive literature on this insecticide and contain information which is particularly valuable towards understanding a pesticide which is of very great ecological significance. Useful and extensive summaries of the toxicological literature are provided in the handbooks of Negherbon (1959) and Rudd and Genelly (1956). Among review papers the following are particularly useful introductions to their subjects: Ripper (1956) on the effects of pesticides on arthropod populations, Stern et al. (1959) on integrated control, Cope (1966) on the special problems of pesticides in fresh water, and Butler (1966) on pesticides in the marine environment. The problems of interaction ("potentiation") are reviewed by Dubois (1961) and pesticide resistance by Brown (1961) and Georghiou (1965a).

Much valuable information is presented in the handbooks produced annually by such organizations as the British Insecticide and Fungicide Council and the British Weed Control Council. Many valuable papers on pesticides, including some of the reviews already mentioned, have been published in the annual series entitled *Advances in Pest Control Research* which is edited by R. L. Metcalf. The annual circulars of the United States Fish and Wildlife Service (United States Department of the Interior) provide useful progress reports of pesticide/wildlife research in that country. British work on this subject is described in the reports of Monks Wood Experimental Station.

Several symposia have been held on the effects of pesticides and their proceedings published. The following are particularly relevant from the ecological point of view:

1. *The Ecological Effects of Biological and Chemical Control of Undesirable Plants and Animals.* I.U.C.N. Symposium, Warsaw 1960. Edited by D. J. Kuenen. (1961) Leiden.
2. *Chemicals and the Land in Relation to the Welfare of Man.* Proceedings of a Symposium held at the Yorkshire (W.R.) Institute of Agriculture, York 1965. Edited by F. M. Baldwin. (1965) Bradford and London.
3. *Research in Pesticides.* Proceedings of the Conference on research needs and approaches to the use of agricultural chemicals from a public health viewpoint. Held at the University of California, Davis, California, 1–3 October, 1964. Edited by C. O. Chichester. (1965) New York and London.
4. *Pesticides in the Environment and Their Effects on Wildlife.* Proceedings of an Advanced Study Institute sponsored by the North Atlantic Treaty Organization, Monks Wood, England 1965. Edited by N. W. Moore. (1966) *J. appl. Ecol.* 3 (Suppl.).

Finally there is *Silent Spring* by Rachel Carson (1962): this is not a balanced view of the pesticide problem; it is an impassioned advocate's plea for a more critical appraisal of pesticide use. In that it has influenced the minds of many people, this famous book is itself an important "ecological factor" which has to be taken into account in assessing the effects of pesticides on living organisms.

The Nature of the Pesticide Problem and the Types of Research Necessary for Studying It

The word pesticide is a useful omnibus term which covers all chemical agents used by man to kill or control living organisms. It includes weed killers (herbicides), fungicides, insecticides, acaricides, nematicides, molluscicides and rodenticides, and is often held to include chemosterilants and growth retarders.

Since no plant or animal lives in isolation, an application of a pesticide never results in a total reaction of the form

pesticide → pest

all pesticide applications in fact consist of the reaction

pesticides → ecosystem in which the pest occurs

This is the only reality; it consists of a man-made factor impinging on a complex system made up of physical elements and interacting populations. The essence of all pesticide problems, therefore, is one of population ecology. All other approaches are simplifications imposed for the purpose of scientific analysis. This has the important corollary that the results of analysis are likely to be misleading unless they are related to the system as a whole: in other words resynthesis is essential. Of course this concept applies to all ecological work, but there is abundant evidence to show that it is frequently forgotten in pesticide research. For example, it is often thought that effects in the field can be reliably predicted from simple toxicity tests.

Misunderstanding about the nature of pesticide problems is probably reinforced by the use of the term "side effects." For commercial and pest control reasons emphasis has to be made on the pesticide and the pest; as a result all other effects are thought of as "outside" the essential pesticide/pest relationship. It is then very easy to fall into the error of thinking that the actual reaction is

pesticide → pest ($\pm$ side effects)

I believe that many of the misunderstandings about the effects of pesticides stem from this way of thinking. It might be countered by the use of the term "biocide," since this emphasizes the whole effect of the chemical agent, not just the intentional effect. On the other hand it could be argued that many factors other than pesticides are biocidal.

Once it is recognized that all pesticide applications consist of a toxic factor impinging on a highly complex system, useful deductions can be made.

1. In studies on the effects of pesticides, a scientist is always dealing with the effects of a factor (which can be accurately described) on a system whose nature is scarcely understood at all. This is because the complexity of ecosystems is such that in no case have their workings been demonstrated. At best we have crude models illustrating what are believed to be salient features.

2. At any one time the population size of each species in an ecosystem is the result of a number of intrinsic and extrinsic factors. Therefore, even if the exact physiological response of a species to a pesticide can be predicted this will not necessarily demonstrate

whether the species will increase or decrease as a result of the application of the chemical.

3. The pesticide will impinge at all levels of organization: at levels of the cell and the organ, and at the levels of the individual, the population and the ecosystem. Therefore the most adequate explanation of pesticide effects consists of a synthesis of information from a wide range of scientific disciplines, notably from toxicology and population dynamics.

4. All possible effects cannot be forseen from toxicological and autecological studies, therefore it is often desirable to carry out experiments on whole systems: to compare sprayed areas with unsprayed controls. Nevertheless, experiments of this type can never produce absolutely reliable predictions on effects on constituent species, because ecosystems change and so alterations in the system may result in alterations in the effects on constituent species.

From all these considerations it will be seen that while we can accurately describe the pesticide we cannot accurately describe the system it affects. Its impact is on ill-defined complex interactions and on phenomena that are frequently continuous; therefore effects can only be predicted in terms of probability.

Most ecological situations are multifactorial; when discussing the increase or decrease of a population we should expect several causes. Relatively simple cause and effect relationships may be found but we should not expect them.

A full study of the effects of the given pesticide on a species requires assessment of all of the following available information:

1. The properties of the pesticide.
2. The manner of application of the pesticide.
3. Information on the extent to which the species comes in contact with the pesticide. This can be obtained from:

   (*a*) Details of application, see 2 above (e.g. amount of active ingredient per unit area).

   (*b*) The behaviour of the species in the sprayed area.
4. Response of the species to the chemical (data on acute and chronic toxicity).
5. Information on the principal factors which control the population of the species in situations in which no pesticide is used.

6. Toxicological information on the effects of the pesticide on other species which are principal controlling factors of the species studied (see 5 above).

Consideration of information of these types will suggest an hypothesis that the pesticide is likely to have a certain effect on the species (e.g. cause a decrease). This can be tested to a limited extent by performing a field experiment of the type mentioned above.

In no case known to the author has all the necessary information been available in any assessment of pesticide effect. The main gaps in information are in toxicology and population ecology. Frequently no toxicological data exist for the species concerned, and the assumption has to be made that it reacts in the same way as a related standard laboratory animal. Only very broad conclusions can be drawn from such assumptions because organisms react very differently to the same pesticide.

One of the most valuable by-products of the pesticide controversy is that it has shown how very few elementary ecological data are available, even in Britain where the flora and fauna have been more thoroughly studied than in most countries. In this country only the distribution of flowering plants has been systematically studied and recorded (Perring and Walters, 1962). Until recently the distribution of lower plants, vertebrates and the more conspicuous invertebrates has not even been studied systematically. For most animal species even crude distribution maps could not be drawn.

Except for a very few species of birds no attempt has been made to assess population numbers, but even in this group which provides more scope for census work than any other, population data are largely lacking: of the 200 or so species of breeding birds in Britain, population estimates based on total censuses or censuses of extensive sample areas have only been made for fifteen species. In only four have population estimates been made in three or more years and in only two of three were the first censuses made before the large scale introduction of pesticides (Moore, 1965b). Therefore there is extremely little systematic knowledge about population fluctuations in the past, and so much discussion on the effects of pesticides rests on *a posteriori* arguments. Man is the only species for which population data are adequate.

Many animals ingest pesticides through their food, therefore knowledge about their feeding habits is essential in assessing the effects of pesticides upon them. Lists of recorded foods are available for a number of organisms, mainly birds and the larvae of Lepidoptera. Enough is known to show that most animals are not specific in their choice of food but are to a greater or lesser extent opportunists within varying limits. This is not to say that some species do not have preferences or that some do not show distinct patterns. For example several herbivorous birds initially feed their young on insects. In a few cases feeding patterns of local populations have been studied systematically, for example, the Tawny Owls of Wytham Wood by Southern (1954). But for large populations of species which feed on several types of food, statistical information about the proportions of different types eaten is not available. Again we lack essential basic data. In assessing the effects of a pesticide on a species, data on populations, residues, toxicology and feeding habits are required. A list of vertebrate species recently or currently being studied in Britain in order to determine the effects of organochlorine insecticides is given in Table 9.1. A reasonable amount is known about the feeding habits of all these species. It will be seen that in no case do available data approach what is required. For most other species both in Britain and elsewhere even less information is available.

It might be thought from what has been said in this section that the study of pesticide effects contains so many unknown factors that evaluation is pointless. On the other hand the use of pesticides raises so many practical problems that we have to attempt to understand their effects. In the present state of ignorance we rarely expect to do more than bring a little order into chaos and to provide tentative answers.

Ecological Effects on Single Species

Introduction

As shown above, each application of a pesticide results in a pesticide-ecosystem reaction however much it may be aimed at one species. Nevertheless, to understand the whole we need to understand the parts, that is the effect or effects of a pesticide on individual species, and for many practical purposes information on

154   N. W. Moore

Table 9.1 Information Available on the Effects of Organochlorine Insecticides on Some Vertebrate Species Studied in Britain

Species	Population				Residue Analysis		Toxicology		Recent Population Changes	Authorities
	Systematic national studies made before 1940	Systematic national studies made after 1940	Local studies made before 1940	Local studies made after 1940	Systematic national	Incidental national	Acute oral toxicity data available	Sublethal effects on reproduction studied		
Heron (*Ardea cinerea*)	+	+	+	+	0	+	0	0	No change	Stafford, in prep.
Great Crested Grebe (*Podiceps cristatus*)	+	+	+	+	0	+	0	0	Increase	Prestt (1966)
Peregrine (*Falco peregrinus*)	0	+	0	+	0	+	0	0	Large decrease	Ratcliffe (1963, 1965)
Sparrow-Hawk (*Accipiter nisus*)	0	+	0	+	0	+	0	0	Large decrease	Prestt (1965 and in prep.)
Buzzard (*Buteo buteo*)	0	+	0	+	0	+	0	0	Slight decrease	Prestt (1965)
Pheasant (*Phasianus colchicus*)	0	0	+	+	0	+	+	+	No change	Genelly and Rudd (1956), DeWitt (1956), Azevedo et al. (1966), Taylor and Ash (1964)
Feral Pigeon (*Columbus livia*)	0	0	0	0	0	+	+	0	?	Turtle et al. (1963)
Shag (*Phalacrocorax aristotelis*)	0	0	0	+	0	+ (limited systematic)	0	0	Increase	Potts (unpub.)
Fox (*Vulpes vulpes*)	0	0	0	0	0	+	+	0	?	Blackmore (1963), Taylor and Blackmore (1961)

155 Synopsis of the Pesticide Problem

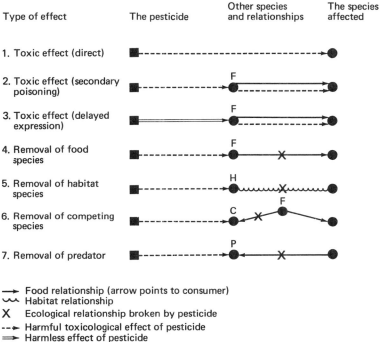

Figure 9.1 The types of effect of pesticides on one species.
Note: C = competing species. F = food species. H = habitat species. P = predator species.

the effects on one species is all that is required. There is now a very extensive literature on the effects of pesticides on single species. Many papers on very narrowly defined problems raise questions of fundamental interest. The aim of this section is to provide an ecological classification to help in assessing the literature.

The principal ways in which a pesticide can affect a species are shown in diagrammatic form in Figure 9.1. Each type is discussed below. It is obvious that in nature more than one type of effect is likely to be operating at any one time.

Toxic Effect on a Species—Direct Effects

A direct toxic effect is one in which the pesticide is either taken up percutaneously (including respiration) or ingested by drinking or by eating food directly contaminated by the pesticide. The extent to which the pesticide has a direct effect on the species depends first on the intrinsic reaction of the species to the chemical —this can be determined by toxicological experiments; secondly it depends on the extent to which a species comes in contact with the pesticide, and this depends on the amount of pesticide used, its properties and manner of application, and on the behavioural characteristics of the species.

Most of the recorded direct effects are of pesticides on pests. There are very few systematic descriptions of the effects on non-pest species.

A pesticide may have a direct effect on the individual or on its reproduction—in other words on the next generation. In many cases it is not obvious which type of effect is the more important. In a laboratory study on pheasants by Azevedo et al. (1965) the overall effect of DDT on reproduction was probably not significant at dietary levels which could be tolerated by adults. In a field study of residues in pheasants in California they found high levels of DDT in the fat (0–2 930 ppm, mean 740 ppm). Any effects on the population in this case could be attributed to direct poisoning rather than to an indirect effect on reproduction.

Toxic Effect on a Species—Delayed Effects

A species may be affected by eating plants or animals which contain residues of a pesticide previously ingested. As shown above, organochlorine insecticides become stored in body fat and this enables an animal to concentrate these substances. When one animal eats another the effects depend on the amount of pesticide that is ingested from the prey and the reaction of the animal to it. A number of situations may occur and special terms have been coined to describe them (Rudd, 1964). Secondary poisoning occurs whenever one animal dies from eating another animal containing a pesticide. Rudd makes the distinction between secondary poisoning which occurs shortly after consuming poisoned prey and "delayed toxicity" when contamination results in death at a later date. When an animal eats contaminated prey it may not die itself but it may kill another animal which eats it—Rudd calls this "delayed expression." This situation usually occurs when the intermediate species has the characteristic of being able to concentrate the pesticide to levels which would cause death in its predator. Such species have been called "biological concentrators."

157 Synopsis of the Pesticide Problem

Earthworms appear to be an example of biological concentrators (e.g. Barker, 1958; Davis, 1966). Stickel et al. (1965) show that normal field use of heptachlor is likely to result in concentrations of heptachlor epoxide in earthworms which are likely to affect American woodcock feeding upon them.

Secondary poisoning has been recorded among the organochlorine insecticides DDT, aldrin and dieldrin, the organophosphorus insecticides demeton methyl (metasystox) and dimethoate, and among the rodenticides zinc phosphide, strychnine, thallium sulphate and sodium fluoroacetate (1080). "Delayed toxicity" and "delayed expression" have only been observed among the organochlorine insecticides and among the organomercury fungicides.

Concentration of organochlorine insecticides in food chains has been demonstrated in many different types of environment. Since the process results in concentration at the end of food chains, predators are potentially at greater risk than species at lower trophic levels. There is insufficient evidence to say categorically that predators are declining at a greater rate than other species, but the evidence suggests that this is happening. For example, much work in agricultural entomology shows that many insect predators are severely affected by pesticides. There has almost certainly been an unprecedented decline of predatory birds in the Northern Hemisphere in recent years (Prestt, 1965; Ratcliffe, 1963). Some of the declines may be due to other factors, but since so many of the avian predators contain residues at, or approaching levels of toxicological significance (Moore, 1965b), the suspicion that indirect poisoning by organochlorine insecticides is an important cause is well based. The effects observed may be due to birds eating a few highly contaminated specimens (Jefferies and Prestt, 1966) or to consuming many less contaminated specimens over a long time. The effects may be direct and/or due to effects on reproduction (Ratcliffe, 1965). As shown above, delayed effects rarely operate through more than five stages in a food chain. Nevertheless considerable concentration of pesticides can occur.

Delayed effects are not restricted to animal/animal relationships or to unintentional effects. For example, aphids are poisoned by sucking the juices of plants which have taken up systemic organophosphorus insecticides from the soil. Recently C. H.

Walker (pers. comm.) has shown that sufficient BHC can be taken up by brussels sprouts from the soil to kill lepidopterous larvae feeding on them. Little is known about the ecological effects of intentional and unintentional systemic action.

Reduction of Food Species

In all countries where herbicides are widely used there has been an extensive reduction in certain weed species. Animal species dependent on these weeds must have suffered a considerable loss of potential food. For example, all insects dependent on poppies (*Papaver* sp.) have suffered an enormous reduction in their food supply in East Anglia, where poppies were once common in many fields but are now banished from most of them. Similarly predators and parasites which are dependent on herbivores living on plants now reduced by herbicides must have been affected. For example, D. J. Cross and T. R. E. Southwood (pers. comm.) have shown that young partridges, which are dependent on insect foods in the first few weeks of their lives, find less food in sprayed crops than in unsprayed. Accordingly they have to use more energy in finding the necessary food, and under wet conditions this involves a proportionally higher loss of energy through cooling. As a result partridges are likely to do less well where herbicides are used extensively. The decline in many partridge populations in Britain recorded in recent years may be partly due to the increase in the use of herbicides.

Extremely few studies have been made of this type of effect, and it may be much more important to non-pest species than is generally believed. Similarly the extensive use of insecticides must have greatly reduced the food of certain species, especially of insectivorous species with narrow food ranges. There is little evidence that pesticides reduce total production, therefore species with very wide food tolerances are less likely to be affected by pesticide applications on their food supply—they merely feed more extensively on those species which are unaffected or which have increased as a result of pesticide use.

Reduction of Habitat

Plants, in addition to providing food, are important to animals and other plants because they provide shelter. Pesticides, by destroying a habitat, may have a direct effect on numbers. Herbicides are the main agents of this sort. The simplification of crop ecosystems lessens the range of microhabitat for many species. The use of 2,4-D and 2,4,5-T to control scrub by roads and in woods reduces the essential habitat of almost all British land birds, which, because they are survivors of the original forest fauna, are still dependent on trees and bushes. Again very little work has been done to demonstrate the floristic and faunal changes resulting from habitat changes due to herbicides. But it is safe to assume that many species are being affected in this way.

However, it should be noted that herbicide treatments rarely destroy a habitat completely: they alter it. For example, scrub may be changed to grassland, or an aquatic flora consisting of submerged flowering plants may be changed to one consisting entirely of green algae. The new habitat may be more rather than less suitable for certain species. Again the loss is more likely to be one of diversity than of biomass.

Removal of a Competitor

Many plants and animals appear to be limited in their numbers by competition with other species which make use of the same basic requirements. There are many examples where it is known that plant species have increased as a result of herbicide use. For example Yemm and Willis (1962) showed that treatments of maleic hydrazide caused the progressive replacement of the tufted grasses *Arrhenathrum elatius* and *Dactylis glomerata* by the rhisomatous species *Festuca rubra* and *Poa pratensis*. The widespread use of MCPA and 2,4-D in crops has caused a great increase in weed species which previously were relatively rare. If effects on competition are extensive they may lead to the virtual substitution of one species for another. For example Aitken and Trapido (1961) have shown that when the mosquito *Anopheles labranchiae* was greatly reduced by DDT spraying in the anti-malarial campaign in Sardinia, 1947–48, four other species of *Anopheles* greatly increased. An explanation of the substitution was available in the case of *Anopheles hispaniola*: *A. labranchiae* rested in houses much more than did *A. hispaniola,* and houses were the habitat most thoroughly sprayed. Further, when water was sprayed the larvae of *A. hispaniola* were found to dive and avoid the surface for much longer periods than were the larvae of *A. labranchiae,* and so they had less contact with DDT.

Most of the cases of substitution of species have been discovered among invertebrates. Perhaps it is less likely to occur in animals with more highly developed nervous systems, since the latter may be able to adapt more quickly to new situations. The carrion crow has probably increased during the period when the sparrowhawk and kestrel have been virtually eliminated from Eastern England (Prestt, 1965). It is possible that this is a case of substitution caused by pesticides; one in which a highly adaptable species of omnivorous habits has filled the niche of species with more selective habits and which are, perhaps, more sensitive to organochlorine insecticide poisoning.

The effects of pesticides on ecological "buffering" should be noted. The effects of predators are often modified and reduced in severity because, when they begin to cause a severe reduction in one species, they turn their attention to another and so reduce pressure on the first. This system of buffering is almost certainly the cause of the greater stability of complex ecosystems. If a pesticide reduces the numbers of an alternative prey it may cause a predator to feed proportionally more upon another species which hitherto it had scarcely affected. For example, Edwards et al. (1964) showed that an application of TDE led to a great reduction of culicids, and as a result the trout in the experimental pond fed on crustacea more than before. If the crustacea were the competitors of the culicids the immediate effect of the spray may have been advantageous to the crustacea by removing competitors, but by indirectly causing the trout to feed more extensively upon them the eventual result may have been, theoretically at least, a reduction in their numbers. The effects of changes in predator pressure are discussed in the next section.

Removal of a Predator

For many years it has been known that resurgences of pests may occur after an application of an insecticide. Theoretically (see, for example, Kuenen and Post, 1958) these may be due to:
1. A stimulatory effect of the pesticide on the pest's reproduction.
2. A stimulatory effect of the pesticide on the food of the pest.
3. A reduction in competition.
4. A reduction in predators.

In most cases there is strong circumstantial evidence that resurgences have been caused mainly by a reduction of predators. In many other cases the use of a pesticide has resulted in the rise of another pest, i.e. substitution (see above) has occurred.

The most studied case is that of the tetranychid mites, in particular *Panonychus ulmi*. These animals have become important economic pests as a result of the use of pesticides (see, for example, Collyer, 1964 and Muir, 1965).

Of the forty-five species of predator found in orchards with *P. ulmi*, two species, the mirid *Blepharidopterus angulatus* and the phytoseiid mite *Typhlodromus pyri*, appear to be important in controlling the red spider mite. The work of Collyer (1964) shows that whereas the univoltine mirid requires more than one season to build up numbers large enough to control *P. ulmi* the predaceous mite breeds fast enough to control it in most cases. In an insectary experiment another mite, *Aculus fockeri*, was introduced into the system. Despite the presence of this alternative prey, *T. pyri* continued to control *P. ulmi*. Spray programmes directed against other pests of orchards reduce the populations of the predators including *T. pyri* and so cause eruption of *P. ulmi*. Kuenen and Post (1958), however, point out that low populations of *P. ulmi* can occur when predator populations are very low and so other factors may sometimes operate. They suggest that the reduction of competitors, changes in chemical composition in the leaves due to fertilizers applied at the same time as the pesticides, and stimulation of egg production by DDT at low levels, may also be important.

Effects on Ecosystems

Introduction

The ways in which a pesticide can affect a species or simple relationships between species have been outlined above. Again it must be emphasized that studies of this sort are likely to be misleading unless they are related to the whole situation. In this section the effects of the pesticide are described in terms of total effect—on changing the characteristics of the whole ecosystem on which the pesticide impinges.

Ecosystems can be described and so compared in quantitative terms. Measurements of species diversity, supported by knowledge of feeding habits, provide patterns showing the relative proportions of species in different trophic levels. If information about species is extended to cover information on populations and biomass, a further step can be taken: patterns of energy exchange can be assessed. Eventually it should be possible to assess the effects of pesticides in terms of production as well as diversity. Ecosystems tend to develop in time until a relatively stable state is reached: by affecting diversity and production a pesticide may also affect succession.

Effects on Diversity

Few studies have been made to determine the effects of pesticides on species diversity as such. Menhinick (1962) compared the animal populations of soil and litter of mown grassland in areas treated with fifteen insecticides and fungicides and those in unsprayed areas. The sprayed areas contained fifty-three taxa, the unsprayed eighty-two. Phenoxyacetic acid herbicides applied to pastures rich in species result in a decline in species because they kill off most of the dicotyledonous species which make up a large proportion of the total (see Yemm and Willis, 1962). It is reasonable to suppose that the loss of many plant species results in a corresponding loss of animal species. Yet herbicidal treatment does not necessarily result in a decline in species. For if a herbicide is used to kill a dominant species which is growing in a virtually pure stand, the result may be an increase since the deduction of the dominant allows colonization by a wide range of new species. For example, hawthorn scrub at a site in Kent was given a basal bark treatment with a 2,4-D/2,4,5-T mixture. Before treatment the scrub was so dense that there was virtually no ground flora beneath it. As a result of chemical treatment the canopy was broken and light reached the ground flora which was quickly colonized by nettles, sow thistles and other species (Davis, Moore and Way, unpublished). In general, however, pesticides are applied to complex ecosystems and so it can be assumed that they normally cause a decrease in diversity. The decline of diversity is not random—some groups are affected more than others. The differences between the sprayed and unsprayed areas referred to above (Menhinick, 1962) were as shown in Table 9.2, columns 1 and 2. He concluded that there was a decrease in the number of taxa and that the decline was most notable in the case of large organisms, especially in predaceous species. Unfortunately no records were

Table 9.2 Effects of Pesticides on Diversity, Density, Number of Individuals and Biomass: Comparisons Between the Fauna of Sprayed and Unsprayed Grassland

	Diversity			Density		Numbers		Biomass	
	No. of taxa in		Common to both	No. of taxa in which density significantly more abundant		No. of individuals per 3×10^7 m^2		Biomass per 3×10^7 m^2 in gms	
Category	Treated	Untreated		Treated	Untreated	Treated	Untreated	Treated	Untreated
Phytophagous	8	**14**	8	4	**5**	3,548	**2,311**	11.0	2.0
Fungivorous-Saprophytic	37	**42**	27	13	**22**	**94,800**	15,159	146.6	**175.6**
Predaceous	14	**21**	14	2	**21**	**15,085**	8,602	1.2	**4.1**
Total	53	**82**	51	17	**44**	**113,453**	26,591	154.0	**183.0**

Source: Data from Menhinick (1962).
Note: The higher value in each set of figures is printed in bold.

made of the pre-spray situation on the treated areas so the differences observed could, theoretically, have been due to factors other than pesticides. However, the results are consistent with other studies, e.g. of Edwards et al. (1964), and there is little doubt that the pesticide treatments caused the differences observed. The results are also in accordance with what is known about the effects of other forms of pollution. They are also in accordance with probability: most ecosystems consist of species which range from the very abundant to the very rare. If a pesticide had deleterious effects on all the species present it would be likely to exterminate some of the rarer species and so reduce diversity. Predators are necessarily rarer than their prey and hence are always present in relatively small numbers in any ecosystem. Therefore it can be stated as a general law that any deleterious density-independent factor is likely to have a particularly severe effect on predators. In the case of ecosystems affected by organochlorine insecticides, the effect on predators is likely to be enhanced because they receive higher doses of organochlorine insecticides than do other species. Therefore a differential decline in predators is likely to occur both because they are relatively rare and because they are at the end of food chains. These facts are of course interrelated.

It was shown above that pesticides have differential effects on species, therefore superimposed on the patterns outlined above is the pattern determined by differences in the susceptibility to pesticides. Menhinick (1962) recorded a greater decline in large herbivores compared with small ones in the chemically treated areas. This could have arisen because the larger animals had lower population densities but other explanations are possible, for example, a slower rate of recovery because of low reproductive rate. Differential toxicity may run parallel to taxonomic groups and feeding habit categories. For example, many crustacea, insect larvae and spiders appear to be more susceptible to most insecticides than are molluscs; worms and adult insects are intermediate. Algae and flowering plants are more susceptible than bacteria to herbicides. The innate "resistance" or adaptability of bacteria to pesticides is a fact of enormous practical importance; if it were not so, pesticides would greatly reduce the fertility of farmland.

In general, pesticides reduce diversity and this is likely to produce less stable systems (Elton, 1958). The extent to which such a decrease in diversity persists will depend on the rate at which the sprayed area can be recolonized, and this depends on the size of the sprayed area and the extent to which neighbouring ones have been or are being treated. It is probable that organochlorine insecticides have caused the extinction of certain birds of prey over large areas in which their original population density was relatively low (e.g. Ratcliffe, 1963). However, I do not know of any case where pesticide treatment has caused the total extinction of dense populations of non-pest species over large areas. Extinction is only likely when pesticides are applied to isolated communities, for example on oceanic islands. Where extinction occurs the pattern of diversity will be permanently affected, but in most cases the ecosystem is likely to return to a pattern similar to that which occurred before the application of the pesticide.

Few studies have been made which could indicate the long-term effects of pesticides on diversity or other features of ecosystems, whether the pesticides were applied on one occasion or as repeated treatments. The study of the effects of TDE and other pesticides on the fauna of Clear Lake by Hunt and his colleagues is exceptional. Catastrophic declines in bird populations caused by spraying forests with organophosphorus insecticides appear to be made good by the following year (e.g. Schneider, 1966). When DDT was applied over a period of four years in a forest in Maryland, U.S.A., there was a decline of 26 percent in the bird population; but a significant decline was only observed in three of the twelve common species present (Robbins et al., 1951). Since birds are known rapidly to recolonize vacant habitat, the true effects of pesticide treatment can only be discovered by marking all individuals in the experimental population and recording their fate. Studies on the effects of pesticides on aquatic invertebrates show a much slower return to normal (see, for example, Ide, 1956 and Hitchcock, 1960). This is not surprising in view of three special characteristics of aquatic animals. First they are more likely than terrestrial animals to concentrate pesticides from their medium. Secondly they are often unable to escape from the sprayed area, and thirdly the rate of recolonization tends to be slower owing to the slower dispersal rates of most aquatic species. It can be concluded that in general pesticides are more likely to reduce species diversity in aquatic ecosystems than in terrestrial ones.

Effects on Production

Menhinick's data (see Table 9.2) suggest that pesticide treatment has considerable effects on the number of individuals and on the biomass of the animals at different trophic levels. He concludes that there has been a decrease in the numbers of large organisms and a compensating increase in the numbers of small tolerant organisms. He also noted a shift in trophic structure in favour of lower trophic levels. He suggests that increased decomposition rates could result from chemical treatments. Much work will have to be done before generalizations can be made, but it is clear from this work that effects on diversity, population density and biomass can vary between trophic levels. Such effects are bound to affect production.

A number of studies have been made (e.g. Laverton, 1962) on changes in economic production resulting from pesticide use. But these are no substitutes for studies on total biological production. Studies on total biological production would be in the interests of agriculture since they would provide a better basis for understanding the long-term requirements of crop production.

Effects on Succession

In the majority of cases succession of animal populations depends on botanical succession. Therefore herbicides are more likely to affect succession than insecticides.

Total herbicides have a similar effect to a strong fire: they may wipe the area clean and allow a new succession to begin. The effects of more selective herbicides are of a different nature. Seral development is frequently arrested intentionally at the scrub stage. For example, 2,4,5-T is often used to prevent the invasion of grassland by bushes and to control woody competitors of planted trees in forests. There is likely to be an increasing use of herbicides for this purpose. Since the scrub stage is often particularly rich in plant and animal species, increased control of scrub may have considerable repercussions on wildlife. Succession is rarely affected unintentionally because herbicidal drift is rarely severe enough to kill shrub species. Increase in his own population is forcing man to reclaim more natural and semi-natural habitats, both on a small scale, e.g. hedge destruction, and on a large scale, e.g. drainage of large marshland areas. In general man favours vegetation which is equivalent to the early and late stages of ecological succession—field crops and high forest. Relatively few crops represent the intermediate scrub stage. Wild scrub has far less economic value than wild forest. Therefore, there is likely to be an increasing pressure on land in middle stages of succession throughout the world. The invention of herbicides (2,4,5-T, etc.) makes such reclamation easier than hitherto. Therefore the scrub habitat is likely to be reduced yet further as a result of the invention of herbicides. The scale of scrub reduction is already so great that there may be significant declines of plant and animal species dependent on it, and this could eventually affect climax vegetation.

Few studies have been made on the effects of pesticides on seral development. Laird (1958) studied those of DDT on succession of microflora and microfauna of infusions containing larvae of *Anopheles maculipennis atroparvus*. He found that the pattern of succession was similar in the cultures containing DDT and the controls, but that the rate of change was greater in the sample containing DDT. He showed that this was due to the DDT killing many of the mosquito larvae which resulted in increased bacterial activity, which in turn caused an increase in the growth and multiplication of protozoa. The surviving mosquitoes fared better in the treated samples than in the controls because more food was available for them at an early stage of development.

Summary

The pesticide problem is essentially ecological and an interdisciplinary approach to it is essential. Pesticides always affect ecosystems although they are usually applied to control single or very few pest species. Conclusions about pesticide effects on populations must generally remain tentative owing to the lack of information on population dynamics, feeding habits and toxicology. Most pesticides are non-specific in their action and their effects are density-independent. Pesticides vary greatly in their toxicity and persistence and solubility. There are differences of response to one chemical between species, sexes, age groups and individuals. Sublethal effects on reproduction and behaviour are likely to be ecologically significant. Persistence and fat solubility cause organochlorine insecticides to become widely dispersed and concentrated in food chains. Little is known about the ecological effects of pesticide interaction (potentiation). Pesticides have been used for over two hundred years but they have only become an important ecological factor during the past twenty years. Existing statistics are inadequate in assessing the amount of pesticides in the environment. A pesticide may affect a species directly by delayed effect and by affecting food species, habitat, competitors and predators. Pesticides usually reduce diversity and since they have differential effects on taxa at different trophic levels they may affect production. Pesticides, like all deleterious density-independent factors, are likely to have a particularly severe effect on predators. This effect is enhanced in the case of organochlorine insecticides by the food chain effect (see above). Pesticides are likely to affect freshwater ecosystems more severely than terrestrial ones. Modifications of succession by pesticides are described. The natural selection of resistant strains by pesticides can cause qualitative changes in species. Pest control and pesticide control are both ecological factors. Some reasons for not taking an ecological approach to pesticide problems are discussed. It is suggested that future trends in pest control will polarize towards two types of system, one involving complete control of very simple systems and others involving integrated control of complex systems. The use of pesticides is to some extent self-regulatory. Pesticides, particularly the more selective ones, have value as tools in ecological research.

Acknowledgments
I should like to thank Dr. J. P. Dempster, Dr. D. J. Jefferies, Dr. K. Mellanby, Dr. F. Moriarty and my wife Dr. Janet Moore for valuable criticisms of the manuscript. Discussions with my colleagues in the Nature Conservancy's Toxic Chemicals and Wildlife Division were most helpful in preparing this paper—I am most indebted to all members of the Division. I should also like to thank Miss Philippa Nathan who helped with the figures and Miss Jackie Potter for typing the manuscript.

References

Aitken, T. H. G. and Trapido, H. (1961). *In* "Symp. 8th Technical Meeting I.U.C.N. Warsaw" 1960, Leiden.

Albert, T. F. (1962). *Auk.* **79**, 104–107.

Ames, P. L. (1956). *In* "Pesticides in the Environment and Their Effects on Wildlife." *J. appl. Ecol.* 3(Suppl.), 87–97.

Arrington, L. G. (1956). *World Survey of Pest Control Products.* U.S. Dept. Commerce.

Ash, J. S. and Sharpe, G. I. (1964). *Bird Study* **11**, 227–239.

Azevedo, J. A., Jr., Hunt, E. G. and Woods, L. A., Jr. (1965). *Calif. Fish Game* **51**, 276–299.

Barker, R. J. (1958). *J. Wildl. Mgmt.* **22**, 269–274.

Bernard, R. F. (1963). *Publs. Mich. St. Univ. Mus.* **2**, 155–192.

Blackmore, D. K. (1963). *J. comp. Path. Therapeutics* **73**, 391–409.

Borg, K., Wanntorp, H., Erne, K. and Hanko, E. (1966). *In* "Pesticides in the Environment and Their Effects on Wildlife." *J. appl. Ecol.* 3 (Suppl.), 171–172.

Boyd, C. E. and Ferguson, D. E. (1964). *Mosquito News* **24**, 19–21.

Boyd, C. E., Vinson, S. B. and Ferguson, D. E. (1963). *Copeia* 1963, 426–429.

Boyle, C. M. (1960). *Nature. Lond.* **188**, 517.

Brown, A. W. A. (1951). "Insect Control by Chemicals." New York: Wiley.

Brown, A. W. A. (1958). *Adv. Pest Control Res.* **2**, 351–414.

Brown, A. W. A. (1961). *In* "Symp. 8th Technical Meeting, I.U.C.N. Warsaw" 1960, Leiden.

Burdick, G. E., Harris, E. J., Dean, H. J., Walker, T. M., Skea, J. and Colby, D. (1964). *Trans. Am. Fish. Soc.* **93**, 127–136.

Bushland, R. C. (1960). *Adv. Pest Control Res.* **3**, 1–25.

Butler, P. A. (1966). *In* "Pesticides in the Environment and Their Effects on Wildlife." *J. appl. Ecol.* 3 (Suppl.), 253–259.

Carson, R. (1962). *Silent Spring.* Boston: Houghton Mifflin. (English ed. 1963, London: Hamish Hamilton.)

Collyer, E. (1964). *Acarologia* **6**, 363–371.

Cope, O. B. (1966). *In* "Pesticides in the Environment and Their Effects on Wildlife." *J. appl. Ecol.* 3 (Suppl.), 33–44.

Davis, B. N. K. (1966). *In* "Pesticides in the Environment and Their Effects on Wildlife." *J. appl. Ecol.* 3 (Suppl.), 133–139.

DeBach, P. (1964). "Biological Control of Insect Pests and Weeds." London: Chapman and Hall.

DeWitt, J. B. (1955). *J. agric. Fd. Chem.* **3**, 672–676.

DeWitt, J. B. (1956). *J. agric. Fd. Chem.* **4**, 863–866.

Dubois, K. P. (1961). *Adv. Pest Control Res.* **4**, 117–151.

Edwards, R. W., Egan, H., Learner, M. A. and Maris, P. J. (1964). *J. appl. Ecol.* **1**, 97–117.

Elton, C. S. (1958). "The Ecology of Invasions by Animals and Plants." London: Oxford University Press.

Genelly, R. E. and Rudd, R. L. (1956). *Auk* **73,** 529–539.

George, J. L. and Frear, D. E. H. (1966). *In* "Pesticides in the Environment and Their Effects on Wildlife." *J. appl. Ecol.* **3** (Suppl.), 155–167.

Georghiou, G. P. (1965a). *Adv. Pest Control Res.* **6,** 171–230.

Georghiou, G. P. (1965b). *J. econ. Ent.* **58,** 58–62.

Giles, R. H., Jr. and Peterle, T. J. (1963). *In* "Radiation and Radioisotopes Applied to Insects of Agricultural Importance." pp. 55–84. I.A.E.C. Vienna.

Grolleau, G. and Giban, J. (1966). *In* "Pesticides in the Environment and Their Effects on Wildlife." *J. appl. Ecol.* **3** (Suppl.), 199–212.

Heath, R. G. and Stickel, L. F. (1965). *Circ. Fish Wildl. Serv., Wash.* **226,** 18–24.

Hickey, J. J., Keith, J. A. and Coon, F. B. (1966). *In* "Pesticides in the Environment and Their Effects on Wildlife." *J. appl. Ecol.* **3** (Suppl.), 141–154.

Higgins, E. (1951). *J. Wildl. Mgmt.* **15,** 1–12.

Hitchcock, S. W. (1960). *J. econ. Ent.* **53,** 608–611.

Howard, L. O. and Fiske, W. F. (1911). *Bull. U.S. Bur. En.,* **91,** 169, 183.

Huffaker, C. B. and Kennett, C. E. (1956). *Hilgardia* **26,** 191–222.

Hunt, E. G. and Bischoff, A. I. (1960). *Calif. Fish Game* **46,** 91–106.

Ide, F. P. (1956). *Trans. Am. Fish. Soc.* **86,** 208–219.

Jefferies, D. J. and Prestt, I. (1966). *Brit. Birds* **59,** 49–64.

Keith, J. A. (1966). *In* "Pesticides in the Environment and Their Effects on Wildlife." *J. appl. Ecol.* **3** (Suppl.), 57–70.

Keith, J. A. (1966). *In* "Pesticides in the Environment and Their Effects on Wildlife." *J. appl. Ecol.* **3** (Suppl.), 71–85.

Knutson, H. (1955). *Ann. ent. Soc. Am.* **48,** 35–39.

Koeman, J. H. and van Genderen, H. (1966). *In* "Pesticides in the Environment and Their Effects on Wildlife." *J. appl. Ecol.* **3** (Suppl.), 99–106.

Kuenen, D. J. and Post, A. (1958). *In* "Proc. Xth Int. Congr. Ent." **4,** 611–615.

Laird, M. (1958). *Can. J. Microbiol.* **4,** 445–452.

Laug, E. P., Kunze, F. M. and Prickett, C. S. (1951). *A.M.A. Arch. Ind. Hyg. occup. Med.* **3,** 245–246.

Laverton, S. (1962). "The Profitable Use of Farm Chemicals." London: Oxford University Press.

Lu, F. C., Jessup, D. C. and Lavallee, A. (1965). *Fd. Cosmet. Toxicol.* **3,** 591–596.

Ludwig, H. F. (1965). (Engineering Science Inc.) Toxicant-induced behavioural and histological pathology: a quantitative study of sublethal toxication in the aquatic environment. Final Report.

Martin, H. (1963). "Insecticide and Fungicide Handbook for Crop Protection." Oxford: Blackwell.

Martin, H. (1964). "The Scientific Principles of Crop Protection." London: Arnold.

Menhinick, E. F. (1962). *Ecology* **43,** 556–561.

Metcalf, R. L. (1955). "Organic Insecticides, Their Chemistry and Mode of Action." New York and London: Interscience Publ.

Moore, N. W. (1956). *Terre et Vie* 3, 220–225.

Moore, N. W. (1965a). In "Ecology and the Industrial Society," *5th Symp. of the Brit. Ecol. Soc. Oxford.*

Moore, N. W. (1965b). *Bird Study* 12, 222–252.

Moore, N. W. (1966). Ed. "Pesticides in the Environment and Their Effects on Wildlife." *J. appl. Ecol.* 3 (Suppl.).

Moore, N. W. and Tatton, J. O'G. (1965). *Nature, Lond.* 207, 42–43.

Moore, N. W. and Walker, C. H. (1964). *Nature, Lond.* 201, 1072–1073.

Muir. R. C. (1965). *J. appl. Ecol.* 2, 31–41 and 43–57.

Mulla, M. S. (1966). In "Pesticides in the Environment and Their Effects on Wildlife." *J. appl. Ecol.* 3 (Suppl.), 21–28.

Müller, P. (1955 and 1959). "DDT. The Insecticide Dichlorodiphenyltrichloroethane and Its Significance." Vol. I and II. Basle and Stuttgart: Birkhauser Verlag.

Murton, R. K. and Vizoso, M. (1963). *Ann. appl. Biol.* 52, 503–517.

Negherbon, W. O. (1959). "Handbook of Toxicology, Vol. III Insecticides." Washington: National Academy of Sciences.

Nicholson, A. J. (1939). Verhandl. Intern. Ent. 7th Kongr. Berlin, 4, 3022–3028.

Odum, E. P. (1959). "Fundamentals of Ecology." Philadelphia and London: Saunders.

Ogilvie, D. M. and Anderson, J. M. (1965). *J. Fish Res. Bd. Can.* 22, 503–512.

Ozburn, G. W. and Morrison, F. O. (1962). *Nature, Lond.* 196, 1009–1010.

Peakall, D. B. (1962). *Bird Study* 9, 198–216.

Pendleton, R. C. and Hanson, W. C. (1958). In "2nd U.N. Geneva International Conference on the Peaceful Uses of Atomic Energy." London: Pergamon.

Perring, F. H. and Walters, S. M. (1962). "Atlas of the British Flora." London: Nelson.

Prestt, I. (1965). *Bird Study* 12, 196–221.

Prestt, I. and Mills, D. H. (1966). *Bird Study* 13, 163–203.

Ratcliffe, D. A. (1958). *Brit. Birds* 51, 23–26.

Ratcliffe, D. A. (1963). *Bird Study* 10, 56–90.

Ratcliffe, D. A. (1965). *Bird Study* 12, 66–82.

Ripper, W. E. (1956). *A. Rev. Ent.* 1, 403–438.

Robbins, C. S., Springer, P. F. and Webster, C. G. (1951). *J. Wildl. Mgmt.* 15, 213–216.

Rudd, R. L. (1964). "Pesticides and the Living Landscape." Madison: University of Wisconsin Press. (English edition 1965, London: Faber & Faber.)

Rudd, R. L. and Genelly, R. E. (1956). *Calif. Dept. Fish Game, Game Bull.* 7, 5–209.

Satchell, J. E. and Mountford, M. D. (1962). *Ann. appl. Biol.* 50, 443–450.

Schneider, F. (1966). *In* "Pesticides in the Environment and their Effects on Wildlife." *J. appl. Ecol.* **3** (Suppl.), 15–20.

Sheldon, M. G., Mohn, M. H., Ise, G. A. and Wilson, R. A. (1964). *In* "Pesticide-Wildlife Studies 1964." *Circ. Fish Wildl. Serv., Wash.* 199.

Southern, H. N. (1954). *Ibis* **96**, 384–410.

Stern, V. M., Smith, R. F., van der Bosch, R. and Hagen, K. S. (1959). *Hilgardia* **29**, 81–101.

Stickel, W. H., Hayne, D. W. and Stickel, L. F. (1965). *J. Wildl. Mgmt.* **29**, 132–146.

Street, J. C. (1964). *Science, N.Y.* **146**, 1580–1581.

Strickland, A. H. (1966) . *In* "Pesticides in the Environment and Their Effects on Wildlife." *J. appl. Ecol.* **3** (Suppl.), 3–13.

Taylor, A. and Ash, J. S. (1964). *Game Research Assoc. 4th Annual Report.*

Taylor, A. and Brady, J. (1964). *Bird Study* **11**, 192–197.

Taylor, J. C. and Blackmore, D. K. (1961). *Vet. Rec.* **73**, 232–233.

Templeton, W. L. (1965). *In* "Ecology and the Industrial Society." 5th Symp. of the Brit. Ecol. Soc. Oxford.

Thomas, A. S. (1960). *J. Ecol.* **48**, 287–306.

Turner, N. (1965). *Conn. Agr. Exp. Sta. Bull.* **672**, 3–11.

Turtle, E. E., Taylor, A., Wright, E. N., Thearle, R. P., Egan, H., Evans, W. H. and Soutar, N. M. (1963). *J. Sci. Food Agric.* **14**, 567–577.

Vinson, S. B., Boyd, C. E. and Ferguson, D. E. (1963). *Science, N.Y.* **139**, 217–218.

Wain, R. L. (1958). *Adv. Pest Control Res.* **2**, 263–306.

Warner, R. E., Peterson, K. K. and Borgman, L. (1966). *In* "Pesticides in the Environment and Their Effects on Wildlife." *J. appl. Ecol.* **3** (Suppl.), 223–247.

Wheatley, G. A. and Hardman, J. A. (1965). *Nature, Lond.* **207**, 486–487.

Wheatley, G. A., Hardman, J. A. and Strickland, A. H. (1962). *Plant Pathology* **11**, 81–90.

W.H.O. Report (1964).

Winteringham, F. P. W. and Hewlett, P. S. (1964). *Chem. Ind.* 1964, 1512–1518.

Woodford, E. K. (1964). *Proc. 7th Brit. Weed Control Conf.* 944–62.

Woodford, E. K. and Evans, S. A. (1965). Weed Control Handbook. Oxford: Blackwell.

Wynne-Edwards, V. C. (1962). "Animal Dispersion in Relation to Social Behaviour." London: Oliver and Boyd, Edinburgh.

Yemm, E. W. and Willis, A. J. (1962). *Weed Res.* **2**, 24–40.

Part III Climatic Change and
Terrestrial Ecosystems

During the month of the Study of Critical Environmental Problems, there were three major Work Groups, each concerned with either climatic effects of man's activities, ecological effects, or the monitoring and measurement of these effects. Although there was considerable interaction between the monitoring and the other two groups, SCEP did not consider in detail the effect on ecosystems of changes in climate which might be caused by man.

In this section, the summary of the SCEP Reports on climatic effects and related monitoring provides the background for three papers that present analyses of ecological responses to postulated climatic changes and serve as important supplements to the SCEP Report. The first two papers, prepared for SCEP by scientists in the U.S. Department of Agriculture, discuss the effects of climate on crops and forests, respectively. With respect to crops, the areas of primary concern are the effects on photosynthesis and on other physiological and behavioral processes of plants. In addition to the effects on forests of changes in CO_2 concentration, sunlight intensity, temperature, and precipitation, the next paper prepared by a U.S. Forest Service team discusses information needs and possible programs for monitoring.

The effect of climatic change on forest diseases is an area of narrow scope but one with profound implications in forest ecology. The review paper by Dr. Hepting is reprinted here not only because it treats fully the role of diseases in the effects of climate on forests, but also because it develops in an eloquent fashion the ecosystem concept as a valuable tool in evaluation. This is a well-documented case that demonstrates how whole ecosystems can be remarkably sensitive to change and how seemingly subtle effects can prove to be much more dominant than a review of first-order effects would indicate.

10
Climatic Effects of Man's Activities
Summary of SCEP Report

Introduction

There is geological evidence that there have been five or six glacial periods (ice ages); the most recent (the Pleistocene) lasted 1 to 1.5 million years. In the past century there has been a general warming of the atmosphere of about 0.4 °C up to 1940, followed by a few tenths degree cooling. It seems clear that our climate is subject to a wide variety of fluctuations, with periods ranging from decades to millennia, and that it is changing now.

We know that the atmosphere is a relatively stable system. The solar radiation that is absorbed by the planet and heats it must be almost exactly balanced by the emitted terrestrial infrared radiation that cools it; otherwise the mean temperature would change much more rapidly than just noted. This nearly perfect balance is the key to the changes that do occur, since a reduction of only about 2 percent in the available energy can, in theory, lower the mean temperature by 2 °C and produce an ice age.

That there have not been wider fluctuations in climate is our best evidence that the complex system of ocean and air currents, evaporation and precipitation, surface and cloud reflection and absorption form a complex feedback system for keeping the global energy balance nearly constant. Nonetheless, the delicacy of this balance and the consequences of disturbing it make it very important that we attempt to assess the present and prospective impact of man's activities on this system.

The total mass of the atmosphere and the energy involved in even such a minor disturbance as a thunderstorm (releasing the energy equivalent to many hydrogen bombs) should convince us immediately that man cannot possibly hope to intervene in such a gigantic arena. However, in reality man does intervene, because he can—without intending to do so—reach some leverage points in the system.

All the important leverage points that this Study has identified control the radiation balance of the atmosphere in one way or another, and most of them control it by changing the compo-

Reprinted from Study of Critical Environmental Problems, 1970. *Man's Impact on the Global Environment* (Cambridge, Massachusetts: The M.I.T. Press), pp. 10–19.

sition of the atmosphere. For example, man can change the temperature of the atmosphere by introducing a gas such as CO_2 or a cloud of particles that absorbs and emits solar and terrestrial infrared radiation, thereby altering the delicate balance we have described. He can also affect the heat balance by changing the face of the earth or by adding heat as a result of rising energy demands.

A thorough understanding and reliable prediction of the influence of atmospheric pollutants on climate requires the mathematical simulation of atmosphere-ocean systems, including the pollutants. At present, computer models successfully simulate many observed characteristics of the climate and have significantly advanced our knowledge of atmospheric phenomena. They have, however, a number of drawbacks that become serious when modeling new states of equilibrium or changes of climate in its transition toward these new states. Unless these limitations are overcome, it will be difficult, if not impossible, to predict inadvertent climate modifications that might be caused by man.

Recommendations
1. We recommend that current computer models be improved by including more realistic simulations of clouds and air-sea interaction and that attempts be made to include particles when their properties become better known. Such models should be run for periods of at least several simulated years. The effects of potential global pollutants on the climate and on phenomena such as cloud formation should be studied with these models.
2. We recommend that possibilities be investigated for simplifying existing models to provide a better understanding of climatic changes. Simultaneously, a search should be made for alterternative types of models which are more suitable for handling problems of climatic change.

Carbon Dioxide from Fossil Fuels
All combustion of fossil fuels produces carbon dioxide (CO_2), which has been steadily increasing in the atmosphere at 0.2 percent per year since 1958. Half of the amount man puts into the atmosphere stays and produces this rise in concentration. The other half goes into the biosphere and the oceans, but we are not

certain how it is divided between these two reservoirs. CO_2 from fossil fuels is a small part of the natural CO_2 that is constantly being exchanged between the atmosphere/oceans and the atmosphere/forests.

A projected 18 percent increase resulting from fossil fuel combustion to the year 2000 (from 320 ppm to 379 ppm) might increase the surface temperature of the earth 0.5°C; a doubling of the CO_2 might increase mean annual surface temperatures 2°C. This latter change could lead to long-term warming of the planet. These estimates are based on a relatively primitive computer model, with no consideration of important motions in the atmosphere, and hence are very uncertain. However, these are the only estimates available today.

Should man ever be compelled to stop producing CO_2, no coal, oil, or gas could be burned and all industrial societies would be drastically affected. The only possible alternative for energy for industrial and commercial use is nuclear energy, whose by-products may also cause serious environmental effects. There are at present no electric motor vehicles that could be used on the wide scale our society demands.

Although we conclude that the probability of direct climate change in this century resulting from CO_2 is small, we stress that the long-term potential consequences of CO_2 effects on the climate or of societal reaction to such threats are so serious that much more must be learned about future trends of climate change. Only through these measures can societies hope to have time to adjust to changes that may ultimately be necessary.

Recommendations
1. We recommend the improvement of present estimates of future combustion of fossil fuels and the resulting emissions.
2. We recommend study of changes in the mass of living matter and decaying products.
3. We recommend continuous measurement and study of the carbon dioxide content of the atmosphere in a few areas remote from known sources for the purpose of determining trends. Specifically, four stations and some aircraft flights are required.
4. We recommend systematic scientific study of the partition of

carbon dioxide among the atmosphere, the oceans, and the biomass. Such research might require up to twelve stations.

Particles in the Atmosphere

Fine particles change the heat balance of the earth because they both reflect and absorb radiation from the sun and the earth. Large amounts of such particles enter the troposphere (the zone up to about 12 km or 40,000 feet) from natural sources such as sea spray, windblown dust, volcanoes, and from the conversion of naturally occurring gases into particles.

Man introduces fewer particles into the atmosphere than enter from natural sources; however, he does introduce significant quantities of sulfates, nitrates, and hydrocarbons. The largest single artificial source is the production of sulfur dioxide from the burning of fossil fuel that subsequently is converted to sulfates by oxidation. Particle levels have been increasing over the years as observed at stations in Europe, North America, and the North Atlantic but not over the Central Pacific.

In the troposphere, the residence times of particles range from six days to two weeks, but in the lower stratosphere micron-size particles or smaller may remain for one to three years. This long residence time in the stratosphere and also the photochemical processes occurring there make the stratosphere more sensitive to injection of particles than the troposphere.

Particles in the troposphere can produce changes in the earth's reflectivity, cloud reflectivity, and cloud formation. The magnitudes of these effects are unknown, and in general it is not possible to determine whether such changes would result in a warming or cooling of the earth's surface. The area of greatest uncertainty in connection with the effects of particles on the heat balance of the atmosphere is our current lack of knowledge of their optical properties in scattering or absorbing radiation from the sun or the earth.

Particles also act as nuclei for condensation or freezing of water vapor. Precipitation processes can certainly be affected by changing nuclei concentrations, but we do not believe that the effect of man-made nuclei will actually be significant on a global scale.

Recommendations

1. We recommend studies to determine optical properties of fine particles, their sources, transport processes, nature, size distributions, and concentrations in both the troposphere and stratosphere, and their effects on cloud reflectivity.

2. We recommend that the effects of particles on radiative transfer be studied and that the results be incorporated in mathematical models to determine the influence of particles on planetary circulation patterns.

3. We recommend extending and improving solar radiation measurements.

4. We recommend beginning measurements by lidar (optical radar) methods of the vertical distribution of particles in the atmosphere.

5. We recommend the study of the scientific and economic feasibility of initiating satellite measurements of the albedo (reflectivity) of the whole earth, capable of detecting trends of the order of 1 percent per ten years.

6. We recommend beginning a continuing survey, with ground and aircraft sampling, of the atmosphere's content of particles and of those trace gases that form particles by chemical reactions in the atmosphere. For relatively long-lived constituents about ten fixed stations will be required, for short-lived constituents, about 100.

7. We recommend monitoring several specific particles and gases by chemical means. About 100 measurement sites will be required.

The Role of Clouds

The importance of clouds in the atmosphere stems from their relatively high reflectivity for solar radiation and their central role in the various processes involved in the heat budget of the earth-atmosphere system.

Recommendations

1. We recommend that there be global observations of cloud distribution and temporal variations. High spatial resolution satellite observations are required to give "correct" cloud population

counts and to establish the existence of long-term trends in cloudiness (if there are any).
2. We recommend studies of the optical (visible and infrared) properties of clouds as functions of the various relevant cloud and impinging radiation parameters. These studies should include the effect of particles on the reflectivity of clouds and a determination of the infrared "blackness" of clouds.

Cirrus Clouds from Jet Aircraft

Contrail (condensation trail) formation, which is common near the world's air routes, is more likely to occur when jets fly in the upper troposphere than in the lower troposphere because of the different meteorological conditions in these two regions.

There are very few, if any, statistics that permit us to determine whether the advent of commercial jet aircraft has altered the frequency of occurrence or the properties of cirrus clouds. We do not know whether the projected increase in the operation of subsonic jets will have any climate effects.

Two weather effects from enhanced cirrus cloudiness are possible. First, the radiation balance may be slightly upset, and, second, cloud seeding by falling ice crystals might initiate precipitation sooner than it would otherwise occur.

Recommendations

1. We recommend that the magnitude and distribution of increased cirrus cloudiness from subsonic jet operations in the upper troposphere be determined. A study of the phenomenon should be conducted by examining cloud observations at many weather stations, both near and remote from air routes.
2. We recommend that the radiative properties of representative contrails and contrail-produced cirrus clouds be determined.
3. We recommend that the significance, if any, of ice crystals falling from contrail clouds as a source of freezing nuclei for lower clouds be determine.

Supersonic Transports (SSTs) in the Stratosphere

The stratosphere where SSTs will fly at 20 km (65,000 feet) is a very rarefied region with little vertical mixing. Gases and particles produced by jet exhausts may remain for one to three years before disappearing.

We have estimated the steady-state amounts of combustion products that would be introduced into the stratosphere by the Federal Aviation Agency projection of 500 SSTs operating in 1985–1990 mostly in the Northern Hemisphere, flying seven hours a day, at 20 km (65,000 feet), at a speed of Mach 2.7, propelled by 1,700 engines like the GE-4 being developed for the Boeing 2707-300. We have used General Electric (GE) calculations of the amount of combustion products because no test measurements exist. In our calculations we used jet fuel of 0.05 percent sulfur. We have been told that a specification of 0.01 percent sulfur could be met in the future at higher cost.

We have compared the amounts that would be introduced on a steady-state basis with the natural levels of water vapor, sulfates, nitrates, hydrocarbons, and soot in the stratosphere. We have also compared these levels with the amounts of particles put into the atmosphere by the volcano eruption of Mount Agung in Bali in 1963.

Based on these calculations, we have concluded that no problems should arise from the introduction of carbon dioxide and that the reduction of ozone due to interaction with water vapor or other exhaust gases should be insignificant. Global water vapor in the stratosphere may increase 10 percent, and increases in regions of dense traffic may be 60 percent.

Very little is known about the way particles will form from SST-exhaust products. Depending upon the actual particle formation, particles from these 500 SSTs (from SO_2, hydrocarbons, and soot) could double the pre-Agung eruption global averages and peak at ten times those levels where there is dense traffic. The effects of these particles could range from a small, widespread, continuous "Agung" effect to one as big as that which followed the Agung eruption. (The analogy between the SST input and that by the Mount Agung eruption is not exact.) The temperature of the equatorial stratosphere (a belt around the earth) increased 6° to 7°C after the eruption and remained at 2° to 3°C above the pre-Agung level for several years. No apparent temperature change was found in the lower troposphere.

Clouds are known to form in the winter polar stratosphere.

Two factors will increase the future likelihood of greater cloudiness in the stratosphere because of moisture added by the SSTs: the increased stratospheric cooling due to the increasing CO_2 content of the atmosphere and the closer approach to saturation indicated by the observed increase of stratospheric moisture. Such an increase in cloudiness could affect the climate. The introduction of particles into the stratosphere could also produce climatic effects by increasing temperatures in the stratosphere, with possible changes in surface temperatures.

A feeling of genuine concern has emerged from these conclusions. The projected SSTs can have a clearly measurable effect in a large region of the world and quite possibly on a global scale. We must, however, emphasize that we cannot be certain about the magnitude of the various consequences.

Recommendations

1. We recommend that uncertainties about SST contamination and its effects be resolved before large-scale operation of SSTs begins.

2. We recommend that the following program of action be initiated as soon as possible:

   a. Begin now to monitor the lower stratosphere for water vapor, cloudiness, oxides of nitrogen and sulfur, hydrocarbons, and particles (including the latter's composition and size distribution).

   b. Determine whether additional cloudiness or persistent contrails will occur in the stratosphere as a result of SST operations, particularly in certain cold areas, *and* the consequences of such changes.

   c. Obtain better estimates of contaminant emissions, especially those leading to particles, under simulated flight conditions and under real flight conditions, at the earliest opportunity.

   d. Using the data obtained in carrying out the preceding three recommendations, estimate the change in particle concentration in the stratosphere attributable to future SSTs *and* its impact on weather and climate.

3. We recommend implementation now of a special monitoring program for the lower stratosphere (about twenty km or 60,000 to 70,000 feet) to include the following activities:

a. Measurement by aircraft and balloon of the water vapor content of the lower stratosphere. The area coverage required is global, but with special emphasis on areas where it is proposed that the SST should fly.

b. Sampling by aircraft of stratospheric particles, with subsequent physical and chemical analysis.

c. Monitoring by lidar (optical radar) of optical scattering in the lower stratosphere, again with emphasis on the region in which heavy traffic is planned.

d. Monitoring of tropospheric carbon monoxide concentration because of its potential effects on the chemical composition of the lower stratosphere.

Atmospheric Oxygen: Nonproblem
Atmospheric oxygen is practically constant. It varies neither over time (since 1910) nor regionally and is always very close to 20.946 percent. Calculations show that depletion of oxygen by burning all the recoverable fossil fuels in the world would reduce it only to 20.800 percent. It should probably be measured every 10 years to be certain that it is remaining constant.

Surface Changes and the Climate
The most important properties of the earth's surface that have a bearing on climate and are likely to be affected by human activity are reflectivity, heat capacity and conductivity, availability of water and dust, aerodynamic roughness, emissivity in the infrared band, and heat released to the ground.

Since the amount of carbon dioxide in the atmosphere is dependent on the biomass of forest lands which serves as a reservoir, widespread destruction of forests could have serious climatic effects. Population growth or overgrazing that increases the arid or desert areas of the earth creates conditions that allow the introduction of dust particles to the atmosphere.

Other important surface changes are from man's activities that modify snow and ice cover, particularly in polar regions, and from some possible projects involving the production of new, very large water bodies. Increased urbanization is of possible global importance only as it produces extended areas of contiguous cities. Still, it is not certain whether effects of urbanization extend far beyond the general region occupied by the cities.

Recommendation

We recommend that before actions are taken which result in some of the very extensive surface changes described mathematical models be constructed which simulate their effects on the climate of a region or, possibly, of the earth.

Thermal Pollution

Although by the year 2000 global thermal power output may be as much as six times the present level, we do not expect it to affect global climate. Over cities it does already create "heat islands," and as these grow larger they may have regional climatic effects. We recommend that these potential effects be studied with computer models.

11
Potential Effects of Global Temperature Change on Agriculture

Sherwood B. Idso

Introduction

The nature and distribution of earth's vegetation is primarily controlled by climate (Good, 1931; Barghoorn, 1953); and of the climatic elements, temperature ranks as perhaps the most important factor (Gleason and Cronquist, 1964). Indeed, as Polunin (1960) has stated it, "temperature is vitally important as it conditions the speed of the chemical actions and activities comprising life. The great world vegetational zones, like altitudinal ones, depend primarily on temperature." Thus, in view of man's complete dependence upon plant life for sustenance, it becomes important to consider possible effects of global temperature change on vegetation. Preliminary to this investigation, however, it is reasonable first to dwell briefly on the possibilities for temperature change per se, so that inferences with respect to vegetational adjustments can be made in a realistic framework.

Temperature and Photosynthesis

Most biological activity, including photosynthesis, is confined to the temperature range 0 to 60°C (Gates, 1963). Over this range, each plant species will have a characteristic photosynthetic dependency upon temperature, starting with a poor efficiency at low temperatures, rising to maximum production at some optimum temperature, and then decreasing again at higher temperature (Gates, 1965; Idso and Baker, 1967).

For different plant species, the optimum temperature for photosynthesis will be different. Scott and Billings (1964) have investigated this relationship for several plants of alpine tundra origin. For two of them the optimum temperature was near 0°C, while for four others it was as high as 15°C.

For the temperate region plant *Lupinus,* Thompson (1942) found the optimum temperature to be near 27°C; and for a warm region plant *Zea,* he obtained a value of 32°C. Similar values for corn have also been obtained by Lehenbaur (1914), Lundegardh (1931), and Wilson (1966). Recently, however, Win-

Prepared for SCEP.

Effects of Global Temperature Change on Agriculture

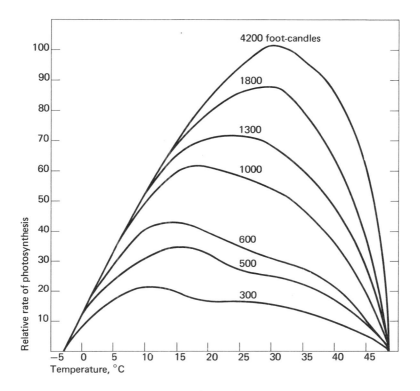

Figure 11.1 Photosynthesis in *Mimulus cardinalis* as a function of light intensity and leaf temperature
Source: Adapted from Milner and Hiesey, 1964.

ter and Pendleton (1970) have obtained optimum leaf temperatures for corn of approximately 40°C. Other warm region plants and their respective optimum temperatures for photosynthesis are cotton (35°), sunflower (35°), and sorghum (42°) (El-Sharkawy and Hesketh, 1964).

Sometimes, however, the situation is not so simple as has been just presented. Milner and Hiesey (1964), for instance, have found that the photosynthetic temperature dependence of *Mimulus* plants is also a function of light intensity, as shown here in Figure 11.1. This realization forces one to abandon analyses based solely on one variable and to consider in detail the dynamic environmental system as a whole in its relation to plant productivity.

The Holocoenotic Environment

Billings (1938) and Allee and Park (1939) have emphasized that factors of the environment act collectively and simultaneously and that the action of any one factor is qualified by other factors, using the term "holocoenotic environment" to denote this situation. Cain (1944) has elucidated the concept:

Physiological processes are multi-conditioned, and an investigation of the effects of variation of a single factor, when all others are controlled, cannot be applied directly to an interpretation of the role of that factor in nature. It is impossible, then, to speak of a single condition of a factor as being the cause of an observed effect in an organism . . . quite definitely, the environment *is* holocoenotic. . . .

Several models of net photosynthesis in plant communities have been developed which adopt, to varying degrees, the holocoenotic environment principle. The model of Monteith (1965) and Duncan's group (1967) concentrate chiefly on light intensity. Gates (1965), Idso and Baker (1967, 1968), and Idso, Baker, and Gates (1966) deal with light intensity and leaf temperature. DeWit (1965) considers light intensity and CO_2; and Paltridge (1970a, b) includes light intensity, CO_2, and soil water. Idso (1968, 1969) probably considers more factors than others: light intensity, leaf temperature, CO_2 concentration, and soil and atmospheric moisture. By means of these models it is possible to make calculations of net photosynthesis in characteristic environmental situations from basic plant parameters.

Photoynthesis Predictions

Idso and Baker (1967, 1968) have made calculations of net photosynthesis for maize and sorghum with their simplified light intensity-leaf temperature model that are useful in demonstrating the effects of warming and cooling trends on different types of vegetation. The photosynthetic dependencies they used for these two plants are shown in Figure 11 2, as constructed from data of Gates (1965), Waggoner, Moss and Hesketh (1962), Lundegardh (1931), El-Sharkawy and Hesketh (1964), Hesketh and Moss (1963), and Moss (1963).

In one study (Idso and Baker, 1968), the total daily net photosynthesis of a single horizontal leaf over bare soil was calculated for each plant for four characteristic types of days: clear-cool

187 Effects of Global Temperature Change on Agriculture

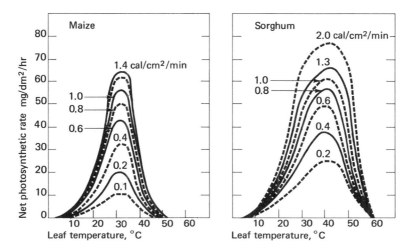

Figure 11.2 Net photosynthesis in maize and sorghum as a function of leaf temperature
Source: Idso and Baker, 1967.

Table 11.1 Total Daily Net Photosynthesis (mg/dm^2) as Computed for a Horizontal Sunlit Leaf over Bare Soil

Day	Maize	Sorghum
August 28	355	256
July 22	305	685
September 6	20	30
August 24		
Clear	440	392
Cloudy	364	236
Average	402	314

Source: Idso and Baker (1968). Reprinted by permission of *Agronomy Journal*.

($T_{max} = 19°C$ on August 28), clear-warm ($T_{max} = 33°C$ on July 22), cool-cloudy ($T_{max} = 11°C$ on September 6), and intermittent clear-cloudy ($T_{max} = 25°C$ on August 24). The results are contained in Table 11.1 and Figure 11.3.

It is evident from these results that when light is not limiting, maize is slightly more productive than sorghum during cool weather while sorghum is more than twice as productive as maize during warm weather. Since Figure 11.2 indicates that these two plants have similar low-temperature cutoff points for photosynthesis, it is evident that a cooling trend starting at the tempera-

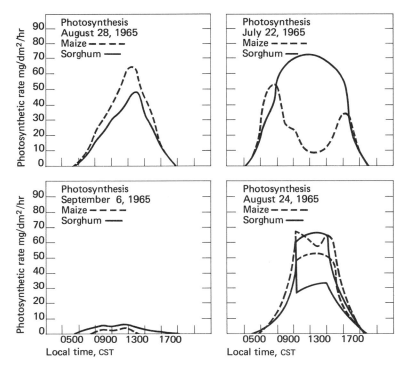

Figure 11.3 Net photosynthesis in maize and sorghum on four different days
Source: Idso and Baker, 1967.

tures of the clear-cool day will affect both plants nearly identically. Any warming trend starting from that point, however, would be quite disastrous for maize, yet perhaps beneficial to sorghum, depending upon how great a temperature rise occurred.

In another study (Idso and Baker, 1967), the photosynthesis of maize and sorghum was calculated for a complete crop as it would develop through the growing season for a constant daily set of environmental conditions (clear-warm). These results are contained in Table 11.2. They demonstrate that although individual leaves of different species may have quite different photosynthetic capabilities, when they are aggregated together into a plant canopy producing mutual shading, and so on, very similar results may be obtained. Thus, conclusions drawn from basic plant properties per se, as in the previous section for individual leaves, may not necessarily be valid. More detailed models of the

Table 11.2 Average Total Daily Net Photosynthesis (mg/dm^2) in Maize and Sorghum for Different Periods throughout the Growing Season

Stage of growth	Maize	Sorghum
Less than 10 leaves	305	685
50% cover, north-south row orientation	169	179
50% cover, east-west row orientation	177	181
100% cover, any row orientation	129	117

Source: Idso and Baker (1967). Reprinted by permission of *Agronomy Journal*.

type developed by Idso (1968, 1969) are required to improve our understanding of the photosynthetic response of a crop to its environment.

In this regard, one of the most important problems encountered in evaluating the temperatures of the differently oriented and located leaves of the plant canopy is to specify the light intensity incident on these several leaves. Of the models mentioned herein, the only ones that differentiate between differently illuminated leaves within individual horizontal canopy layers are those of DeWit (1965) and Idso (1968, 1969). Together, Idso and DeWit (1970) have presented this theory in detail.

Nonphotosynthetic Considerations

Although photosynthesis is the basic plant process that must proceed to supply us with food and fiber, there are other plant processes that are equally important. Consider, for instance, seed germination, establishment, flowering, fruit set, translocation of photosynthates, fruit development, and maturation. Each of these component physiological functions has its own temperature dependency (Polunin, 1960; Schimper, 1903); and if any one of them is inhibited or upset, it is equally disastrous for the end result.

In this respect, it is often the extremes of climatic conditions that are most important for cultivated crops (Mason, 1936), as compared to the mean conditions for total growing season photosynthesis (Clements, 1916). A severe freeze at the beginning or end of an otherwise very successful growing season, for instance, may completely destroy an entire citrus crop. In many areas, a decrease in the mean temperature of only 2 or 3°C may be sufficient to create extreme temperatures low enough to prohibit the successful production of citrus for this reason. Similarly, in other areas, a temperature increase of like amount could easily prohibit the successful production of cool weather crops.

With man on the scene, we even have the possibility of a small temperature *decrease* prohibiting successful production of *cool* weather crops. This is a very real possibility where scientists have developed hardy strains of wheat and other cereals and have pushed the areas of their production farther and farther north in Canada and Siberia (Polunin, 1960).

References

Allee, W. C., and Park, T., 1939. Concerning ecological principles, *Science, 89:* 166–169.

Barghoorn, E. S., 1953. Evidence of climatic change in the geologic record of plant life, *Climatic Change: Evidence, Causes, and Effects* (Cambridge, Massachusetts: Harvard University Press), pp. 235–248.

Billings, W. D., 1938. The structure and development of old field short-leaf pine stands and certain associated physical properties of the soil, *Ecological Monograph, 8:* 437–499.

Cain, S. A., 1944. *Foundations of Plant Geography* (New York: Harper and Brothers).

Clements, F. E., 1916. Plant succession, *Carnegie Institution of Washington Publication, 242*.

DeWit, C. T., 1965. Photosynthesis of leaf canopies, *Verslagen van Landbouwkundige Onderzoekingen, 663*.

Duncan, W. G., Loomis, R. A., Williams, W. A. and Hanau, R., 1967. A model for simulating photosynthesis in plant communities, *Hilgardia, 38:* 181–205.

El-Sharkawy, M. A., and Hesketh, J. D., 1964. Effects of temperature and water deficit on leaf photosynthetic rates of different species, *Crop Sciences, 4:* 514–518.

Gates, D. M., 1965. Energy, plants, and ecology, *Ecology, 46:* 1–13.

Gleason, H. A., and Cronquist, A., 1964. *The Natural Geography of Plants* (New York: Columbia University Press).

Good, R. D'O., 1931. A theory of plant geography, *New Phytology, 30:* 149–171.

Hesketh, J. D. and Moss, D. N., 1963. Variation in the response of photosynthesis to light, *Crop Sciences, 3:* 107–110.

Idso, S. B., 1968. A holocoenotic analysis of environment-plant relationships, *Technical Bulletin, 264* (Saint Paul: University of Minnesota).

Idso, S. B., 1969. A theoretical framework for the photosynthetic modeling of plant communities, *Advancing Frontiers of Plant Science, 23:* 91–118.

Idso, S. B., and Baker, D. G., 1967. Method for calculating the photosynthetic response of a crop to light intensity and leaf temperature by an energy flow analysis of the meteorological parameters, *Agronomy Journal, 59:* 13–21. Tables 11.1 and 11.2 and Figures 11.2 and 11.3 are reprinted by permission of *Agronomy Journal*.

Idso, S. B., Baker, D. G., and Gates, D. M., 1966. The energy environment of plants, *Advances in Agronomy, 18* (New York: Academic Press), pp. 171–218.

Idso, S. B., and DeWit, C. T., 1970. Light relations in plant canopies, *Applied Optics, 9:* 177–184.

Lehenbaur, R. A., 1914. Growth of maize seedlings in relation to temperature, *Physiological Research, 1:* 247–288.

Lundegardh, H., 1931. *Environment and Plant Development,* translated and edited by E. Ashby (London: Edward Arnold and Company).

Mason, H. L., 1936. The principles of geographic distribution as applied to floral analysis, *Madrono, 3:* 181–190.

Milner, H. W., and W. M. Hiesey, 1964. Photosynthesis in climatic races of *Mimulus.* I. Effect of light intensity and temperature on rate, *Plant Physiology, 39:* 208–213. Figure 11.1 is reprinted by permission of the American Society of Plant Physiologists.

Monteith, J. L., 1965. Light distribution and photosynthesis in field crops, *Annals of Botany New Series, 29:* 17–37.

Moss, D. N., 1963. The effect of environment on gas exchange of leaves, *Stomata and Water Relations in Plants,* Bulletin 664 (New Haven: Connecticut Agriculture Experimental Station), pp. 87–101.

Paltridge, G. W., 1970a. A model of a growing pasture, *Agricultural Meteorology, 7:* 93–130.

Paltridge, G. W., 1970b. Experiments on a mathematical model of a pasture (unpublished)

Polunin, N., 1960. *Introduction to Plant Geography and Some Related Sciences* (New York: McGraw-Hill).

Schimper, A. F. W., 1903. *Plant Geography Upon a Physiological Basis* (Oxford: Clarendon Press).

Scott, D., and Billings, W. D., 1964. Effects of environmental factors on standing crop and productivity of an alpine tundra, *Ecological Monograph, 34:* 243–270.

Thompson, D'A., 1942. *On Growth and Form* (New York: Cambridge University Press).

Waggoner, P. E., Moss, R. A., and Hesketh, J. D., 1963. Radiation in the plant environment and photosynthesis, *Agronomy Journal, 55:* 36–39.

Wilson, J. W., 1966. Effect of temperature on net assimilation rate, *Annals of Botony New Series, 30:* 753–761.

Winter, S. R., and Pendleton, J. W., 1970. Results of changing light and temperature regimes in a corn field and temperature effects on the apparent photosynthesis of individual leaves, *Agronomy Journal, 62:* 181–184.

12
Potential Effects of Global Atmospheric Conditions on Forest Ecosystems

Karl F. Wenger,
Carl E. Ostrom,
Philip R. Larson, and
Thomas D. Rudolph

Effects of an Increase in the Carbon Dioxide Content of the Atmosphere

According to the President's Science Advisory Committee (1965), the carbon dioxide content of the atmosphere increased about 7 percent between 1860 and 1960. They estimate further that, with the projected rate of fossil-fuel consumption, atmospheric carbon dioxide would increase about 25 percent by the year 2000 over the amount present during the nineteenth century, which was about 0.03 percent of the atmosphere. If all known recoverable fossil fuels were consumed, which could be approached within the next 150 years, atmospheric carbon dioxide could increase by nearly 170 percent. It seems reasonable to assume a 50 percent increase in atmospheric carbon dioxide by the year 2050.

An increase in carbon dioxide content would affect productivity but probably not species composition of forest ecosystems. In laboratory studies of photosynthesis, the photosynthetic rate increased linearly with an increase in carbon dioxide concentration up to 0.04 to 0.05 percent. Above that concentration, photosynthesis continued to increase but at a slower rate (Hellmers, 1964).

However, the growth of a forest would not increase to the same degree because of limitations of moisture and nutrients. Furthermore, the air in a closed forest is already enriched in carbon dioxide. The soil atmosphere may have carbon dioxide concentrations as high as 13 percent (Voigt, 1962). This carbon dioxide is released continuously through the day and night, but at night the forest vegetation is also releasing carbon dioxide since photosynthesis is not in progress. Consequently, the forest atmosphere becomes enriched with carbon dioxide at night. This higher carbon dioxide concentration results in a greater photosynthetic rate and increases dry matter production substantially over what would be produced in normal air. Hellmers (1964) presents data which show that more than 25 percent of the dry matter production of a European beech stand was attributable

Prepared for SCEP.

to this increased carbon dioxide concentration. Consequently, a general increase in atmospheric carbon dioxide probably would not result in a proportionate increase in forest productivity.

A 50 percent increase in atmospheric carbon dioxide would mean an increase in concentration from 0.03 to 0.45 percent, about the upper limit of the linear increase in photosynthetic rate. But because of the limitations in moisture and nutrients, and the effect of forest carbon dioxide enrichment, a corresponding 50 percent increase in forest productivity would be unlikely. A more reasonable expectation would be an increase of 20 to 25 percent.

Effects of Changes in Sunlight Intensity, Temperature, and Precipitation

Increases in CO_2 concentration, amounts of particulate matter, water vapor, concentration of ozone and other gaseous pollutants, and the addition of waste heat to the atmosphere will affect the intensity of incoming solar radiation, the temperature, and the rainfall.

A Reduction in Intensity of Sunlight

A decrease in the intensity of sunlight caused by the increase in the atmospheric factors and pollutants listed earlier would be accompanied by a related decrease in productivity of the earth's forests. Seedlings of broad-leaved species reach their maximum photosynthetic rate at about one-third of full sunlight. However, a forest of broad-leaved species photosynthesizes at a rate proportional to the light intensity because of the mutual shading and light diffusing effect of large masses of leaves. At low light intensity, many leaves are receiving less light than required for maximum photosynthesis; as light increases, more and more leaves receive this light requirement.

In needle-leaved species, the photosynthetic rate of individual needles follows the same trend in relation to light intensity as individual seedlings of broad-leaved species, reaching maximum rate at about one-third of full sunlight. But the photosynthetic rate of seedlings, and presumably of larger trees also, follows a trend similar to that of broad-leaved forests, reaching its maximum only at full sunlight. The needle arrangement results in mutual shading and light diffusion. Therefore, forests of needle-leaved species could be expected to follow the same trend as broad-leaved forests.

The reduction in productivity of the earth's forests would not bear a 1-to-1 relation to the reduction in light intensity; that is, a 10 percent reduction in light intensity would not result in a 10 percent reduction in productivity. This conclusion follows from the relation of photosynthetic efficiency to light intensity. Photosynthetic efficiency is the efficiency with which plants use a unit of light energy to produce plant material. Photosynthetic efficiency increases as light intensity decreases from full sunlight (Hellmers, 1964). The photosynthesizing mechanism approaches saturation between 0.1 percent and 0.2 percent of full sunlight. At full sunlight a smaller percentage of the total incident light is used than at lower light intensities. Efficiencies as high as 9 to 19 percent have been obtained at 0.1 or less of full sunlight (Hellmers, 1964). Thus, a 10 percent decrease in sunlight intensity would result in somewhat less than a 10 percent reduction in productivity because the lower light intensity would be used more efficiently.

A reduction in the intensity of solar radiation would very likely be accompanied by an overall reduction in temperature. Consequently, the effect of reduced light needs to be considered in relation to temperature effects.

Changes in Temperature at the Earth's Surface

The net effect on temperature at the earth's surface of changes in atmospheric factors and pollutants is unpredictable at present. According to the American Chemical Society (1969), "It is generally agreed that the 10-year mean global temperature rose about 0.4°C between 1880 and 1940 and then reversed itself, falling nearly 0.2°C by 1967." Meteorologists believe that the decline since 1940 is the result of increased particulate matter and cloudiness, which override the effect of increasing CO_2 concentration (James, 1970). Increasing CO_2 content and water vapor tend to raise temperature by the "greenhouse effect," while particulate matter, clouds, ozone, aerosols, and other substances tend to scatter, absorb, or reflect incoming sunlight, decreasing the amount of heat that reaches the earth's surface.

In addition to changing the mean global temperature, increases in CO_2, water vapor, and pollutants probably will affect day-night temperature differences, which strongly influence growth of many tree species. The most likely effect would be to raise night temperatures, because all the substances on the increase tend to suppress reradiation at night. Thus, a drop in mean global temperature would be the result mainly of a drop in day temperature, while a rise in the global mean would be the combined effect of rising day and night temperatures. Thus, any change in day-night temperature differences is most likely to be in the direction of reducing this difference.

According to Hellmers (1962),

Heat requirements between the upper and lower lethal limits for various representatives of a species may affect (1) the range of the species, (2) the distribution of the species within the range, (3) the site classification for the species, (4) the growth rate for individuals during a growing season, and (5) the induction of summer dormancy.

Determining temperature effects on tree growth and species occurrence is a complex problem because species vary widely in their response to temperature variations. Hellmers (1962) found that loblolly pine, northern red oak, and Douglas fir grow best with large day-night temperature differences, 10° to 13°C. Redwood growth was affected mainly by day temperature, while Digger pine responded most to night temperature. Jeffrey pine, erectcone pine, and eastern hemlock responded mainly to the total daily heat, regardless of when it was received, up to an optimum.

The effect of a rise in global mean temperature, assuming that it would change the day-night difference very little, would be to displace tree ranges poleward, with a widening of the tropical zone. On a long-term basis, productivity perhaps would increase, mainly because of the widened tropical zone and the forest area added by the poleward extension. However, the poleward extension of tropical forests conceivably could leave a less productive equatorial band if conditions there became less favorable for existing tropical species.

During the shorter term future, productivity might decrease while range shifts were taking place. Temperature patterns would become less favorable toward the equatorial side of ranges while the poleward extension remained incomplete.

A drop in global mean temperature with a decrease in day-night differences would be unfavorable for most species. Hellmers (1962) and others have shown the heat requirements of different species. Even within species, particular geographic strains likely have characteristic requirements in day-night temperature differences. McMillan (personal communication) found that high-elevation strains of sweet gum did not grow as well in a simulated low-elevation, higher latitude climate as they did with a day-night temperature range characteristic of their native habitat. To some degree, species would tend to migrate toward localities with more favorable day-night differences, particularly toward higher elevations. But, in general, a drop in global mean temperature with a reduction in day-night difference would tend to reduce productivity because it would reduce the photosynthetic-respiratory ratio. Photosynthesis might not be greatly reduced by the lower day temperature, but higher night temperatures would increase respiration, leaving less photosynthate available for growth and reproduction. Increased susceptibility to insects and diseases might also result.

Changes in Precipitation

Changes in temperature would undoubtedly be accompanied by changes in precipitation. These relationships are extremely complex and beyond the authors' area of competence. However, some evidence has been accumulated that particulate matter and water vapor tend to increase rainfall (James, 1970).

If we assume a general increase in precipitation without a change in geographic or seasonal distribution, the effect would likely be a significant increase in productivity of the earth's forests. Throughout forested regions of the earth, with the exception of swampy areas and possibly rain forests, moisture is the major limiting factor in tree growth. In temperate regions, annual growth of trees is so dependent on moisture that dating by tree ring analysis is possible. Forests characteristically exhaust available soil moisture before other growing season factors, mainly temperature and day length, become limiting.

Another effect of a general increase in rainfall that would increase productivity is an expansion of forested areas part way into presently drier regions, such as the Great Plains of North America. Such boundaries are characteristically elastic, advancing into the dry area during wet periods and retreating during dry periods. A permanent increase in precipitation would result in shrinking these nonforested regions. It would also facilitate afforestation of vast areas of savanna and scrub with productive timber trees such as pines.

A third effect would be on species composition of forests. More mesophytic species of both conifers and hardwoods would be favored and would expand their ranges where other factors were favorable. Generally, hardwood species would be favored over conifers, since there are a greater number of mesophytic hardwood species. Xerophytic species would react unfavorably and might disappear entirely from some areas. Also, they would suffer more in competition with mesophytic species.

A general decrease in precipitation would have opposite effects, reducing growth, shrinking forested regions, and favoring xerophytic over mesophytic species. Pines generally might fare better than hardwoods and other conifers.

If an increase or decrease in precipitation were accompanied by changes in seasonal and geographic distribution, the effects on forest ecosystems would be much more complex. Conceivable changes of that kind are so numerous that they cannot be considered individually here. Information available is probably not adequate for realistic assessments, in any event.

General Considerations

While some of the temperature and precipitation changes discussed earlier would apparently be favorable, other factors need to be considered. Migrations of species, such as might occur with temperature changes, depend on the ability of species to reproduce themselves in the invaded areas. Flowering and fruiting of many species are controlled by photoperiod. Consequently, north-south migrations would be limited by this factor even though temperature and precipitation patterns are favorable.

The additions to the atmosphere of CO_2, water vapor, and pollutants have drawn attention and are causing concern because of the rate at which they are occurring. Estimates of future conditions are usually cast in terms of 25, 50, or 100 years. For example, a 25 percent increase in the CO_2 content of the atmosphere by the year 2000 over that in 1965 has been estimated (President's Science Advisory Committee, 1965). This increase without other atmospheric changes could result in a rise in the global mean temperature of as much as 1°C. Most trees that will compose the forests of the year 2000 are already in existence. Consequently, little adaptation to this temperature change could occur in forest ecosystems through range displacements. Changes in composition would be in the direction of decreasing diversity as species that were less tolerant of the new temperatures failed in competition or reproduction.

In general, any of the climatic changes that might occur from atmospheric additions are likely to be deleterious. Tree species have become adapted to the climate within their ranges through long-term evolution. With few exceptions, research has shown that forest species are best fitted to the climate in which they are found. Such adaptations exist even within species. Thus, evolutionary adaptations to a rapidly changing climate could not be expected. Consequently, any appreciable permanent change in worldwide climate is likely to reduce productivity for a long time.

If climatic trends could be reliably predicted, or a change verified soon after occurrence, the reduced productivity of present forest ecosystems might be partially compensated for by planting species selected for their adaptation to the new conditions. For example, increased rainfall would permit planting of some species in areas presently nonforested because of low rainfall. A temperature rise might permit some species to be moved to higher elevations. In short, changes in natural forest ecosystems might be anticipated by planting. It is very doubtful, however, that such steps would go very far to compensate for the worldwide reduced productivity of forests resulting from climatic change.

Because of the enormous complexity of the relationships and interactions and the great number of species involved, it would seem virtually impossible to predict reliably the vegetational changes likely to occur by constructing models from individual species data. A more promising method would be base predictions on ecosystem-climate analogies. This method would require detailed description of currently existing ecosystem-climate associations. These records would then be the basis for foreseeing the kind of ecosystem that would develop in a particular locality with a specific predicted or verified climatic change. Analogies may not currently exist for all conceivable climates that might come about from atmospheric additions, but it seems likely that a high proportion of the possibilities could be provided for with this approach.

Effects of an Increase in Radioactivity and Toxic Pollutants
A significant increase in radioactivity in the atmosphere resulting from a catastrophe would have a considerably greater impact on coniferous forest than on deciduous, as indicated by the work of Woodwell (1970) and others at Brookhaven. Miksche (1968) has shown that in *Pinus banksiana* and *Picea glauca* the DNA content varies within the species according to geographic origin of the tree. One would expect that this would mean that the origins having the higher DNA content would be more sensitive to damage by radioactivity. However, this is probably not the case because Miksche's work indicates that conifers are an exception to the general rule that radiosensitivity is directly related to DNA content of the species.

Ionizing radiation would also have certain aftereffects on forest ecosystems, and particularly on the conifers. Sublethal exposure could reduce reproduction potential and could result in genetic effects that would show up only in future generations.

Gross changes in forest composition can be expected from harmful concentrations of air pollutants. For example, it is said that essentially all of the conifers in Saint Louis are less than about thirty odd years in age, because all of the conifers were killed by earlier severe air pollution.

In Tokyo many of the conifers in the heavily polluted sections are dead or dying and will have to be replaced by other ornamentals until air pollution is brought under control.

In the Rhine Valley a similar change in forest composition is taking place, and attention is being given to the types of trees which may have to be introduced as replacements.

Experiments by Hagestead at Beltsville indicate that plants are growing much better in pure air than in city air from Washington. This result indicates that the growth rate of trees and forests in cities may be significantly reduced even without visible symptoms of air pollution.

It is reported that recent reduction of air pollution in London has permitted a considerable increase in tree growth and that certain trees are again invading the city, according to Fred Last of Edinburgh University.

Some scientists are concerned about the possible effects on forest ecosystems of the silver in silver iodide released into the atmosphere in weather modification programs. However, there is no agreement as to the probable effects, and in certain areas silver is already present in larger quantities than those resulting from weather modification.

The effects of radiation and air pollutants on forest ecosystems have recently been discussed by Woodwell (1970). He holds that the effect of air pollutants and other adverse influences are similar to those of radiation. Trees are eliminated at low exposures; as exposure increases the tall shrubs go next, then the lower shrubs, then herbs, and finally mosses and lichens. He cites evidence that other adverse influences, such as high-altitude exposure, salt spray, water availability, repeated fire, and SO_2, produce the same trends. He believes that other chronic disturbances would have similar effects. He reasons that large plants, which have a relatively smaller photosynthetic area and a larger respiratory demand are more sensitive to chronic disturbance, because photosynthetic capacity is likely to be reduced without a reduction in respiratory demand. Smaller plants can sustain relatively more damage to photosynthetic capacity, because they have a relatively larger photosynthetic area and smaller respiratory demand.

Monitoring Forest Ecosystems
The President's Science Advisory Committee (1965) has recommended base-line studies of natural populations in diverse relatively unpolluted habitats to establish a basis for comparison with populations under pollution stress. A framework already exists for such base-line studies in the Research Natural Areas already set aside on publicly and some privately owned lands. These natural areas were set aside because they are undisturbed representatives of distinctive forest ecosystems. They are intended to remain undisturbed. Most are distant from population centers and very likely among the least polluted ecosystems in the United States.

The Committee also suggests that the general environment

be monitored by means of 100 small drainage areas selected to be representative of the general environment. Air, soil, water, and living organisms would be sampled periodically to determine general pollution levels and the trends in populations of living organisms. The Research Natural Areas (RNA) and these sample drainages would constitute a fairly comprehensive sample of all kinds of environments, especially if the RNA array were rounded out to insure that all identifiable natural ecosystems were included. The lands of the drainage areas would vary from highly cultivated farm and forest through rangelands to desert, so some might be selected to include established RNA. Such an arrangement should provide excellent comparisons of pollution effects on managed forests with those on otherwise undisturbed forest ecosystems.

Somewhat the same approach might be extended to worldwide monitoring, especially since the countries most concerned are the more industrialized. A number of these already have pollution control and monitoring programs in progress.

Monitoring population densities in forest ecosystems is comparatively easy, since it can be done by periodic sample counts. Effects on productivity are much more difficult to determine because of the longevity of trees and the large number of species, which respond differently to environmental changes. One approach might be to select fast-growing, short-lived species, such as cottonwood in the southern United States, and grow them in a replicated design on a short rotation, perhaps only five years, in a continual sequence with annual plantings. Uniform soil conditions would be necessary within the experimental area and soil management measures would have to be devised to maintain soil productivity. Careful and complete measurement at harvest of each year's planting might then show trends related to atmospheric and climatic changes. An array of such plantings in critical places around the world would be needed. Such an approach, although fraught with complications, would still be about the simplest and quickest way to detect possible changes in productivity resulting from atmospheric and climatic changes. The productivity of such plantings would be only an index, rather than a direct evaluation, of forest ecosystem productivity.

References

American Chemical Society, 1969. *Cleaning Our Environment: The Chemical Basis for Action* (Washington, D.C.: American Chemical Society).

Federal Committee on Research Natural Areas (Compilers), 1968. *A Directory of Research Natural Areas on Federal Lands of the United States of America* (Washington, D.C.: U.S. Government Printing Office).

Hellmers, Henry, 1962. Temperature effect on optimum tree growth, *Tree Growth*, edited by T. T. Kozlowski (New York: Ronald Press), pp. 275–287.

Hellmers, Henry, 1964. An evaluation of the photosynthetic efficiency of forests, *Quarterly Review of Biology, 39:* 249–257.

James, Richard D., 1970. Changing climate, *Virginia Journal of Education, 63*(6): 11, 22.

Miksche, Jerome P., 1968. Quantitative study of intraspecific variation of DNA per cell in *Picea glauca* and *Pinus banksiana, Canadian Journal of Genetics and Cytology, 10*(3): 590–600.

President's Science Advisory Committee, 1965. *Restoring the Quality of Our Environment*, Report of the Environmental Pollution Panel (Washington, D.C.: U.S. Government Printing Office).

Voigt, G. K., 1962. The rate of carbon dioxide in soil, *Tree Growth*, edited by T. T. Kozlowski (New York: Ronald Press), pp. 205–220.

Woodwell, G. M., 1970. Effects of pollution on the structure and physiology of ecosystems, *Science, 168:* 429–433.

13
Climate and Forest Diseases George H. Hepting

Changes in climate, which we may consider long-term weather trends, can influence known diseases or create new disease problems. Examples have been given (1) of how weather influences many tree diseases including blister rust, fusiform rust, Phytophthora root rot, Dutch elm disease, black root rot in southern nurseries, pole blight of western white pine, sweetgum blight, ozone damage, and others.

In general we might consider that our field crops and other annual or short-lived vegetation reflect weather changes; whereas trees, soil flora, and other perennial flora will reflect, in addition, climate changes. The main climatic factors that pathologists are concerned with are precipitation, temperature, humidity, fog and dew, wind, and radiation. If a pathogen is well equipped for reproduction and means of spread, we find that weather factors are usually paramount in determining disease severity. Most of our forest diseases are caused by fungi. Weather factors will affect the growth of a fungus, its sporulation, its rate of spread, spore germination and infection, and the response of the host to infection. In fact, for most plant diseases a rather special concomitance of circumstances is necessary to produce an epidemic (Yarwoood [2], Humphrey [3]). Waggoner (4) expressed the normally small chances of infection in a curious but interesting way—"A severe outbreak is a rare removal by the weather of obstacles that ordinarily restrain the pathogen."

In order to explore how climatic change can influence tree diseases, we must look first at what pathologic processes might be influenced by changes of a given magnitude, and then review some known cases of limitations put on the development of some diseases by climatic factors.

There is no intent here to review the voluminous literature on specific effects of meteorological factors on plant diseases, since so many of these are common knowledge, and since this subject has been covered elsewhere (2, 3). I will, however, present evidence on the instability of our climate, and then on the basis of what we know about weather limitations governing some forest diseases, show how recent climatic changes could be influencing these

Reprinted from *Annual Review of Phytopathology*, 1: 31–50 (1963). Survey of literature pertaining to this review was concluded in December 1962.

diseases. I should like to avoid becoming involved with the many effects of climatic changes on such normal physiological processes as photosynthesis and respiration, and would rather hold as close as possible to climate change as it influences the impact of pathogens and such physiogenic disturbances as air pollutants and prolonged moisture stress.

Our Climate Is Changing
The evidence is clear that one of the features that characterizes the world's climate is its instability. Going back millions of years, there have been long periods of moderate climate called "normal times" by the geologist. Man has never known the "normal" periods because he arrived here after the last one. The brief intervals of comparative climatic violence the geologist calls "revolutions." As Russell (5) points out, two outstanding features characterize the revolutions: (a) There is unrest in the crust of the earth—earthquakes, volcanoes, upthrusts of high mountains, and retreat of oceans. He points out that we are now in a period of as violent crustal unrest as the earth has ever known, with 325 volcanoes active. (b) There are ice caps, frozen seas in polar regions, and ice extending down to warm regions (glaciers). The meeting of cold air and warm air masses results in violent weather changes. During the geologist's "normal" periods, there was no ice, but warm seas and an equable climate year-round.

In the last ice age five advances and five retreats of the ice probably occurred. We are now living in a period of ice retreat that many geologists feel may be marking the end of our last great ice age.

Coming to modern times, Russell (5) traced climate change in periods of one-half to several centuries duration. In the first century Europe was wet, and from 180 to 350 it was wet. The fifth century was very dry in Europe and Asia, and so dry in North America that many western American lakes dried up completely.

In the seventh century Europe was so warm and dry that glaciers retreated permitting heavy traffic over Alpine passes now completely closed with ice. The study of tree rings shows that this period was also dry in the United States. In the ninth century precipitation became heavy, lakes rose, and the rings of trees dating this period were wider. The 10th and 11th centuries were again warm and dry. In fact, the Arctic ice cap may have disappeared entirely then. Greenland was settled in 984 and abandoned to the ice in 1410.

Early in the 17th century Europe was very wet. The glaciers extended, and there were disastrous floods in northern Italy. The glaciers receded from about 1640 to 1770 and then advanced again until the mid-19th century. Since then they have receded to about their 16th century positions. The recent warming appears to Russell (5) to be a worldwide condition. It suggests that the 19th century had higher temperatures than the 18th and, in its first half, the 20th had higher temperatures than either (6).

Through all the changes listed above there is little evidence of cycles. The type of weather and the duration of that weather seem to most investigators little short of random. The reader should bear in mind that the duration of many of the above climatic periods runs from 100 to 300 years, since this point will take special meaning in challenges made by some to the validity of the "climax forest" concept, and in support of Holloway's (7) concept of unstable climate vitiating the climax idea at least for the South Island of New Zealand, and probably elsewhere.

We now come to the present century and find general agreement that during the first forty or more years the warming trend has not only been continuing but possibly accelerating. The world's mean annual temperature increased 1°F from 1885 to 1945 (6). In most of the United States, summers are 1°F hotter and winters 2 to 4°F warmer. In the Iceland, Greenland, Spitzbergen area the mean annual temperature has risen 6 to 12°F. Temperature data obtained in Antarctica during the International Geophysical Year seem to confirm the world's long-term warming trend. In 1951 Holloway (7) mentioned that the eastern glaciers of New Zealand's South Island were in general and rapid retreat. During the next ten years, according to the New Zealand Geological Survey, the South Island's Franz Josef Glacier receded another 3500 feet.

The warming trend, especially notable since 1900, has meant about a 2°F rise in mean annual temperature in the North Temperate zone and about a 5°F rise for the Arctic. A 2°F change is equivalent to about 100 miles of latitude. If one wonders whether a 2°F general rise in temperature would have much effect on the world's complex of biological activity, it is interesting to note that Russell (5) maintains that a rise of 2°F in the temperature of the earth now would be sufficient to clear the polar seas of all ice. It would also materially increase evapotranspiration.

Glaciers have been receding in Alaska, the tree line in Finland has been migrating northward, and there has been a northward migration of snow lines, freeze-free seasons, and of animals and plants. The weather over most of the eastern United States has been becoming more maritime, which means less severe winters, cooler springs, longer summers, lingering autumns, and a generally more equable climate [Brean (8)]. Landsberg (9) and Baum and Havens (10) have presented effective reviews of the evidence of a warming trend.

Some climatologists feel that in the early 1940s the United States again entered a cooling phase (6). Appraisals of the effects of the warming trend in the United States discussed in this paper, therefore, cover mainly the period from 1885 to 1945.

As we explore what these changes in long-term weather mean with respect to forest diseases, we must realize that while mean rises or drops in temperature or moisture are important, the extremes which vary with these means may be more important. Daubenmire (11), in stressing the kinds of climatic data that the ecologist needs to interpret his problems, questions whether temperature and precipitation means themselves are worth the trouble of computing. He has developed special methods of expressing climate for the biologist.

The time of year when weather changes are greatest and the shifts in seasons such as the warmer autumns and later, cooler springs being experienced in southeastern United States will influence diseases. Also, the shifts in storm paths and other accompanying changes, such as the time and duration of periods of fog or dew, must be of great significance in the behavior of disease organisms and the vectors that carry them.

Why Our Climate Is Changing

There are many theories as to why we have had variable climate over the ages and, also, within an age such as the current one. Russell (5) summarizes these theories and the principal ones are based upon the following phenomena: change in the angle between the earth's axis and the ecliptic plane; sunspots; variation in the amount of energy emitted from the sun; changes in the sun's radiant energy reaching the earth due to changes in the earth's atmosphere through which this energy travels; effects of volcanic dust; changes in the rate of escape of the earth's heat through the atmosphere; variations in activity of the earth's crust; and atmospheric accumulations and dissipations of carbon dioxide.

The CO_2 theory is widely held as at least part of the explanation for climate change, and particularly the recent warming period. Since the middle of the 19th century, man has put into the atmosphere an enormous quantity of CO_2 mainly through the combustion of coal and oil. It has been estimated that in another fifty years, 47 billion tons will be added annually. Junge (12) reported a 10 percent increase in atmospheric CO_2 since 1900. He says the oceans have absorbed much of this 10 percent rise, but by the year 2000 the accumulations of CO_2 could be very serious.

Plass (13) mentions the postulate that CO_2 regulates the temperature of the earth and that much of the reradiation of energy away from the earth is most intense at wavelengths very close to the principal absorption band of the CO_2 spectrum. He says that because the CO_2 blanket prevents its escape into space, the trapped radiation warms the atmosphere. He presents a theory of climate change through the ice ages based on CO_2 being absorbed by the oceans thus cooling the earth and the resulting sea ice accumulations shrinking the seas, concentrating the absorbed CO_2 and causing it to be given off again, during the formation of sea ice, leading to a warming trend, when the oceans expand again, reabsorb the CO_2, and the process starts over again.

Plass computes that the approximately 13 percent increase in CO_2 that has taken place in the past 100 years would theoretically raise the earth's temperature 1°F, and that is almost exactly what it has risen. He estimates that by the year 2000 the earth's average temperature would rise 3.6°F if the present trend continues, but there is evidence that his estimates may be high (14). Such a change would surely affect the balance between forest trees and their environments and their parasites.

The view that establishing more tree growth would mean more CO_2 absorption, reducing the problems arising from CO_2 excess has been expressed by Leake (15). We could count on some excess of CO_2 being utilized by existing forests and reflected in increased growth [Plass (13), Kramer and Kozlowski (16)]. However, a tremendous increase in the volume of green vegetation would be required to use a large proportion of the CO_2 increase resulting from fuel consumption. Anyway, if Plass' (13) CO_2 balance sheet is approximately correct, then the consumption of CO_2 by biological activity, mainly photosynthesis, is exactly countered by CO_2 release through respiration and decay, and this would mean *no* net reduction in CO_2 by increasing green vegetation.

Is the Climax Forest Concept Valid?
Much of the forester's long-range planning is based upon the concept of the climax forest. Thus, his cutting methods may appear to him to be favoring climax, subclimax, or temporary species. With reference to the climax forest Zon (17) says, "Any forest if left undisturbed for long periods of time will become permanent in character as a result of attaining perfect balance with its environment. This final stage in forest development is usually known as the climax type."

On this topic Hayes and Buell (18) write, "Note that moisture seems to control this natural succession, and that over the centuries, or the millenniums, all forest if undisturbed will approach a condition of medium moisture, as nearly as is possible within the limits imposed by climate. That is, they approach the climax forest. . . ."

Now, Zon's ultimate adaptations of the forest, leading eventually to some permanent composition, and Hayes and Buell's millennia of undisturbedness leading to a permanent climax forest require that the local climate settle down long enough so that the forest can finally seek a permanent composition appropriate to this climate. Daubenmire (11) indicates such stability for vegetation boundaries for one region in stating: "This question is adequately answered for eastern Washington and northern Idaho by fossil pollen studies which indicate very little changes in the positions of ecotones since the beginning of post-Wisconsin time." With respect to the present century, however, Keen (19) shows decided drops in tree ring width of ponderosa pine in eastern Oregon beginning about 1908 and through 1930, with a critical drop from 1912 until the study was concluded in the early 1930s. His tree ring calendar, starting in the 14th century, shows that the diameter growth reduction period from 1912 to 1931 was of longer duration and of greater magnitude than any other such period in almost 700 years of ring history (Figure 13.1). He relates this drop largely to moisture deficit. While he considers that cumulative drought has led to the drastic growth drop since 1917, there remains the possibility that other factors—biotic as well as climatic—may have played a part.

Marshall (20) working in northern Idaho made annual-ring analyses of large western white pines that showed definite periods of climatic change (Figure 13.2). Marshall says his curves of growth over time "reveal one outstanding point. In every age class of trees (280, 230, 180, 140, and 75 years old) . . . there is a rapid decrease in growth during the past 40 years (1885 to 1925). . . . The only possible solution seems to lie in a deficiency of precipitation. The deficit . . . appears unprecedented in the 250-year period covered by these growth curves." His work was done in the general area in which the pole blight disease of western white pine has occurred, being reported first in 1929.

Accepting the position that climate is dynamic and not static, and that periods of weather lasting 50 to several hundred years have occurred that differed markedly in temperature and precipitation from those preceding or following, we know that the impact of these changes on our forest vegetation must be great. As Holloway (7), writing of New Zealand, puts it, "Readjustments in response to one particular change in climate may still be in progress when the next change occurs. . . . Thus, for example, the forests may still be responding to the onset of an era of desiccation for some considerable time after climates have again tended to become wetter. . . . This is a possibility wholly in line with developments in the northern hemisphere."

Holloway (7) was able to study ecological changes in areas of New Zealand's South Island spared by fire and the disturbance of animals and man. He had at hand a remarkable natural laboratory for his studies, and presents some penetrating new ideas on forests in relation to climate. The crux of his hypothesis is that climate changes could have been so recent that existing forests bear the imprint of old forests to this date, that existing forests

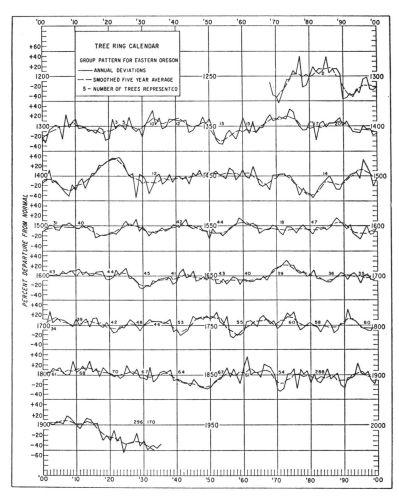

Figure 13.1 Keen's "tree ring calendar," expressing radial wood growth of ponderosa pine in eastern Oregon in terms of percent departure from normal, annually, 1268–1935

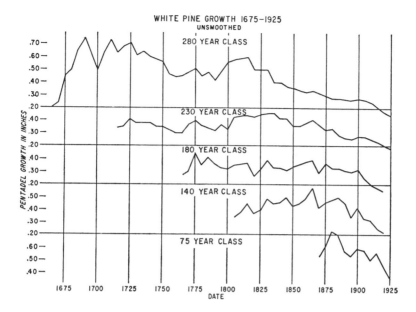

Figure 13.2 Marshall's radial wood growth of western white pines in northern Idaho of 5 age classes
Each point represents 5 years of accumulated growth (pentadel). There are 5 trees in each age class.

are in a plastic and unstable condition, and that with respect to forest types and species distribution existing forests may be largely out of phase with respect to present climates.

In countering the position that a mean annual climatic shift of approximately 2°F (with appropriate shifts in extremes) or a few inches of rainfall can cause a major recession of a species, the argument is sometimes put forth that such changes could affect a species only near the limits of its range, but not through the main range. This argument ignores the existence of provenance, or geographic races. Parts of New York State have a mean annual temperature just 2°F lower than Asheville, North Carolina, about 700 miles to the southwest, yet the flora of the two areas are quite different. New York has gray birch, aspen, and other boreal species, while the Asheville area sustains shortleaf pine, southern red oak, and other warm-climate species. Many other species thrive in both localities. Yet a native black oak or hemlock growing on

Long Island has different requirements and different responses than a native black oak or hemlock in North Carolina.

Squillace and Silen (21) deal with provenance in ponderosa pine, and cite much of the literature on racial variation in forest species. This review presents evidence that when seeds of certain tree species from different geographic sources were planted at a single site, the trees they produced have been found to vary significantly in survival, height growth, frost resistance, weevil and tip moth damage, needle color, and other characteristics. Zobel and Goddard (22) reported a survival test of loblolly pine of different seed sources planted in Texas, with drought the main factor involved, in which 8 percent of the seedlings raised from a Florida source survived as compared with 95 percent survival of seedlings from seed from a droughty area in Texas.

Raup (23) gives evidence of recent climatic changes in New York and New England, and states that forest managers should take climatic trends into consideration in their silviculture. A consistent change in the weather over many years will affect trees to some degree throughout the range of a species. The general effect of such a change might be illustrated by taking two identical rulers and placing them side by side so that the numbers and markings match. Let one ruler represent a species with each inch-mark a gradation of latitude, and the other represent habitat conditions with each inch-mark the appropriate environment for its counterpart ecotype on the other ruler. Now, slide one ruler along the other and the fit of ecotype and environment is off *all along the scale*—not just at the ends.

The "provenance" or local adaptation characteristic of a strain of trees and the fact that long-lived trees will keep their genetic makeup throughout their life while local climate may change could be reasons why some general declines of tree species have been taking place over wide areas.

Zon (17) describes the great diversity in the forest vegetation of the United States and attributes most of it to variations in temperature and moisture. He shows an interesting chart of the forests along the 39th parallel of north latitude depicting forest development in relation to elevation, rainfall, and prevailing west-to-east winds. He states, "The smallest difference in elevation (a depression or a hill) or in exposure (north or south slope) is immediately reflected in the character of the vegetation. . . ."

In an example from a locality in New Mexico, Pearson (24) shows how species composition varies with temperature and precipitation through altitude:

Forest type	Precipitation (Inches)	Maximum air temp. June to September (°F.)
Pinyon juniper woodland	12–16	78–81
Ponderosa pine	20	70
Douglas-fir	24–26	60–62
Engelmann spruce and associates	30	57–58

The effect of climate on species distribution and character of vegetation has often been noted. Beilmann and Brenner (25), writing of the eastern and northern portions of the Ozark region of Missouri, give evidence supporting their position that,

> Within historic times this vast region was a prairie, or at least park-like in that the trees were widely spaced and confined to the water-courses and drainage ways. The logging operations which are now so much a part of the industry of the Ozark region are but little more than a century old. The loggers have been, and still are, cutting the first crop of trees to mature there.

Although conceding that fire control has probably had a role in the change from prairie to forest, these authors feel that climate change has been a major factor and have recorded "pertinent observations—which indicate that the Ozarks today enjoy a much milder and wetter climate."

If climate change alters tree species distribution, the next questions would be how these weather changes influence a tree's normal functions, and how they influence the biota—the insects, fungi, bacteria, viruses, and mistletoes that prey upon trees. In this connection Paine and O'Regan (26) have shown that the optimum temperatures for the growth of isolates of the heart rot fungus *Echinodontium tinctorium* from central California were 1.2°C higher than the optimum temperatures for isolates from northern California.

Diseases and Climate Change

Many of our worst tree pathogens have caused great damage over a wide range of climatic factors, being exceptionally well equipped to spread, infect, cause disease, and reproduce. These include the fungi causing chestnut blight, Dutch elm disease, mimosa wilt, and red heart. Other killing diseases with a high potential for damage find conditions so rarely suitable for their development that they only occasionally or never have become serious. In this group we might place Phytophthora root rot of Douglas-fir in the Pacific Northwest, sapstreak of maple and yellow-poplar in the Appalachians, pine pitch canker in the South, some rusts and leaf blights, and probably many other "sleepers" that a consistent change in temperature and moisture could make very damaging. Wagener (27) refers to several such maladies as "sporadic diseases of infectious origin." He cites an area of pinyon pine and ribes that was hit hard by the pinyon rust in the 1920s and not again from then through 1960. Another area of ponderosa pine was hit hard by comandra rust in 1932 to 1933 and almost not at all since. A disease of Douglas-fir caused by *Phomopsis lokoyae* adjoining the Napa Valley in California frightened observers by its severity in 1930 when weather conditions especially favored attack. It then subsided and has caused little concern since.

Between the pathogen-caused diseases that are relatively insensitive to a wide range in weather factors and those that might be considered oversensitive to them, we have many of our important tree diseases. In this last group are the threshold diseases—those which have demostrated that they often can be very damaging but only under certain climatic conditions. In this group are most of the leaf diseases and most of the rusts, including blister rust, fusiform rust and southern cone rust, the powdery mildews, Rhabdocline needle blight of Douglas-fir, sycamore anthracnose, and *Keithia thujina* on western red cedar.

Woods (28) proposed the view that climate change could be a major factor in the accelerated activities of certain forest pathogens, and that the recession of some species following a warming climate may be through effect on disease organisms. He cited the heavy losses of American chestnut in the Southeast from the soilborne fungus *Phytophthora cinnamomi* prior to the chestnut blight. This same usually tropical to subtropical fungus later has played a role in the destruction of shortleaf pine from little leaf in the Southeast.

There are also physiogenic diseases not resulting from attack by living organisms, but from such causes as prolonged moisture stress, water excess, and toxic gases. While direct impacts of weather extremes over short periods produce effects such as drought, flooding, or sunscald, such conditions are usually obvious. Hepting et al. (29) described the disastrous effects of a severe freeze on a large number of woody plants, mostly well-established exotic or subtropical trees in the Southeast in 1951. Sometimes, however, as in the case of ozone damage to white pine (6), effects directly related to weather may not be obvious.

Occasional low temperatures are regarded by Wagener (30) as restricting two species of leafy mistletoe (*Phoradendron*) to their present limits in California and account for their absence over other parts of their host ranges. He also determined that severe cold almost completely eliminated the comandra rust (*Cronartium comandrae*) from a ponderosa pine plantation in Idaho by killing almost all of the cankered branches (31). Unusual heat has caused conspicuous injury to some California broadleaf trees (32). Ice damage is undoubtedly a major limiting factor in restricting the northern limits for slash pine.

It is probable that weather changes of many years duration have had a role in sweetgum blight, pole blight of western white pine, pitch streak of turpentine pines, and littleleaf of shortleaf pine; and it is possible that such changes have had a part to play in the upsurge of birch dieback, diebacks of sugar maple, ash, black walnut, and other species in the Northeast, and a notable decline or dieback of many hardwood species in several southeastern states that is causing grave concern among many foresters in this region today.

In New Zealand, Gilmour (33) stresses several destructive sequences following weather aberrancies, including frost followed by the canker fungi *Diplodia pinea* and *Phomopsis strobi* on *Pinus radiata,* wood rots following freezing of bark on this species, and hail followed by *D. pinea* on *P. radiata* and *P. ponderosa.* Warm wet summers there led to fungus diseases of the foliage and to *Rhizoctonia solani* damage to nursery seedlings. Excessive soil moisture together with species of *Phytophthora* kill *P. radiata* on heavy soils in New Zealand (34).

Climate Change and Pathogen Behavior

The limitations imposed on pathogens by weather are great. In Minnesota and the Dakotas, epidemics of stem rust of wheat did not occur in seasons when the average temperature during the critical period was below 61°F (3). At Quincy, Florida, and Tifton, Georgia, 28 years of records showed that every one of the four severe tobacco blue-mold years was preceded by a well-above-normal mean January temperature close to or above the 62°F optimum for infection (35).

Van Arsdel et al. (36), working on sporidial formation and germination in white pine blister rust, stress that even when periods of moisture-saturated air occurred, a prior period of two weeks with no three consecutive days over 28°C became important to provide fertile teliospores; and for subsequent infection on pine another period of 48 hours of moisture-saturated air required temperatures under 20°C.

Dew is a function of temperature and moisture. In the southern Appalachian region, even though white pine and ribes occur together in many areas and the rust has occurred throughout the region for decades on ribes, only at elevations approaching or above 3000 feet, where dew and fog prolong free moisture on the foliage well into the mornings, does infection occur on pine. The rust is minor in the southern Appalachian region wherever the mean July temperature exceeds 70°F, but becomes major in the 68° zones.

The southward spread of blister rust in the West into the sugar pine region of California is strongly limited by climatic factors. Comparing conditions for pine infection at Prospect Point, Oregon (latitude 42°44′), with those at Giant Forest, California (latitude 36°34′), Kimmey and Wagener (37) reported 92 "precipitation periods" in the late summer and fall between 1927 and 1956 at the Oregon station and only 28 at the California station. They report that 18 seasons out of 30 were too dry to allow any possibility of pine infection at the former, and no such dry season occurred at the latter in the same 30 years.

The incidence of fusiform rust in the South has been on the increase for many years. This is partly because fire control has resulted in a greater preponderance of the rust-susceptible slash and loblolly pines over the resistant longleaf, and in much more oak, which harbors the alternate stages of the rust. However, any change in spring weather that would advance the dates of budbreak in oaks, aecial fruiting, and pine shoot growth could increase rust incidence, especially if followed by a cool and damp late spring. Siggers (38) laid great stress on these relations of early spring temperature and moisture to this important disease.

Although oak wilt has been known at points near the 35° north parallel of latitude from northern Arkansas to North Carolina and northward since 1951, it has never been found south of this line in 11 years of surveys since 1951, although susceptible oaks occur abundantly to the Gulf of Mexico. Kuhlman's work (39) indicates that in spite of the high susceptibility of Douglas-fir seedlings to *Phytophthora cinnamomi* and the occurrence of the root rot in many small areas in the Northwest, the disease will probably not become important in the great Douglas-fir forests of that region because in the wetter months the weather is usually too cold for effective zoospore production and infection, and in the warmer months the soil is too dry.

If the temperature range of Minnesota and the Dakotas consistently rose 2°F, we could expect severe wheat rust years there frequently (3). If the Quincy, Florida-Tifton, Georgia area went up 2°F in mean January temperature, tobacco growers there could expect the incidence of severe blue-mold years to about double (35). On the other hand, such a general rise could greatly *reduce* the area of severe blister rust in the East, the Lake States, and the Northwest, but might mean putting the Douglas-fir forests of the Northwest in a temperature bracket favorable to *P. cinnamomi* during the wetter months. If oak wilt spread or development is limited by temperature, a mean 2°F rise would probably cause a cessation of the disease in at least North Carolina, Tennessee, and Arkansas.

Soil temperature can usually be expected to follow air temperature closely down to a 2-inch depth, within which a high percentage of the feeding roots of many forests trees occur. In the upper inch soil temperature may exceed air temperature. Hodges (40) working with the fungus-caused black root rot of southern pine nursery seedlings found that even when the two causal fungi occur in a nursery no damage occurs unless or until the soil temperature reaches 90°F and remains there or higher for many hours. A warming trend could thus expose the more southerly nurseries to root rot longer and could bring nurseries further north into the temperature range for this disease.

These examples concern some easily comprehended phenomena. Yet some of the most profound effects of temperature and moisture changes would be on the soil microflora and anywhere else in or on a tree where a complex of organisms live in relationships bordering pathogenesis, symbiosis, or saprogenesis. Tree roots are covered and surrounded by a multitude of fungi, bacteria, actinomycetes, insects, protozoa, and nematodes. The balance among them can be upset, perhaps turning a normal mycorrhizal association to pathogenesis, or through shifts in the ascendency of competing organisms due to their different temperature or moisture optima, permitting a pathogen to take dominance from a saprogen. In my earlier paper (1), I cited the work of Redmond (41), who put heating cables in the soil around birch trees in the field and found that raising the temperature an average of 3.6°F for 100 days raised root mortality from a normal of 6 percent to 60 percent. Since the heat could have practically no direct lethal effect on the roots themselves, Redmond explored the temperature relations of two birch root fungi and found that a *Cylindrosporium* inhibited a mycorrhizal associate at temperatures above optimum for root growth.

Ward and Henry (42) found that on agar the saprophyte *Trichoderma viride* grew rapidly at 25–30°C with a distinct growth advantage over the slower-growing parasite *Ophiobolus graminis* at that temperature, but at 10°C *O. graminis* grew as fast as *T. viride*. Henry (43) had found that in unsterilized soil wheat seedling infection by *O. graminis* was severe at 13°C but absent at 27°C, and this difference is attributed to antagonism to the parasite by other soil flora at the higher temperature.

While weather can affect rhizosphere organism interactions in an annual plant, a far different and more persistent set of factors comes into play around a tree's roots. Here the mycelia of fungi growing on them, the other organisms surrounding them, and the mycorrhizal fungi and nematodes in them must remain in association for scores or hundreds of years, subject to modification largely by the soil and root environment.

Climate and Physiogenic Diseases

Leaphart (44), working on pole blight of western white pine, reports that this decline, first reported in 1929, involves root deterioration and physical soil characteristics. He cites the observations of Canadian (45) and American workers that suggest an edaphic-physiologic cause of pole blight, and reports that climatic records for the pole blight area show that the western white pine stands went through a period of low precipitation and high temperature from 1917 to 1940. Leaphart and A. R. Stage are now studying climatic fluctuations and their implications in the pole blight problem and will soon be reporting their interesting findings. Tree ring analyses showed the onset of the disease followed a general and pronounced reduction in ring width of large white pines in the pole blight area.

Parker (46) describes an unexplained decline in lodgepole pine in southern British Columbia that has many of the characteristics of pole blight in western white pine. If pole blight is brought about partly as a result of climate change, Parker's lodgepole decline, which occurs partly in stands containing western white pine, may have a similar genesis.

Berry and Ripperton (47) present evidence that oxidant injury to eastern white pine foliage, called emergence tipburn, results from a build-up of oxidant, probably ozone, associated with unusual spring weather patterns. Any lengthening of the period of cold spring weather, which has been a feature of the recent climate trend in the Southeast, might reflect itself in increased ozone damage to white pine and to bright-leaf tobacco (a spotting called weather fleck) (48).

Air pollution from many sources and involving a variety of gases can cause serious damage to trees (49). The extent of damage and where it takes place with reference to the pollutant source are often governed by meteorological factors. Wind direction, wind speed, temperature, humidity, and conditions leading to air inversions are critical factors in the air pollution picture. A consistent change in weather would influence damage to forests from smelter fumes (50), stack gas from coal consumption (49), smog (51), or from virtually any source.

Toole (52) regards moisture deficit since 1950 as the cause of the wide-spread dying of *Liquidambar styraciflua,* called sweetgum blight, in the South. Consistent small moisture deficits over long periods bring about higher soil imbibitional water values, high concentrations of sodium, potassium, and other metal salts, and result in less water and more toxic ions being taken up by trees.

Rusden (53) refers to the "gradual tropicalization" of the northeastern United States and the maritime provinces of Canada, and is concerned about this trend with respect to the health of trees. He cites Pomerleau of Canada as relating the death of birch to this warming trend, and himself relates the dying of shallow-rooted red and scarlet oaks in West Virginia, the death of sweetgum in Delaware and Maryland, and the death of individuals and groups of many species in western New England on quick-draining sites to unusual heat and drought in recent years. While weather changes may have played a role in birch dieback, and have been given intensive study, the current views as summarized by Clark and Barter (54) lean toward an infectious disease, with a virus [Berbee (55)] possibly triggering the decline.

Pitch streak of turpentine pines, which may be regarded as an extended and serious form of dry face, has been reported only during the two very prolonged periods of drought in this century in the southern United States, the 1930s and the 1950s [Cobb (56)].

MacDonald (57), after describing many ways that climate restricts silvicultural aspirations in Britain, states, "All we can say with any confidence at the present time is that the climatic factor which will exert the dominating influence is neither rainfall nor temperature, but exposure to wind." Climatic influences on forests have thus taken many different forms aside from their effects on microorganisms or the obvious *acute* effects of "unusual" weather.

Climate Change and Disease Vectors

A great many tree diseases are carried by insects or by wind. Elm bark beetles carry the Dutch elm disease. There is good reason to believe that the unprecedented series of hurricanes that swung through the northeast in the early 1950s resulted in carrying great numbers of infested beetles into Canada from southern New England, causing the ensuing surge in the spread of this disease through Maine and into New Brunswick. In this case a change in storm paths affected the distribution of a disease.

Wellington (58, 59) has contributed greatly to our knowledge of insect behavior and climatic factors, and his own work and that of others indicate clearly that the shifts in storm paths that accompany or bring about climatic changes can have a great influence on those diseases dependent upon insects for their dissemination.

Elm phloem necrosis, a virus disease and a catastrophic killer that has been moving eastward, is carried by the elm leaf hopper. Its spread eastward with only slight penetration in the South (in the Chattanooga, Tennessee, area) could well be related to the distribution of the leafhopper mentioned by Baker (60) as occurring mainly in several states east and north of the known distribution of the disease.

The overland spread of oak wilt is regarded by most investigators (True et al. [61]) as being the result of ascospores and conidia being carried by insects, with Nitidulids and wood borers prime suspects. This disease spreads overland with some difficulty and in spite of its wide distribution in the East, economic damage there has been small. A climatic change that consistently increased the formation of the spore-bearing mats upon which Nitidulids and other insects feed (Boyce [62]), or caused a great multiplication of the insect vectors or increased their efficiency as carriers, could transform oak wilt from a worry in much of the East to a disaster worse than the chestnut blight.

Over 100 species of leafhoppers carry plant viruses (63), including the elm phloem necrosis virus. Aphids, also, are among the most common virus vectors. Broadbent (64) brings out that plant viruses spread fastest under conditions optimal for insect multiplication and that in temperate climates temperature is their most important regulating factor. Aphids will not fly at temperatures below 55°F. Activity increases to about 85°F. They may be soon killed if the temperature rises only a few degrees above the optimum for activity.

In assessing the impact on forest diseases of any established or continuing change in our climate, we must deal with the fact that each tree is genotypically fixed for its lifetime, and that these individuals cannot change genetically to meet climatic changes favorable to their parasites or to the carriers of these parasites. Acquired resistance as it takes place in animals, where they have been exposed to sublethal infection by disease organisms, is yet to be demonstrated for plants.

Summary and Discussion
The changes in climate taking place in parts of North America and elsewhere in the world during the past century, and particularly from 1900 to about 1945, are very likely affecting the distribution and severity of many of our tree diseases. I have stressed the general warming trend during this period and its probable effects because of the strong documentation of this trend. Along with this temperature change go changes in many other meteorological elements; in evapotranspiration, the direction of storm paths, and in the length and character of seasons. I have also cited evidence of a marked decrease in tree ring widths in some areas, during the present century, that the investigators attribute largely to precipitation decreases. Since a tree's genetic constitution is fixed, an individual is limited in what adjustments it can make to weather changes, and since there has not even been enough of a time lapse in the current period of change for provenance adaptations to take place in progeny, it is not surprising that we are having to face many enigmatic declines of trees not readily attributable to pathogens.

Considering the diseases that *are* caused by aggressive specific pathogens, such as the rusts, the many foliage disease fungi, and the Phytophthoras and Pythiums, we have clear-cut evidence that certain years of disease severity are directly correlated with special weather conditions. Thus, depending upon the parasite's optima with respect to weather factors, a consistent rise or fall of 1° to 3°F, together with accompanying changes in extremes in temperature and accompanying meteorological factors, would undoubtedly mean a reduction or increase in the number of "rust years" or "damping-off years" or "leaf blight years" per decade.

A continuation of the recent warming trend would doubtless accentuate the already established northward movement of tropical or subtropical organisms, such as *Phytophthora cinnamomi*, a fungus now causing diseases of many woody plants in temperate regions. A reversal of this warming trend (6) could cause the disappearance of many troubles of recent origin, but would probably bring us others.

Climatic instability has already caused some observers to revise the concept of the climax forest—a concept predicated on forests ultimately reaching a permanent composition with climate holding constant except for ups and downs of short duration. In some areas of the world Holloway (7) considers that climatic change results in the forest never quite catching up in its adaptations to these changes and thus being out of phase with the current climate much of the time. Diseases are among the forces determining forest composition, and particularly in North America diseases have influenced composition in virtually every major region.

If the probability of a relation of climatic influences to certain tree declines is appreciated by foresters and pathologists, as in the case of pole blight (44), sweetgum blight (52), and some tree problems in California (30, 31, 32) and New Zealand (33), we will manage forests accordingly. We at least will know that we can hardly stop a decline of a forest species when climate change is responsible. We are more likely to plan our silviculture to encourage or to plant unaffected species or geographic races of the declining species that are better adapted to the newer climate.

Forest pathology, now largely the study of *fungus* diseases of trees, will doubtless give more attention to long-continued environmental changes as direct and indirect pathologic influences.

Dr. H. E. Landsberg, Director of the United States Weather Bureau's Office of Climatology (9), has said that "The greatest advances of climatology are destined to lie in the border field of biology, provided an adequate cooperative program is started. The interactions between the physical changes in the atmosphere and living organisms are too great a challenge to scientific curiosity to remain in a relatively unexplored state."

References

1. Hepting, G. H., Climate change and forest diseases, *Proc. 5th World Forestry Congr.*, 2, 842–47 (Univ. Washington, Seattle, 2066 pp., 1962).

2. Yarwood, C. E., Microclimate and infection, *Plant Pathol. Probl. and Progr. 1908–1958*, 548–56 (Univ. Wisconsin Press, Madison, 588 pp., 1959).

3. Humphrey, H. B., Climate and plant diseases, *Yearbook Agr. 1941*, 499–502 (U.S. Dept. Agr., Washington, 1248 pp., 1941).

4. Waggoner, P. E., Weather, space, time, and chance of infection, *Phytopathology*, **52**, 1100–08 (1962).

5. Russell, R. J., Climatic change through the ages, *Yearbook Agr. 1941*, 7–9, 67–97 (U.S. Dept. Agr., Washington, 1248 pp., 1941).

6. Mitchell, J. M., Jr., Recent secular changes of global temperature, *Ann. N.Y. Acad. Sci.*, **95**, 235–50 (1961).

7. Holloway, J. T., Forests and climate in the South Island of New Zealand, *Trans. Roy. Soc. New Zealand*, **82**, Part 2, 329–410 (1954).

8. Brean, H., Our new weather, *Life*, **41**, 117–30 (1956).

9. Landsberg, H. E., Trends in climatology, *Science*, **128**, 749–58 (1958).

10. Baum, W. A., and Havens, J. M., Recent climatic fluctuations in maritime provinces, *Trans. 21st North Am. Wildlife Conf.*, 436–53 (1956).

11. Daubenmire, R., Climate as a determinant of vegetation distribution in eastern Washington and northern Idaho, *Ecol. Monographs*, **26**, 131–54 (1956).

12. Junge, C. E., Continental and global aspects of air pollution (Abstr.), *Syllabus for the Third Air Pollution Research Seminar*, p. 2 (U.S. Dept. Health, Education, and Welfare, 81 pp., 1960).

13. Plass, G. N., Carbon dioxide and climate, *Scientific Am.*, **201**, 41–47 (1959).

14. Kaplan, L. D., The influence of carbon dioxide variations on the atmospheric heat balance, *Tellus*, **12**, 204–08 (1960).

15. Leake, C. D., Social aspects of air pollution, *Proc. Natl. Conf. on Air Pollution 1958*, 20–24 (U.S. Dept. Health, Education, and Welfare, Washington, 526 pp., 1959).

16. Kramer, P. J., and Kozlowski, T. T., *Physiology of Trees*, 84–85 (McGraw-Hill Book Co., Inc., New York, 642 pp., 1960).

17. Zon, R., Climate and the nation's forests, *Yearbook Agr. 1941*, 477–98 (U.S. Dept. Agr., Washington, 1248 pp., 1941).

18. Hayes, G. L., and Buell, J. H., Trees also need water at the right time and place, *Yearbook Agr. 1955*, 219–28 (U.S. Dept. Agr., Washington, 751 pp., 1955).

19. Keen, F. P., Climatic cycles in eastern Oregon as indicated by tree rings, *Monthly Weather Rev.*, **65**, 175–88 (1937).

20. Marshall, R., Influence of precipitation cycles on forestry, *J. Forestry*, **25**, 415–29 (1927).

21. Squillace, A. E., and Silen, R. R., Racial variation in ponderosa pine, *Forest Sci. Monograph No. 2*, 27 pp. (1962).

22. Zobel, B. J., and Goddard, R. E., Preliminary results on tests of drought hardy strains of loblolly pine (*Pinus taeda* L.), *Texas Forest Serv. Res. Note 14*, 22 pp. (1955).

23. Raup, H. M., Recent changes of climate and vegetation in southern New England and adjacent New York, *J. Arnold Arboretum*, **18**, 79–117 (1937).

24. Pearson, G. A., Forest types in the Southwest as determined by climate and soil, *U.S. Dept. Agr. Technical Bull. 247*, 144 pp. (1931).

25. Beilmann, A. P., and Brenner, L. G., The recent intrusion of forests in the Ozarks, *Ann. Missouri Botan. Garden*, **38**, 261–82 (1951).

26. Paine, L. A., and O'Regan, W. G., Growth studies of regional isolates of *Echinodontium tinctorium,* the Indian paint fungus, *Can. J. Botany,* **40,** 13–23 (1962).

27. Wagener, W. W., Sporadic diseases in young stands in California and Nevada, *Western Insect and Forest Disease Work Conf. Rept.,* 14–22 (1960) [Processed].

28. Woods, F. W., Disease as a factor in the evolution of forest composition, *J. Forestry,* **51,** 871–73 (1953).

29. Hepting, G. H., Miller, J. H., and Campbell, W. A., Winter of 1950–51 damaging to southeastern woody vegetation, *Plant Disease Reporter* **35,** 502–03 (1951).

30. Wagener, W. W., The limitation of two leafy mistletoes of the genus *Phoradendron* by low temperatures, *Ecology,* **38,** 142–45 (1957).

31. Wagener, W. W., Severe cold reduces rust in pine plantations, *Plant Disease Reporter,* **34,** 193 (1950).

32. Mielke, J. L., and Kimmey, J. W., Heat injury to the leaves of California black oak and some other broadleaves, *Plant Disease Reporter,* **26,** 116–19 (1942).

33. Gilmour, J. W., The importance of climatic factors in forest mycology, *New Zealand J. Forestry,* **8,** 250–60 (1960).

34. Newhook, F. J., The association of *Phytophthora* spp. with mortality of *Pinus radiata* and other conifers. I. Symptoms and epidemiology in shelterbelts, *New Zealand J. Agr. Res.,* **2,** 808–43 (1959).

35. Miller, P. R., Plant disease forecasting, *Plant Pathol. Probl. Progr. 1908–1958,* 557–65 (Univ. Wisconsin Press, Madison, 588 pp., 1959).

36. Van Arsdel, E. P., Riker, A. J., and Patton, R. F., The effects of temperature and moisture on the spread of white pine blister rust, *Phytopathology,* **46,** 307–18 (1956).

37. Kimmey, J. W., and Wagener, W. W., Spread of white pine blister rust from ribes to sugar pine in California and Oregon, *U.S. Dept. Agr. Technical Bull. 1251,* 71 pp. (1961).

38. Siggers, P. V., Weather and outbreaks of the fusiform rust of southern pines, *J. Forestry,* **47,** 802–06 (1949).

39. Kuhlman, E. G., Survival and pathogenicity of *Phytophthora cinnamomi* rands in forest soils (*Doctoral thesis,* Oregon State University, 95 pp. 1961).

40. Hodges, C. S., Black root rot of pine seedlings, *Phytopathology,* **52,** 210–19 (1962).

41. Redmond, D. R., Studies in forest pathology. XV. Rootlets, mycorrhiza, and soil temperatures in relation to birch dieback, *Can. J. Botany,* **33,** 595–627 (1955).

42. Ward, E. W. B., and Henry, A. W., Comparative response of two saprophytic and two plant parasitic soil fungi to temperature, hydrogen-ion concentration, and nutritional factors, *Can. J. Botany,* **39,** 65–79 (1961).

43. Henry, A. W., Influence of soil temperature and soil sterilization on the reaction of wheat seedlings to *Ophiobolus graminis* Sacc., *Can. J. Res.,* **7,** 198–203 (1932).

44. Leaphart, C. D., Pole blight—how it may influence western white pine management in light of current knowledge, *J. Forestry,* **56,** 746–51 (1958).

45. Wellington, W. G., Pole blight and climate, *Can. Dept. Agr., Sci. Serv. Forest Biol. Div. Bi-monthly Progr. Report,* **10,** (6), 2–4 (1954).

46. Parker, A. K., An unexplained decline in vigor of lodgepole pine, *Forestry Chron.,* **35,** 298–303 (1959).

47. Berry, C. R., and Ripperton, L. A., Ozone, a possible cause of white pine emergence tipburn. [In press].

48. Heggestad, H. E., and Middleton, J. T., Ozone in high concentrations as cause of tobacco leaf injury, *Science,* **129,** 208–09 (1959).

49. Berry, C. R., and Hepting, G. H., Injury to eastern white pine by unidentified atmospheric constituents (In press).

50. Scheffer, T. C., and Hedgcock, G. G., Injury to northwestern forest trees by sulfur dioxide from smelters, *U.S. Dept. Agr. Technical Bull. 1117,* 49 pp. (1955).

51. McDermott W., Air pollution and public health, *Sci. Am.,* **205,** 49–57 (1961).

52. Toole, E. R., Sweetgum blight, *U.S. Dept. Agr. Forest Pest Leaflet* 37, 4 pp. (1959).

53. Rusden, P. L., Trees and the changing weather, *Sci. Tree Topics of Bartlett Tree Res. Lab., Stanford, Connecticut,* **2,** 13–14 (1955).

54. Clark, J., and Barter, G. W., Growth and climate in relation to dieback of yellow birch, *Forest Sci.,* **4,** 343–64 (1958).

55. Berbee, J. G., Birch dieback: present status and future needs (Abstr.), *IX Internal. Botan. Congress Proc.,* **2,** 28–29 (1959).

56. Cobb, F. W., Pitch streak—a disease of turpentined slash pine, *Naval Stores Rev.,* **67,** (9), 4–5 (1957).

57. MacDonald, J., Climatic limitations in British forestry, *Quart. J. Forestry,* **45,** 161–68 (1951).

58. Wellington, W. G., The synoptic approach to studies of insects and climate, *Ann. Rev. Entomol.,* **2,** 143–62 (1957).

59. Wellington, W. G., Atmospheric circulation processes and insect ecology, *Can. Entomologist,* **86,** 312–33 (1954).

60. Baker, W. L., Studies on the transmission of the virus causing phloem necrosis of American elm, with notes on the biology of its insect vector, *J. Econ. Entomol.,* **42,** 729–32 (1949).

61. True, R. P., Barnett, H. L., Dorsey, C. K., and Leach, J. G., Oak wilt in West Virginia, *West Virginia Univ. Agr. Exp. Sta. Bull. 448T,* 119 pp. (1960).

62. Boyce, J. S., Jr., Relation of precipitation to mat formation by the oak wilt fungus in North Carolina, *Plant Disease Reporter,* **41,** 948 (1957).

63. Nielson, M. W., A synonymical list of leafhopper vectors of plant viruses (Homoptera, Cicadellidae), *U.S. Dept. Agr. ARS-33-74,* 12 pp. (1962).

64. Broadbent, L., Insect vector behavior and the spread of plant viruses in the field, *Plant Pathol. Probl. and Progr. 1908-1958,* 539–47 (Univ. Wisconsin Press, Madison, 588 pp., 1959).

Part IV
Pollution and Oceanic Ecosystems

The paper by Dr. Ketchum in Part I outlines some of the general biological effects of pollution in estuaries and coastal waters. In this section the subjects of the introduction of pollutants into the marine environment and the consequences of specific pollutants will be treated.

The transport of natural and man-made substances over land and through the soil by water flow and percolation is a major mechanism through which terrestrial activity affects aquatic systems. The first two papers in this series present in detail the somewhat different problems associated with runoff from croplands and from forests. These discussions and that by Dr. Smith on range ecosystems in the previous section are fundamental to an understanding of estuarine pollution.

Dr. Harrold reviews the recent work on investigation of pollutants in agricultural runoff in his paper. This source of water pollution has become a high priority problem now that soil erosion has decreased through better conservation procedures. High pollution loads are usually associated with careless agricultural practice or with an expanding feedlot technology. The next paper by Dr. Lull summarizes the interactions of forests with the hydrologic cycle. In discussing the characteristics of the runoff, he relates various forestry practices to water quality and quantity. This paper treats three areas given particular emphasis in the SCEP Report: renewable resources of economic value, benefits to man which are free, and influences on aquatic systems.

Coastal urban centers are also sources for large amounts of matter introduced into marine environment. Dredged wastes from urban areas contain not only sediment but also other wastes, including sewage solids and agricultural and industrial wastes deposited in local waterways. In some areas of the United States and Canada, these dredged wastes exceed the riverborne suspended sediment discharge into coastal waters. The sources of these wastes and the impacts and trends of dredging practices are reviewed in the paper by Dr. M. Grant Gross.

The increasing amounts of man-made wastes introduced into the marine environment may have profound effects on the fundamental chemical nature of complex networks of ecosystems. Dr. Goldberg's paper reviews the seven large groups of pollutants produced by man that influence the chemical composition of the oceans and then discusses some of the implications of potential changes in the marine environment. The underlying question throughout this paper is what will be the responses by organisms, including man, to these intrusions.

The most extensively studied group of substances introduced into the oceans by man is that of the halogenated hydrocarbon pesticides and their degradation products. Their deleterious impacts on sea birds and fish, which were predicted when the pesticides were introduced widely into agricultural usage, are now being carefully documented, but our present knowledge is still woefully incomplete. During SCEP, a Task Force composed of SCEP participants and members of a Workshop of the National Academy of Sciences Committee on Oceanography reviewed this knowledge and made several specific recommendations based on it. Their report, included in this section, served as the background for the development of the final two papers in Part V of this volume.

Oil pollution of coastal waters is often associated only with spectacular disasters such as shipwrecks or oil well blowouts and with effects that are limited to beach amenities or localized smothering of birds or shellfish beds. Yet, most of the petroleum hydrocarbons that enter the oceans are byproducts of normal operations in our highly industrialized societies. Four participants in SCEP prepared a paper in which they compiled much of our present knowledge about sources of oil and the effects of oil on oceanic ecosystems. This paper, which was refined somewhat following the Study, is included here.

The final major pollutant of estuaries which was studied by SCEP was phosphorus. This nutrient is already considered to be responsible for the eutrophication of many lakes and some fear that the same process might become widespread in estuaries, effectively removing these breeding grounds for major fisheries. The conclusions and recommendations of the SCEP Work Group on Ecological Effects are reprinted as the last paper in this series.

One area of major importance which was not covered by the SCEP Report is that of radioactive wastes in the oceans. It was felt that this subject had been thoroughly and authoritatively treated in a recent study by a panel of the Committee on Oceanography of the National Academy of Sciences. The report of that study, "Radioactivity in the Marine Environment," will be published by the National Academy of Sciences in the near future.

14
Runoff from Agricultural Land as a Potential Source of Chemical, Sediment, and Waste Pollutants

Lloyd L. Harrold

Water flowing across agricultural cropland and farm animal feedlots is the most important mechanism in the transport of soil particles, plant residues, manure, salts, nutrients, insecticides, and herbicides from the land into surface water bodies. Also, water percolating down through soil is the prime mover of soluble salts, nutrients, and other agrichemicals into groundwater bodies. Both transport forms discussed in this report relate to critical environmental problems.

Runoff from rainfall on agricultural land is, in some cases, desirable and in others, undesirable. Water running off is no longer available for on-site crop production. Furthermore, it removes fertile soil, organic matter, and agrichemicals. Removal of excess water through surface drainage systems is desirable. Agricultural land is sloped and formed into drainage channels, terraces, and waterways engineered so that water flow velocity therein is so slow that its capacity for transport of sediment is minimized.

Treatment and use of agricultural land to reduce soil erosion have been major efforts of the Soil Conservation Service action program for over thirty years. Research, formerly in the Soil Conservation Service and since 1951 in the Agricultural Research Service, has developed and tested farming practices to do this job. Investigations to develop better practices compatible with trends in expanding farm mechanization, continuous row cropping, and increased chemical usage are continuing.

In the 1930s *soil erosion* was the big problem. Over the past forty years the U.S. Department of Agriculture has accumulated a body of data on erosion that has been used in development of a universal soil erosion equation (Wischmeier and Smith, 1965). This equation and other knowledge are helpful in maintaining water quality by reducing agricultural sediments and chemicals in surface waters. Soon thereafter, *water conservation* was added to form a soil- and water-conservation program. *Water quality* next became a major item of concern. Although water quality is a relatively new item to many areas, it is an old story to the arid west

Prepared for SCEP.

where salt in water and soil has been a major problem for many years.

Within the scope of this report, we treat these three topics: pollution, erosion, and water conservation. But the greatest of these is pollution. Any water containing chemicals, sediments, or wastes in excess of minimum amounts as prescribed by established standards for various water uses is considered polluted.

Runoff Transport

Dissolved Nutrients and Pesticides in Surface Flow

Nutrients used in agriculture of most concern in environmental problems are nitrogen and phosphorus. Timmons, Burwell, and Holt (1968) reported that in Minnesota the greatest transport of nitrogen in runoff from agricultural land occurred on fallow and cornland in the sediment removed during rainstorms. But for hay plots, where sediment yield was negligible, total nitrogen dissolved in runoff from snowmelt was much greater than that in runoff from storm rainfall. Where sediment removal was small, nitrate-nitrogen was higher in snowmelt runoff than in storm runoff. After vegetation died or was frozen, rainwater or snowmelt leached organic residues and contributed phosphorous and nitrogen to runoff.

Johnston and his group (1967) observed insecticides in tailwater surface runoff (excess flood irrigation) in the San Joaquin Valley of California. Tail water contained seven to twelve times as much insecticide residue as that in the applied irrigation water and up to eighty-five times as much after lindane was applied— up to 9400 ppt.

Edwards and Harrold (1970) found that concentration of pesticide pollutants in runoff water decreased as the period after application increased. They also observed herb-algal growth in pools of water in the channel of a wooded watershed where no phosphorus fertilizer had been applied for over thirty-five years. In fact, eutrophication was occurring long before man began to use chemical fertilizers.

Taylor (1967) called attention to the fact that phosphorus applied as fertilizer was converted to water-insoluble forms within a few hours. It was adsorbed on the surface of soil particles. Harrold (1969) indicated that persistent organochlorine insecticides were also water-insoluble, clinging strongly to soil particles. Transport of these pollutants by runoff occurred mostly by sediment removal and will be presented later in this chapter.

Researchers at Coshocton, Ohio, found from 0.03 to 0.06 pound of soluble phosphorus per acre per year in storm runoff from agricultural land (Wadleigh and Britt, 1969). Researchers in Minnesota (Timmons, Burwell, and Holt, 1968) corroborated the Ohio results, with amounts ranging from 0.06 to 0.2lb/acre. These authors called attention to a serious source of nutrient in runoff in the Midwest resulting from the practice of spreading livestock manure on snow-covered fields and frozen soil. When rain falls or snow melts rapidly, runoff rich in nutrients runs off into streams, ponds, and lakes. Runoff from farm animal feedlots is also rich in nutrients (Gilbertson et al., 1970).

Dissolved Nutrients and Pesticides in Subsurface Flow
Water percolating down through the soil picks up soluble compounds in varying amounts. These compounds leached out of the soil contribute to the pollution of underground water supplies.

Edwards and Harrold (1970) reported that percolating water at the eight-foot depth of monolith lysimeters at Coshocton, Ohio, contained no measurable phosphorus. Since phosphorus fertilizer is converted to water-insoluble forms within a few hours after application (Taylor, 1967) and percolating water is well filtered, phosphorus pollution to groundwater supplies is not expected. Nitrate leaching from the lysimeters averaged about 5 lb/acre annually. On nitrogen-deficient lysimeters, crop production and water use were low and percolation and leaching high. At reasonably high nitrogen fertilizer rates, crop yields and water use were high and percolation and leaching of nitrates were low. It is, however, possible to increase nitrate contribution to underground water by fertilization and water applications in amounts greater than that needed by the crop.

Stewart, Viets, and Hutchinson (1968) reported that percolating water in the semiarid High Plains of Colorado caused nitrate concentrations at a depth of 20 feet of 79 lb/acre under alfalfa, 90 in native grassland, 261 in irrigated fields with no alfalfa, and 1,436 in corrals. In irrigated fields with no alfalfa, percolating water carried 25 to 30 pounds of nitrate to the water table annually.

Another group (Johnston et al., 1967) observed insecticides in subsurface drainage effluent from irrigation plots in the San Joaquin Valley of California. DDT and parathion were applied to the experimental plot at rates of 2 lb/acre and 0.1 lb/acre, respectively, after which the plots were flooded. DDT and parathion were again applied at 4 and 0.2 lb/acre, respectively, followed by another flooding. After a crop of rice was grown, lindane was applied at the rate of 3.3 lb/acre.

No parathion was found in tile drainage effluent sampled within fourteen days after the first flood irrigation. Parathion decomposes rapidly.

DDT, a persistent chlorinated hydrocarbon of low solubility, was found in tile drains in small amounts—less than that of the irrigation water applied. Lindane, of relatively high solubility, leached readily through the soil and appeared in the drainage water after the third irrigation at a concentration of 834 ppt.

Wadleigh (1968) told that when nitrate is found in groundwater, the sources may be sewage or septic tank effluent, feedlots or barnyards, field fertilization, or the natural accumulations such as found in caliche of semiarid regions. Research evidence in Missouri indicates that virtually none of the nitrate in groundwater was contributed through percolating water from agriculturally applied fertilization. In Hawaii, however, percolating flow from irrigation carried sizable amounts of nitrate from crop fertilization.

Percolating water is the vehicle for moving salts into and through soils of agricultural fields—a serious problem in the arid western states. Water quality for crop irrigation, salt leaching, and salt concentration in groundwater and return flow downstream from these crop lands is of great concern in these areas.

Sediment Pollution

Runoff from agricultural land is the mechanism for transport of soil eroded from crop fields into streams and water bodies. Wadleigh (1968) wrote "Sediment derived from land erosion constitutes by far the greatest mass of waste materials arising from agricultural and forestry operations." Muddy streams, clogged rivers and harbors, and sediment deposited in reservoirs and lakes present strong evidence of the serious problem of waterborne sediment pollution.

On the contrary, sediment transported into phosphorus-rich lakes has been observed to absorb some phosphorus from the lake water—a reversal of the pollution process providing the sediment settles in deep water and is not later disturbed.

Harrold and Dragoun (1969) reported that sediment yield from continuous corn on the contour in the deep loess of western Iowa averaged 30 tons per acre per year. That on residual silt loam in Ohio was 7.7, and on the loessial regions of the Central Great Plains of south-central Nebraska, 8 tons per acre. Water running off these agricultural areas was indeed polluted with soil.

Some soils erode easily and supply clay-sized particles that stay in suspension for weeks. Others resist erosion, yielding coarser material that settles out rapidly. Erosion-control measures, which vary with different soil, physiographic characteristics, and climate, are discussed in the section on erosion-control farm practices. The subject of nutrient transport attached to eroded soil particles is treated in the next section.

Nutrients and Insecticides Attached to Sediment

Publications relating to pollution of runoff from agricultural land invariably point out that phosphorus and persistent organochlorine insecticides are adsorbed on soil particles and have low solubilities. They become pollutants in water because soil particles on which they are attached become suspended in water and are transported off the land by runoff into streams, ponds, and lakes. As some sediment deposited in surface water bodies has a high capacity to adsorb phosphorus, it may adsorb more from the water than it releases. Since there is practically no sediment in water percolating through the soil pores to the groundwater reservoir, these chemicals do not move to this water body. As a rule, they are not considered as groundwater pollutants. Exceptions to this rule are cases where water flows downward to the water table through large channels.

Thomas, Carraker, and Carter (1969) reported that the 8-year average annual soil loss on Tifton loamy sand of 3 percent land slope on rotation of rye, peanuts, and rye was 1.32 tons per acre. Washoff of phosphorus, applied at 16 lb/acre per year, was 0.4 percent, or 0.06 pound. Nutrient transport by soil in the runoff process was linearly related to erosion, quantitatively. Enrichment ratios in the washoff were three to six times the amount added to the plow layer.

On Barnes loam in Minnesota (Timmons, Burwell, and Holt, 1968), the 2-year (1966–1967) total nitrogen (excluding NH_4 and NO_3—N) removed from fallow plots by storm-water runoff was 0.46 lb/acre and by sediment transport, 109 lb/acre. Comparable values for NH_4 were zero and 1.02; for NO_3—N, 1.52 and 0.48; and for P, 0.05 and 0.56 lb/acre.

Gilbertson's group (1970) found that snowmelt runoff from feed lots of high density of beef cattle in Nebraska was 130 to 170 percent of that of low density. Snowmelt runoff when the feedlot ground was wet and frozen carried off much of the animal waste deposited thereon. Total nitrogen concentration in winter-season runoff ranged from 1,430 to 5,760 ppm and for rainstorm runoff in other seasons, 65 to 555 ppm. Total phosphorus removed from the high animal density area was 1.5 to 4 times that from low density.

Erosion-Control Farm Practices

Agricultural research has developed farming practices and other means for stabilizing soil and reducing its movement into streams. The more these practices are applied to the land, the more constraint is exerted on the movement of phosphorus and organochlorine insecticides into water bodies (Harrold, 1969; Edwards and Harrold, 1970). In southeastern Ohio, conservation farming on silt loam reduced average annual sediment yield from small cornland watersheds to 2.0 tons per acre, compared to 7.7 for unimproved farming (McGuinness, Harrold, and Dreibelbis, 1960). On wheatland, comparable values were 0.14 and 0.88 ton per acre. Sediment yield was reduced further by no-tillage corn—reduced to 0.02 ton per acre (Harrold, Triplett, and Youker, 1967).

Harrold and Dragoun (1969) reported that in the deep loess of western Iowa, level terraces on a continuous corn watershed reduced sediment transport 90 percent, compared to that from an unterraced watershed also in continuous corn. Also, trees planted on eroded abandoned silt loam cropland of North Appalachia in Ohio practically stopped surface runoff and reduced sediment transport 100 percent. Grass seeded on cultivated watersheds on the Central Great Plains of south-central Nebraska reduced sediment transport 50 percent.

Wadleigh and Britt (1969) gave words of warning to those who might feel confident that the conservation job is done, "Let's face it, the job of soil conservation and conservation of associated water resources is a long, long way from being done in the United States." Many practicing conservationists readily concede that they could make improvements in farming that would enhance the quality of our environment. The authors wisely call attention to the fact that the soil conservation movement has, however, done far more to improve the quality of the environment over vast expanses of our nation than any other endeavor.

It is evident that phosphorus and pesticide pollution in surface runoff comes mostly by soil erosion and that erosion-control measures on farmland will play an important role in minimizing pollution from agricultural sources (Edwards and Harrold, 1970).

Monitoring

Monitoring to obtain definitive knowledge of pollutants in runoff from agricultural land is performed so as to identify the source of pollutants. Equipment for monitoring and sampling runoff is selected and located to meet the specific research or study objectives (Edwards and Harrold, 1970b).

Experimental watersheds having historic records of hydrology, land use, and treatment are ideal for monitoring runoff water quality. Observations can then be interpreted in light of current and past chemical applications and climatic variations.

Sophisticated monitoring equipment is now available to transmit and record hourly temperature, conductivity, dissolved oxygen, pH, and dissolved chlorides in runoff water. Where information on quantities of pollutant transport is required, a gauge to measure runoff rates and time is also necessary. This monitoring equipment is very expensive and is usually located on large streams having continuous flow. Under these conditions it is not possible to detect the specific agricultural sources of pollutants. Also it is not economically practical to locate such monitoring equipment at numerous gauges where runoff from individual farm fields is measured. Inexpensive sampling equipment at these many sites is practical (Bentz and Edwards, 1970).

Runoff sampling equipment, techniques, sample size and storage, and analysis differ from study to study. Furthermore, where sediment load in runoff is light, pump samplers with electrically operated solenoid valves for distributing discrete samples to individual containers are satisfactory. For heavy sediment loads, these valves are not satisfactory and are replaced with mechanical distribution systems such as that described by Bentz and Edwards (1970).

After a sample of runoff is obtained, it must be protected from chemical and biological change before laboratory analyses. Samples collected during storm runoff are automatically stored in a refrigerator and then taken to the laboratory as soon as practical. Here they are centrifuged to separate solids from the liquids so that determinations can be made of pollutant transport mechanisms—dissolved in liquid or attached to solids.

Objectives of the monitoring program will dictate the storage, processing, and laboratory analytical procedure.

Modeling
Much effort is being expended in many areas on the development of definitive models of watershed hydrologic processes. These efforts will result in an understanding of watershed flow systems related to physical characteristics of the real river basin. The hydrologist will be able to make acceptable predictions of streamflow for ungauged basins for a variety of land-use practices.

It would be wise to make coincident efforts to attain an understanding of the transport system of agrichemicals and to develop models so that reliable predictions could be made of runoff water quality for a variety of chemical inputs along with a number of land-use treatments.

Summary
Within the past five to ten years, many federal and state agencies have become involved in conducting research on environmental problems—especially in water quality. Many of the results are tentative, as the study period was extremely short. By 1975 there should be much more quantitative material available if the present rate of research activity continues.

Runoff from agricultural land carried pollutants dissolved

in the water and attached to soil particles in sediment transport. Nitrate and salts readily moved into solution both across the land into surface streams and downward through soil to groundwater reservoirs. Land-use and water-management practices that reduced soil erosion and sediment yield improved the quality of our environment over vast expanses of our nation.

Nitrate-nitrogen went into water solution readily and was carried off the surface by runoff from agricultural land and was leached out of the soil to groundwater supplies by percolating water. Sources were nitrogen fertilizer, rainwater, legume fixation, and animal waste. There was no apparent relation between runoff rates and concentration of dissolved nutrients.

Phosphorus and insecticides were attached to fine soil particles within a few hours of application. Transport of these pollutants in solution was low. Concentration of insecticides in runoff water diminished as with time after application.

There is a need to develop mathematical models of pollutant transport systems for the place of input to the land on out to the stream or water body—surface or subsurface pathway—under a variety of land conditions.

References

Bentz, W. W., and Edwards, W. M., 1970. Automatic sampling equipment for large volume samples of runoff, U.S. Department of Agriculture ARS 41-171.

Edwards, W. M., and Harrold, L. L., 1970a. Agricultural pollution of water bodies, *Ohio Journal of Science,* 70(1): 50–56.

Edwards, W. M., and Harrold, L. L., 1970b. Instrumentation considerations for studies of quality of runoff, International Association on Scientific Hydrology, Symposium on Hydrometry, Koblenz, Germany, September 1970.

Gilbertson, C. B., McCalla, T. M., Ellis, J. R., Cross, O. E., and Woods, W. R., 1970. The effect of animal density and surface slope on characteristics of runoff, solid wastes, and nitrate movement on unpaved beef feedlots, University of Nebraska College of Agriculture and Home Economics, *The Agricultural Experiment Station Bulletin,* SB508.

Harrold, Lloyd L., 1969. Pollution of water from agricultural sources, *The Ohio Engineer,* 29(4): 10, 16.

Harrold, L. L., and Dragoun, F. J., 1969. Effect of erosion-control land treatment on flow from agricultural watersheds, *Transactions of the American Society of Agricultural Engineers,* 12(6): 857–859.

Harrold, L. L., Triplett, G. B., and Youker, R. E., 1967. Watershed tests of no-tillage corn, *Journal of Soil and Water Conservation,* 22(3): 98–100.

Johnston, W. R., Ittihadieh, F. T., Craign, K. R., and Pillsbury, A. F., 1967. Insecticides in tile drainage effluent, *Water Resources Research,* 3(2): 525–537.

McGuinness, J. L., Harrold, L. L., and Dreibelbis, F. R., 1960. Some effects of land use and treatment on small single crop watersheds, *Journal of Soil and Water Conservation, 15* (2): 65–69.

Stewart, B. A., Viets, F. G., and Hutchinson, G. L., 1968. Agriculture's effect on nitrate pollution of ground water, *Journal of Soil and Water Conservation, 23* (1): 13–15.

Taylor, A. W., 1967. Phosphorus and water pollution, *Journal of Soil and Water Conservation, 22* (6): 228–231.

Thomas, Adrian, W., Carraker, John R., and Carter, Robert L., 1969. Water, soil, and nutrient losses on Tifton loamy sand, University of Georgia, *College of Agriculture Experimental Station Research Bulletin, 64.*

Timmons, D. R., Burwell, R. E., and Holt, R. F., 1968. Loss of crop nutrients through runoff, Minnesota Agriculture Experimental Station, *Minnesota Science, 24* (4): 16–18.

Wadleigh, Cecil H., 1968. Waste in relation to agriculture and forestry. U.S. Department of Agriculture Miscellaneous Publication No. 1065.

Wadleigh, C. H., and Britt, C. S., 1969. Conserving resources and maintaining a quality environment, *Journal of Soil and Water Conservation, 24*(5): 172–175.

Wischmeier, W. H., and Smith, D. D., 1965. Predicting rainfall-erosion losses from cropland east of the Rocky Mountains, U.S. Department of Agriculture, *Agricultural Handbook No. 282.*

15
Runoff from Forest Lands   Howard W. Lull

Abstract
Forests cover about one-fourth of the world's land surface and constitute the areas of greatest precipitation, runoff, and potential sedimentation. Relationships of forests to runoff have been most intensively investigated in the United States. This paper is drawn from an unpublished manuscript, "Forestry and Water" (Lull, Anderson, Hoover, and Reinhart, Northeastern Forest Experiment Station, in preparation) that brings together results of these studies, describing successively the relationships of forest conditions and treatment to water yield, floods, and erosion sedimentation. Sediment, the principal pollutant from forest land, is intimately tied to runoff.

Water Yield
The 650 million acres of forest land in the United States (one-third of the land area, not counting Alaska and Hawaii) receive about one-half of the total precipitation and yield about three-quarters of the total streamflow. Per unit area, forest land receives annually about twice as much precipitation as other lands, 45 inches versus 22. More of the precipitation runs off: 20 inches, or 44 percent from the forest; 3 inches, or 15 percent from other lands (Wooldridge and Gessel, 1966).

Forests predominate in high-elevation, mountainous areas that receive record amounts of rainfall and produce the greatest runoff. The minimum requirement for forest growth is about 20 inches of water annually; here seasonal differences in evapotranspiration permit water surpluses that provide a minimum of 2 to 5 inches of runoff annually. Average annual runoff from major forest regions ranges from 5 to 50 inches.

Fifty to 65 percent of the precipitation appears as runoff in northern latitudes, such as in the northern hardwood, lodgepole-spruce-fir, and Douglas fir regions. Southward, this proportion diminishes to about 30 percent in the Southeast and to less than 10 percent in the arid Southwest.

Within major forest regions, there is some evidence that runoff varies with forest type, particularly between hardwoods and

Prepared for SCEP.

conifers. In Michigan, water yield from a twenty-eight-year-old jack-pine stand was 12.4 inches compared to 15.3 inches from forty- to sixty-year-old deciduous stands nearby (Urie, 1966). In North Carolina, runoff and interception measurements suggest that annual runoff from a mature white-pine plantation may be as much as 12 inches less than runoff from the preceding hardwood stand (Swank, 1968). In a study of 137 watersheds in the Northeast, runoff increased with precentage of forest cover; this was not the result of a forest influence, but because the areas allowed to remain in forest are generally steeper, are more stony, and have shallower soils (and thus more runoff) than those developed for other purposes (Lull and Sopper, 1967).

The heavier the cutting of a forest, the greater the increase in runoff. Maximum first-year increases from clear-cutting are about 18 inches. Greatest increases occur under greatest precipitation. With average annual precipitation of 48 to 90 inches, first-year increase from clear-cutting ranged from 5 to 18 inches; with 21 to 28 inches precipitation, the increase dropped to 1 to 3 inches; with 19 inches of precipitation, no increase was reported. Increased yields diminish with regrowth. In North Carolina, a first-year increase of 14 inches dropped to 8 inches the fifth year, to 6 inches the tenth year, and to an estimated 1 inch the thirty-fifth year (Kovner, 1956). In areas of more shallow soils, duration of increases was limited to ten years. In western studies, where regrowth is slower and water yield increases are derived largely from differentials in snow accumulation, treatment effects have persisted as long as twenty-four years without sign of change.

As runoff is increased by cutting, so is it reduced by planting trees; the degree depending upon the preceding vegetative cover. Eastern studies have shown a reduction in annual water yield of 6 to 9 inches ten to twenty years after planting conifers in old fields.

Under a sustained yield forestry program, clear-cutting 1 to 2 percent of a management area every year (or 10 to 20 percent every ten years) can increase area water yield by about 1 to 2 percent.

As a rough rule of thumb, clear-cutting upland areas in the humid zone, with subsequent establishment of a minimal protective ground cover, will increase water yield an amount equal to about one-half the average annual evapotranspiration as calculated from mean annual precipitation minus runoff. Increases in water yield may be obtained by cutting phreatophytic vegetation that occupies the floodplains of major rivers in the West; cutting areas of dense growth may save 2 to 3 feet of water annually (Robinson, 1967).

Floods

Because forests occupy about one-third of the United States, receive annually about twice as much precipitation as nonforested lands, and produce about seven times as much runoff, they are the principal sources of major floods. Yet because trees, compared to any other kind of vegetation, have greater foliage, are deeper-rooted, and more fully occupy a site, they provide maximum opportunity for controlling runoff from flood-producing rainfalls. Rarely is overland flow produced from an undisturbed forest floor; commonly, infiltration rates exceed maximum rainfall intensities. However, subsurface flow can rapidly move downslope through the highly permeable surface soil to contribute to flood discharge. Most maximum floods have come from well-forested watersheds; again this is not a "forest influence" but rather the integrated influence of the forest-occupied site—high rainfall, steep topography, and shallow soils.

Forest cutting can have a minor effect on flood peak and flood runoff. Seasonal peak flows and storm runoff are increased by only about 10 to 20 percent by clear-cutting. Overland flow from logging roads, landings, and other disturbed areas can be minimized by proper planning and precautions during the logging operation. Also, peak flows and storm runoffs diminish, probably dropping to negligible amounts within five to ten years, in areas where regrowth rapidly reestablishes interception and transpiration. Finally, the small proportion of total area clear-cut under sustained yield management, or as limited by the variation in age of stands under private ownership, could not mount a major flood threat. Nor, based on trend analysis of land-use changes, is it likely that there will be a major conversion of forest to nonforested land that would increase flood damages.

The chaparral of southern California and Arizona—the country's most inflammable forest—presents a special and serious case. Here fire results in disastrous floods and severe erosion.

Planting conifers on abandoned, eroding land can reduce peak flows by 60 to 90 percent. Also, it can slow the rate of snowmelt and prolong snowmelt further into the spring; however, in the event of unseasonable warm weather late in the melt period, excessive peak flows can be generated from rapidly melting snow accumulated beneath the trees.

Erosion and Sediment
Soil fully protected by a cover of litter and humus contributes little or no sediment to the stream. Erosion occurs almost entirely within the stream channel as the cutting and transporting power of streamflow detaches soil particles from channel banks and carries them downstream. In time, an equilibrium develops between the quantity of streamflow and the length, width, and carrying capacity of its channel. When infrequent high flows overtax its capacity, channel cutting resumes. Thus sediment can be expected from any forested watershed in times of extremely high flow.

When portions of the forest floor have been disturbed by logging, grazing, or burning, infiltration is reduced by the compactive effects of equipment, animals, or raindrops. Overland flow and soil erosion result. Soil loss can then become a function of soil erodibility and length and steepness of slope—all factors of importance in predicting soil loss from fallow land (Agricultural Research Service, 1961).

However, logging disturbances are limited to relatively small areas. Likewise, grazing and fire are no longer extensive enough to reduce seriously the forest's high infiltration rates. Therefore, by and large, forested watersheds contribute little sediment to streamflow. From small, undisturbed forested watersheds under 100 acres in area in the eastern United States, sediment carried by streamflow during nonstorm periods amounts to about four to five tons per square mile per year; stormflow contributes another five to ten tons. In terms of volume and depth of soil loss, ten tons per square mile per year would be equivalent to a soil depth of 0.001 inch eroded from the channel area where channels occupy 1 percent of the watershed. For larger watersheds where roadbanks, logging, and channel erosion contribute more sediment, as much as 30 to 300 tons per square mile per year may come from forested watersheds. This is still well below the soil loss tolerance limits of one to three tons *per acre* accepted for agricultural lands.

In semiarid regions, the erosion potential is much greater. Over broad areas in the West, an average of about two tons of soil per acre per year moves from wildlands into the major streams. Sediment yield is at a maximum where annual effective precipitation is about 10 to 14 inches; the yield decreases sharply where precipitation is on either side of this, because of little runoff or greater density of vegetation (Langbein and Schumm, 1958).

Sedimentation from western, forested, snow zone watersheds is usually less than from watersheds at lower elevations. The snowpack both protects the soil surface from erosion and reduces channel erosion by dampening peak flows. To some degree this protection is offset by steeper slopes, more extensive bare areas, and greater soil erodibility at high elevations (Anderson, 1966).

Sources of major erosion and sedimentation are the earth slides prevalent in the Northeast and Northwest. In the forested White Mountains of New Hampshire, their frequent occurrence has been associated with heavy rain on steep slopes; they were not appreciably affected by man's activity (Flaccus, 1958). In Utah, Oregon, and Alaska, landslides have been caused by forest cutting and road construction. For instance, in one winter in Oregon, thirty-four of forty-seven mass soil movements at the H. J. Andrews Experimental Forest were associated with roads; eight of the slides occurred in logged areas (Dyrness, 1967).

The greater the proportion of a watershed in forest cover, the smaller the sediment yield. For example, in fifteen subbasins of the Potomac River that have 20 to 50 percent forest cover, average annual sediment yield ranged from 90 to 500 tons per square mile. In seventeen watersheds that have forest cover of 60 to 90 percent, the yield ranged from 20 to 200 tons. A regression showed that a fivefold increase in forest cover, from 20 to 100 percent, produced an eighteenfold reduction in sediment yield (Wark and Keller, 1963).

Forest Treatment and Sediment
Cutting trees per se causes little or no erosion; erosion stems from soil disturbance associated with the transportation of felled

timber—roads, skid roads, and landings. This, in turn, is affected by the care used in locating and draining and in postlogging maintenance of these areas. For instance, at the Fernow Experimental Forest in West Virginia, the maximum turbidity from a cutover watershed with no road plan and no provision for drainage was 56,000 ppm (parts per million); with no road plan but roads drained, 5,200 ppm; with moderate planning and drainage, 210 ppm; and with careful planning and drainage, 25 ppm. The maximum for an uncut watershed was 15 ppm (Reinhart, Eschner, and Trimble, 1963).

The 5 to 15 percent of a logging area usually devoted to roads produces 90 percent of the sediment. Road density and grade may be minimized by planning. An unplanned road system at the Fernow Experimental Forest occupied 5 to 7 percent of the area with grades of 14 to 24 percent; a planned system occupied 2 to 5 percent with grades of 9 to 15 percent. By careful planning, the area in skid roads can be reduced 40 percent (Mitchell and Trimble, 1959). The degree of logging disturbance also depends on the type of equipment. Under some conditions, overland-flow conditions can be produced by tractor logging on about one-fourth of the logging area, and by high-lead logging on about 10 to 15 percent.

Logging on coarse-textured, permeable soils in the front range of the Rockies, where care was given to road building, has produced little sediment. Also, the stability of the glaciated soils of New England's forested mountains, despite heavy cutting, has long been recognized. In other forest regions (in the southern Appalachians, in Idaho, Arizona, and California's Sierra Nevada), soil losses have been great.

There is considerable evidence however that careful logging can minimize sediment production. On the Waynesville, North Carolina, municipal watershed of 8,000 acres, 13 million board feet of timber were harvested and 50 miles of road were built in ten years of logging, all without damaging the water resource (Vogenberger and Curry, 1959). Sediment yields from a carefully logged watershed on the Fraser Experimental Watershed in Colorado were relatively large during the year immediately after treatment, about 50 to 100 tons per square mile. Thereafter, they dropped to 25 tons or negligible accumulations even though runoff had increased 25 percent since cutting. Over an eight-year period after logging there was no significant difference in mean annual sediment yield from the logged watershed and uncut watersheds (Leaf, 1966).

Planting conifers on eroding areas can sharply reduce erosion and sedimentation. In Mississippi, a twenty-two-year-old pine plantation, planted on fields from which erosion had removed an estimated two feet of the surface and had cut gullies five feet below the level of the remaining soil, reduced the soil loss rate from several tons per acre per year to 0.00 to 0.08 ton (Ursic and Dendy, 1965).

Potential erosion on many forested areas is limited where stony soils quickly halt the erosion process by forming an erosion pavement. As an example, at the Fernow Experimental Forest in West Virginia, turbidity averaged 490 ppm during a logging operation, 38 ppm the first year after, and 1 ppm the second year after (Reinhart, Eschner, and Trimble, 1963). This rapid recovery is typical of many forested areas of stony soils and suggests why they continue to produce high-quality water despite repeated cuttings. However, self-healing becomes effective only after considerable damage to water quality has been done. Hoover (1945) pointed out that a short stretch of logging road can contribute more sediment to a stream draining a forested watershed than can occasional patches of steep land in cultivated crops.

Besides reducing the quality of water for domestic and industrial users, there is abundant evidence that sediment is detrimental to aquatic life in salmon and trout streams. Adult fishes can stand normal high concentrations without harm, but a small amount of sediment can reduce the survival of eggs, reduce aquatic insect forms, and destroy needed shelter (Cordone and Kelley, 1961). Sediment also harms fish environment by carrying organic matter to the stream bottom, thereby establishing anaerobic decomposition that may be lethal to stream bottom fauna and fish eggs and also by restricting oxygenation of water by limiting light absorption. In trout streams that have forested drainage areas, eroding banks may be the most important source of sediment so that their stabilization may be one means of improving trout habitat (Striffler, 1965).

Water Temperature

Major increases in temperatures have followed clear-cutting: in West Virginia growing-season maximums were increased on the average by 8°F, and temperatures over 78° were observed several times; in western Oregon daily temperatures were increased by 14°F in August, and mean monthly maximum temperatures were increased by from 2° in March to 14° in September (Brown and Krygier, 1967). Scattered checks of similar logged and unlogged drainages in Oregon have shown temperatures to be as much as 10° greater in logged areas where riparian vegetation was completely removed (Chapman, 1962). This can affect fishing; best trout fishing is found in streams with maximum temperatures of 60 to 68°F. Water temperatures for warm-water fish should not exceed 93°F at any time or place (Tarzwell, 1960).

Exposure of the stream surface to direct solar radiation is the principal cause of high water temperatures. Surviving understory vegetation after clear-cutting may provide considerable protection. For instance, a clear-cutting in Oregon increased maximum water temperature only 4°F. But the slash burning that followed removed protective stream cover, increasing the mean monthly maximums for June, July, and August by 12 to 14°F (Levno and Rothacher, 1969).

Management to lower temperatures is possible. For instance, routing an open, exposed trout stream in Wisconsin through the shade of a willow grove reduced later afternoon summer water temperature 10 to 11°F (Stoeckeler and Voskuil, 1960). Large increases in temperature after stream exposure may not have serious effects downstream; in the patch-cut logging system, the warm water can be cooled as it passes through shaded areas downstream or is diluted by cooler water from uncut watersheds (Levno and Rothacher, 1969).

Herbicides

Herbicides used to reduce forest cover and increase water yield, mostly on an experimental basis, have caused little contamination.* For instance, hand-spraying streamside vegetation in New

---

* This publication reports research involving pesticides. It does not contain recommendations for their use, nor does it imply that the uses discussed here have been registered.

Jersey, Pennsylvania, and California and basal spraying, stump treatment, and foliage spraying in West Virginia did not contaminate flow downstream (Lull and Reinhart, 1967). However, helicopter-spraying in Oregon resulted in some herbicide being found in all streams. This can be held to an acceptable minimum by avoiding direct application to large, slow-moving streams and marshy areas. Drift of herbicides during aerial spraying has been a major cause of concern in western watersheds.

Nutrient Discharge

Clearing and herbiciding a small forested watershed in New Hampshire increased nitrates in streamflow about fifty times over a three-year period. Major cation levels rose three- to twentyfold. The total net dissolved inorganic substances in stream water from the cleared watershed were about fifteen times greater than from the forested watersheds (Pierce and others, 1970). These results are of interest but have no practical significance in that clear-cutting a forest does not involve detroying all vegetation.

Forest as Filter

The forest can serve as a living filter for polluted water. Waste water from food processing plants and pulp mills and effluent from sewage treatment plants have been disposed of in forest areas with some success for more than twenty years. The original objective was simply waste-water disposal, to get rid of polluted water without dumping it directly into stream channels. The fact that the water is renovated as it makes its way through forest soil is a welcome bonus.

Sprinkler irrigation was first started by Seabrook Farms in the sandy coastal plains of New Jersey in 1949 (Thornthwaite, 1951). To dispose of 10 million gallons of waste water daily, a sprinkler system was set up on seventy acres of woodland. Rate of application was 0.8 inch per hour for an eight-hour day; annually about 400 to 600 inches of water were applied. There has been no overland flow, for even though the trees did not survive, high infiltration rates were maintained by the forest floor augmented by herbaceous debris in the waste water.

Sprinkler systems are now employed extensively in disposal of waste water. In areas of low infiltration capacity, polluted water has been purified by slow overland flow downhill through forest litter, humus, and ground vegetation. Suspended and dissolved solids are filtered out and converted into humus (Mather and Parmalee, 1963).

More recently, effluent from sewage treatment has been successfully sprayed in woodland at the Pennsylvania State University at intensities of 1 to 4 inches per week. Water samples collected at the 12-inch soil depth under a red-pine plantation showed 95 to 98 percent reduction in concentration of alkyl benzene sulfonate, a constituent of detergents, and reductions of nitrate N and organic N by 68 to 86 percent and of phosphorus by 99 percent. Average concentration of all constituents in the percolate at 12 inches was considerably below allowable limits for drinking water set by the U.S. Public Health Service. With a 2-inch weekly irrigation rate from April 8 to November 18, the total weight of constituents applied was equivalent to 2,500 pounds per acre of 10-10-10 fertilizer (Parizek and others, 1967).

References

Agricultural Research Service, 1961. A universal equation for predicting rainfall-erosion losses, Agricultural Research Service Publication X22-66, U.S. Department of Agriculture.

Anderson, Henry W., 1966. Integrating snow zone management with basin management, *Water Research* (Baltimore: The Johns Hopkins Press), pp. 355–373.

Brown, George W., and Krygier, James T., 1967. Changing water temperatures in small mountain streams, *Journal of Soil and Water Conservation*, 22(6): 242–244.

Chapman, D. W., 1962. Effects of logging upon fish resources of the West Coast, *Journal of Forestry*, 60: 533–537.

Cordone, Almo J., and Kelley, Don W., 1961. The influences of inorganic sediment on the aquatic life of streams, *California Fish and Game*, 47(2): 189–228.

Dyrness, C. T., 1967. Mass soil movements in the H. J. Andrews Experimental Forest, U.S. Department of Agriculture Forest Service Research Paper PNW-42, Pacific Northwest Forest and Range Experiment Station, Portland, Oregon.

Flaccus, Edward, 1958. Landslides and their revegetation in the White Mountains of New Hampshire, Ph.D. dissertation, Department of Botany, Duke University.

Hoover, Marvin D., 1945. Careless skidding reduced benefits of forest cover for watershed protection, *Journal of Forestry*, 43: 765–766.

Kovner, Jacob L., 1956. Evapotranspiration and water yields following forest cutting and natural regrowth, *Proceedings of the Society of American Forestry* 1956: 106–110.

Langbein, W. B., and Schumm, S. A., 1958. Yield of sediment in relation to mean annual precipitation, *Transactions of the American Geophysical Union, 39:* 1076–1084.

Leaf, Charles F., 1966. Sediment yields from high mountain watersheds, central Colorado, U.S. Department of Agriculture Forest Service Research Paper RM-23, Rocky Mountain Forest and Range Experiment Station, Fort Collins, Colorado.

Levno, Al, and Rothacher, Jack, 1969. Increases in maximum stream temperatures after slash burning in a small experimental watershed, U.S. Department of Agriculture Forest Service Research Note PNW-110, Pacific Northwest Forest and Range Experiment Station, Portland, Oregon.

Lull, Howard W., and Reinhart, Kenneth G., 1967. Increasing water yield in the Northeast by management of forested watersheds, U.S. Department of Agriculture Forest Service Research Paper NE-66, Northeastern Forest Experiment Station, Upper Darby, Pennsylvania.

Lull, Howard W., and Sopper, William E., 1967. Prediction of average annual and seasonal streamflow of physiographic units in the Northeast, *Forest Hydrology* (New York: Pergamon Press), pp. 507–521.

Mather, J. R., and Parmalee, D. M., 1963. Water purification by natural filtration. C. W. Thornthwaite Associates Laboratory of Climatology, *Publications in Climatology, 16*(4): 481–511.

Mitchell, Wilfred C., and Trimble, G. R., Jr., 1959. How much land is needed for the logging transport system?, *Journal of Forestry, 57:* 10–12.

Parizek, R. R., Kardos, L. T., Sooper, W. E., Myers, E. A., Davis, D. E., Farrell, M. A., and Nesbitt, J. B., 1967. Waste water renovation and conservation, Pennsylvania State University Studies No. 23.

Pierce, R. S., Hornbeck, J. W., Likens, G. E., and Bormann, F. H., 1970. Effect of elimination of vegetation on stream water quantity and quality, Northeastern Forest Experiment Station, Upper Darby, Pennsylvania (unpublished).

Reinhart, K. G., Eschner, A. R., and Trimble, G. R., Jr., 1963. Effect on streamflow of four forest practices, U.S. Forest Service Research Paper NE-1, Northeastern Forest Experiment Station, Upper Darby, Pennsylvania.

Robinson, Thomas W., 1967. The effect of desert vegetation on the water supply of arid regions, *International Conference on Water for Peace, 3* (Washington, D.C.: U.S. Government Printing Office), pp. 622–630.

Stoeckeler, Joseph H., and Voskuil, Glenn J., 1960. Water temperature reduction in shortened spring channels of southwestern Wisconsin trout streams, *American Fisheries Society Transactions, 88:* 286–288.

Striffler, W. David, 1965. Suspended sediment concentrations in a Michigan trout stream as related to watershed characteristics, U.S. Department of Agriculture Miscellaneous Publication 970, pp. 144–150.

Swank, W. T., 1968. The influence of rainfall interception on streamflow, *Proceedings of Hydrology in Water Resources Management,* Clemson University (March): 101–112.

Tarzwell, Clarence M., 1960. Effects and control of stream pollution, *International Union of Nature and Natural Resources, Seventh Technical Meeting, 4:* 242–252.

Thornthwaite, C. W., 1951. Agricultural climatology at Seabrook Farms, *Weatherwise, 4:* 27–30.

Urie, Dean, 1966. Forest cover in relation to water yields on outwash sand soils, *Michigan Academy of Service, Arts, and Letters Papers, 51, Part I: Natural Science,* edited by John Reichert (Ann Arbor, Michigan: University of Michigan Press), pp. 3–11.

Ursic, S. J., and Dendy, Farris E., 1965. Sediment yields from small watersheds under various land uses and forest covers, U.S. Department of Agriculture Miscellaneous Publication, 1970: 47–52.

Vogenberger, R. A., and Curry, J. A., 1959. Watershed protective logging, *Southern Lumberman* (December 15): 93–94.

Wark, J. W., and Keller, F. J., 1963. Preliminary study of sediment sources and transport in the Potomac River Basin, U.S. Geological Survey and Interstate Commission on the Potomac River Basin.

Woodridge, David D., and Gessel, Stanley P., 1966. Forests and the nation's water, *Water Resources Bulletin,* 2(4): 7–12.

16
Waste-Solid Disposal in Coastal Waters of North America

M. Grant Gross

Introduction

North American estuaries and coastal ocean areas receive large volumes of waste solids derived primarily from coastal urban centers. Much of this material is taken from harbors and placed on the adjacent continental shelf, which is often little affected by wastes derived from other sources. These wastes are commonly not considered in regional waste management planning but may well constitute significant potential long-term problems in many areas.

Sources of Waste Solids

This report is concerned with the large volume of waste solids taken to sea by barges or seagoing dredges for disposal. These wastes include deposits dredged from waterways, solids in waste chemicals, sewage sludges, rubble from building construction and demolition, and fly ash (listed in approximate order of volumes dumped at sea) (Brown and Smith, 1969). Excluded are floatable waste solids such as garbage, refuse, and rubbish. Although produced in large quantities in urban areas (Regional Plan Association, 1968), these floatable wastes are not commonly dumped in coastal or estuarine waters.

Some dredged wastes were originally riverborne sediments. Sediments are also moved into harbors by estuarine circulation in the adjacent coastal ocean and by movement of sand along beaches (Meade, 1969). Artificially deepened navigation channels and irregular shorelines formed by slips and basins used for shipping are often effective sediment traps and must periodically be dredged to maintain water depths adequate for navigation.

If clean gravels, sands, and coarse silts were the only materials dredged, there would be little problem in most coastal urban areas. Dredged materials have been used for landfill, especially hydraulic landfills, and for other construction purposes in many areas. In urban areas, however, dredged materials must commonly be classified and handled as wastes because the mixture of sedi-

Prepared for SCEP.

ment, municipal, and industrial wastes have undesirable physical and chemical properties.

Municipal wastes, including sewage solids, are major contributors of materials dredged from navigable waterways. In many coastal areas, untreated (raw) sewage is discharged to the waterways, which then provide primary treatment (gravitational settling of solids) and secondary treatment (biological degradation of organic matter), and the solids settle out in the waterways. Effluents from sewage treatment plants carry suspended solids into adjacent waterways, and some of these solids doubtlessly sediment, adding to the deposits that must be removed by dredging. In addition to the human wastes, many urban areas also discharge industrial wastes and debris washed from streets.

Waste Disposal Sites

In the United States, regulation of dredging and disposal of dredged wastes is handled by the U.S. Army Corps of Engineers. On the Pacific coast, especially in California, regional and state agencies are also involved in regulation of waste discharges (Brown and Smith, 1969). In most areas, these regulatory functions are conducted under the authority of the Refuse Acts (U.S. Congress, 1899), which prohibit waste disposal in navigable waters without prior permission of the local Corps of Engineers District. In New York, Baltimore, and Hampton Roads, Virginia, the regulation is conducted under provision of the Supervisor of Harbor Act (U.S. Congress, 1888). (Dredging data were supplied by Major General F. P. Koisch, Director Civil Works, U.S. Army Corps of Engineers.) In Canada, dredging is carried out by regional offices of the Department of Public Works, and the dredged materials are dumped at sea in sites whose locations are cleared by the Departments of Fisheries and Transport (C. K. Hurst, written communications).

Only those waste disposal sites known to have been used during the past ten years were included in this study. Many sites were excluded because they were apparently inactive; others were excluded owing to a lack of information.

The data indicate that there are at least 145 waste disposal sites in estuarine and coastal ocean waters of the United States

Table 16.1 Location and Number of Waste Disposal Sites in Various Coastal Ocean Areas with Estimated Tonnages of Wastes Dumped in Each Region

Disposal Sites (no. of active sites)	Estimated Tonnage (10^6 tons/year)	
Atlantic Coast (65)		
Maritime Provinces (4)	1.2	
Gulf of Maine (13)	1.4	
Mid-Atlantic coast (29)*	7.8	15.5
South Atlantic (17)	5.0	
Puerto Rico (2)	0.1	
Gulf of Mexico (20+)		
Eastern Gulf (9)	3.6	
Western Gulf (5+)	10.0	29.3
Mississippi River (6)	15.7	
Pacific Coast (60)		
California (3)	1.7	
Oregon-Washington (38)	10.8	13.5
British Columbia, Strait of Georgia (1)	0.9	
Total	58.3	

* Includes Long Island Sound (12) and Chesapeake Bay (2).

and Canada (Table 16.1). Along the U.S. Atlantic coast, there are fifty-nine active waste disposal sites; four more are located in Canadian waters (Figures 16.1, 16.2). Puerto Rico has two active sites. At least twenty sites are actively used along the Gulf of Mexico coast, and at least sixty sites are in use along the U.S. Pacific coast, including Alaska and a site in the Canadian Strait of Georgia. There is no record of active waste disposal operations in the Hawaiian Islands or U.S. territories or possessions. Although this report does not include the Great Lakes, ninety-five waste disposal sites are reported to be active there (U.S. Army Corps of Engineers, 1969). I have no data for Mexico or Central American countries. Sites (Figure 16.1) used for waste chemical disposal (nine sites) and sewage sludges disposal (two sites) are excluded from the data in Table 16.1.

Most waste disposal sites are situated in open estuarine waters or over the continental shelf, relatively close to the source of the wastes. Disposal sites are usually deeper than 20 meters in the United States, 24 meters in Canada, except where the wastes are used for beach replenishment or construction of artificial

255 Waste-Solid Disposal in Coastal Waters of North America

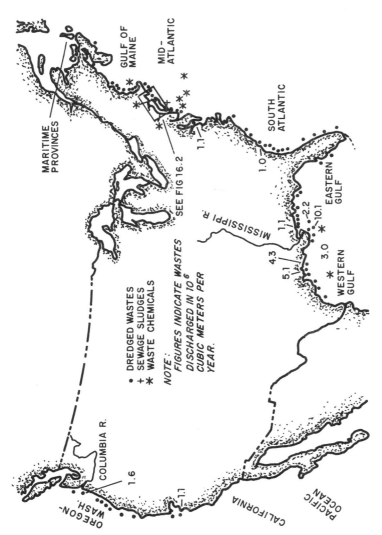

Figure 16.1 Locations of waste disposal sites and the volumes of dredged wastes (10^6 m³/year) discharged in each See Figure 16.2 for locations of disposal sites in the New York Region and Long Island Sound.

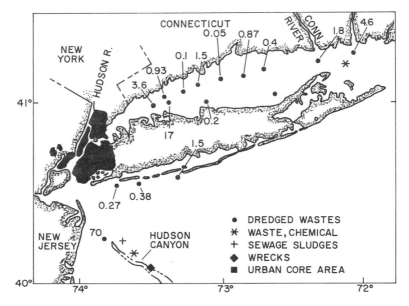

Figure 16.2 Location of active waste disposal sites in the New York Metropolitan Region (*10*)
Volumes (10^5 m³/year) of dredged wastes discharged in each.

fishing reefs. (In the Gulf of Mexico some sites are located in waters as shallow as 4 meters). The data indicate that beneficial uses are the exception rather than the rule.

Estuarine areas receive large volumes of waste solids. Chesapeake Bay, Long Island Sound, and Puget Sound each have several waste disposal sites (Table 16.1). All are used for dredged waste disposal except for Long Island Sound, which receives wastes from the New York Metropolitan Region as well as small volumes of waste chemicals dumped in the eastern end.

In order to avoid interference with navigation and to remove disposal sites as far as possible from beaches and other public areas, many are located more than 5 kilometers from the nearest land, presumably in international waters. Of thirty-one sites along the open Atlantic coast from Maine to Cape Hattaras, North Carolina, for which good location data were available, seventeen were more than 5 kilometers from the coast and eight were more than 10 kilometers from the coastline. A site used for toxic chemicals is approximately 190 kilometers from the entrance to New

York harbor; a similar site used for disposal of arsenic-containing wastes is between 300 and 500 kilometers offshore from Delaware Bay. In general, the more recently established waste disposal areas are farther offshore than older disposal areas.

Most North American continental shelf areas now receive insufficient sediment (Emery, 1968) to dilute or bury waste deposits. Therefore, it is essential to consider the tonnage of sediment brought by rivers to each of the coastal areas in comparison with the tonnage of wastes deposited there. To calculate waste tonnages, I assume a solid content (dry) of 0.9 metric ton per cubic meter of dredged material, which is typical of fine sand when dredged by federal seagoing hopper dredges (R. E. Denning, written communication). Using this assumption, the data indicate an estimated discharge of about 58.3 million metric tons per year.

Waste discharges in the Canadian Maritime Provinces, the U.S. portion of the Gulf of Maine, Long Island Sound, and the mid-Atlantic coast areas greatly exceed the probable suspended sediment discharge by rivers in these areas (Dole and Stabler, 1909). For example, the discharge of riverborne suspended sediment into Long Island Sound is probably about 10^5 metric tons per year, whereas waste discharges exceed 10^6 metric tons per year. Furthermore, the sediment comes primarily from the Connecticut River near the eastern end of the Sound. Waste disposal activities are concentrated in western Long Island Sound because the bulk of the wastes are derived from nearby densely populated urban areas.

In Chesapeake Bay, the suspended sediment load of the Potomac and Susquehanna rivers, about 3.0×10^6 metric tons per year, exceeds the waste discharge of about 1.6×10^6 metric tons per year, and thus could bury or dilute waste deposits. Near the mouth of the Mississippi River the river's suspended sediment load (Holeman, 1968) (about 3.1×10^8 metric tons per year) greatly exceeds the tonnage of wastes dumped there so that wastes are likely to be buried rather quickly. Similar situations probably prevail near the mouths of some rivers in the Gulf of Mexico, near the mouth of the Columbia River (Gross, McManus, and Ling, 1967) and in the Strait of Georgia (Waldichuk, 1953). Elsewhere on the continental shelf, wastes probably accumulate and remain at the water-sediment interface for long periods of time.

Impact of Dredged Waste Disposal

Waste solids pose both immediate and potential long-term problems in the coastal ocean. The immediate effects arise in two ways —from the dredging and from waste disposal operations. Wastes and sediments are disturbed during dredging causing increased turbidity and releasing nitrogen compounds, phosphates, and various reduced substances to the water. Increased oxygen demand arising from introduction of reduced substances (Brown and Clark, 1968) and decomposition of phytoplankton growth resulting from the phosphate and nitrate enrichment of the waters deplete dissolved oxygen concentrations (Pearce, 1969).

Potentially more troublesome is the long-term exposure of unburied waste deposits on the continental shelf or estuary bottom. Since most waste disposal sites are located in relatively shallow water, tidal currents and wave action can act on the deposits to resuspend and move them outside the designated disposal area. Under certain conditions such as upwelling, movement of near-bottom waters is directed landward, toward the estuary or river mouth. Some resuspended wastes may move toward beaches or back into the harbor from which they were removed.

Studies of two waste disposal sites near New York City receiving dredged wastes and sewage sludges indicate that populations of bottom-dwelling organisms in both waste disposal areas were severely reduced over several tens of square kilometers (Pearce, 1969). Among the factors thought to be responsible are low dissolved oxygen concentrations, presence of toxic compounds, and pathogenic organisms in the wastes. In addition to these, it seems probable that physical factors may also play a role inasmuch as the substrate was changed from hard sand or gravel to fine-grained soft bottom. Furthermore, the rapid accumulation of wastes (ranging from 0.4 to 30 cm/year, assuming that the wastes are spread uniformly throughout the affected areas) may make it impossible for many organisms to live in waste disposal sites (Gross, 1970).

Trends in Waste Disposal Operations

Despite proposed changes in governmental regulations (Council on Environmental Quality, 1970), it seems probable that coastal ocean areas will receive even larger volumes of wastes over the next five to ten years. Coastal urban areas are already short of land necessary for present waste disposal operations (Regional Plan Association, 1968) and find waste disposal in the coastal ocean to be an attractive solution for existing and foreseeable waste disposal problems. At present there is no universally acceptable alternative to ocean disposal for large-volume wastes. Hence, there will likely be increasing pressure for new types of wastes to be dumped in ocean waters, at least on an interim basis.

It seems highly probable that there will also be increasing volumes of dredged wastes as certain port facilities are prepared for new generations of deep-draft vessels, requiring 20 to 25 m of water in areas where maximum dredged-channel depths rarely exceed 15 m. The magnitude of this potential problem can be appreciated by considering that the dredging necessary for maintaining depths of the entrances to the Hook of Holland (Rotterdam-Europoort, Netherlands) resulted in the annual dredged-waste discharges increasing from 5.1×10^5 m³ in 1957 to 2.2×10^7 m³ in 1968, a fortyfold increase in eleven years. These figures refer mainly to maintenance dredging. At present the amount of dredged material is substantially increased because of the deepening of the entrance channel for passage of deep-draft tankers. A comparable increase in dredged waste disposal was experienced at the port of Amsterdam during this same period (Director, Rijkswaterstaat, written communication, January 2, 1970). The total volume of material dredged from these ports was 1.3×10^8 m³, about one-third the present annual sediment discharge of the Mississippi River (Holeman, 1968). If comparable port developments occur in North America, the volume of dredged materials will likely increase sharply over present levels.

Considering the widespread use of these disposal sites, their potential long-term effects, and the present inadequate knowledge about the wastes and their effects on the ocean waters and marine organisms, it seems that marine scientists face a challenge to apply their research capabilities in this area. There is also a chance to contribute toward possible solutions for environmental problems as well as to learn more about coastal ocean processes that are otherwise difficult or impossible to study.

Acknowledgments

Financial support was provided Coastal Engineering Research Center, U.S. Army Corps of Engineers, and the Bureau of Solid Waste Management, U.S. Public Health Service. I thank Miss Karen Henrickson for assistance in data reduction.

References

Biggs, R. B., 1968. Overboard spoil disposal, *Journal of the Sanitary Engineering Division, Proceedings of the American Society of Civil Engineers, 94:* 5979.

Brown, C. L., and Clark, R., 1968. Observations on dredging and dissolved oxygen in a tidal waterway, *Water Resources Research, 4* (6): 1381–1384.

Brown, R. P., and Smith, D. D., 1969. Marine disposal of solid wastes: an interim summary, *Marine Pollution Bulletin, 18:* 12–16.

Dole, R. B., and Stabler, H., 1909. Denudation, in papers on the conservation of water resources, *U.S. Geological Survey of Water Supply Paper 234:* 85.

Emery, K. O., 1968. Relict sediments on the continental shelves of world, *Bulletin of the American Association of Petroleum Geologists, 52:* 445–464.

Council on Environmental Quality, Executive Office of the President, 1970. *Ocean Dumping? A National Policy* (Washington, D.C.: U.S. Government Printing Office).

Gross, M. G., 1970. New York Metropolitan Region—a major sediment source, *Water Resources Research, 6* (3): 927–931.

Gross, M. G., McManus, D. A., and Ling, Y-Y., 1967. Continental shelf sediment, Northwestern United States, *Journal of Sedimentary Petrology, 37* (3): 790–795.

Holeman, John N., 1968. The sediment yield of major rivers of the world, *Water Resources Research, 4* (4): 737–747.

Meade, R. H., 1969. Landward transport of bottom sediments in estuaries of the Atlantic Coastal plain, *Journal of Sedimentary Petrology, 39* (1): 222–234.

Pearce, J. B., 1969. The effects of waste disposal in the New York Bight-Interim report for January 1, 1970, Sandy Hook Marine Laboratory, U.S. Bureau of Sport Fisheries and Wildlife, Sandy Hook, New Jersey.

Regional Plan Association, 1968. Waste management, *Regional Plan Association Bulletin, 107.*

U.S. Congress, 1888. Supervisor of Harbor Act, approved June 29, 1888 (33 U.S.C. 441–451), amended July 12, 1952 and August 28, 1958 (Pub. L. 85–802, 72 Stat. 970).

U.S. Congress, 1899. River and Harbor Act, approved March 3, 1899 (30 Stat. 1152; 33 U.S.C. 407); River and Harbor Act, approved March 3, 1905 (33 Stat. 1147; 33 U.S.C. 419).

U.S. Army Corps of Engineers, 1969. *Dredging and Water Quality Problems in the Great Lakes,* Buffalo, New York.

Waldichuk, M. W., 1953. Oceanography of the Strait of Georgia, III. Character of the bottom, *Fisheries Research Board of Canada, Progress Report,* No. 95: 59–63.

17
Chemical Invasion of Ocean by Man

Edward D. Goldberg

Man, a land organism, is influencing the chemical composition of sea water more than any of the species that live within the marine environment. The oceans receive the discharges of an ever-increasing world population in ever-increasing amounts and complexities. People in the United States use annually 7.5 tons of fuel and 5.0 tons of mineral, food, and forest products per person. Nearly all of these materials, used by man in social, agricultural, and industrial activities, are not retained by him but are dispersed, often in degraded forms, about the surface of the Earth. The oceans by intent of man, by their very expanse or by their chemical and physical characteristics, receive a major portion of this discharge. Such metabolic wastes of our civilization can increase the levels of chemical species already a part of the oceanic domain or can introduce substances alien to the marine environment. But most important, in our mining of the Earth's resources to maintain our civilization, we are transferring materials from one site to another and altering natural environments. What are these materials being poured into the oceans through the atmosphere and by the rivers and what will be the responses to them by organisms, including man? Knowledge about the first part of this question provides a necessary background to consider the second part, which still remains primarily in the realm of the unknown.

Some Inputs and Responses

There are seven large groups of pollutants produced by man that influence the chemical composition of the oceans. (1) Mercury and lead form one class of pollutants. Mercury or mercury compounds are used as fungicides and as catalysts in making industrial chemicals, and lead as a component of antiknock gasoline is injected into the atmosphere. (2) A large group is made up of petroleum that is lost during drilling in offshore sites or during transport across the seas. (3) The persistent pesticides used in agriculture are washed off the fields into streams and rivers or are transported to the oceans by the winds. (4) Many volatile industrial chemicals escape to the atmosphere during production, or residues of their

Reprinted from *Yearbook of Science and Technology* (McGraw-Hill, 1970), pp. 65–73. Used with permission of McGraw-Hill Book Company.

manufacture are released into streams. (5) When fossil fuels are burned either for heat or to generate power, the smoke-stack exhaust pollutes the atmosphere and ultimately the oceans. (6) Radioactive species from nuclear-device explosions and from nuclear reactors add to the pollution load of the air and the ocean. (7) The domestic wastes created by man are to a large measure dumped in an untreated form into rivers and streams to contribute to the ocean's changing chemical environment.

Mercury and Lead

These are two elements whose entries into the oceans from the continents appear to be about equally influenced by normal weathering processes and by the discharges of civilization.

Weathering processes result in the transfer of mercury from the continents to the oceans by way of rivers at a rate of about 5000 tons per year. About one-half of the world production of mercury, 9200 tons per year, is utilized by agriculture and industry with a subsequent uncontrolled release to the environment. Since the early decades of this century, organic mercury compounds have been used both as fungicides, primarily as seed dressings, and as catalysts in the production of chemicals such as chlorine and sodium carbonate. Such usages are held responsible for the increases in the mercury contents of birds collected in Sweden, where the levels in the 1940s and 1950s were 10–20 times those from previous years (Figure 17.1) and may explain the covariance of high concentrations of atmospheric mercury and smog such as occurs in the San Francisco Bay region (Figure 17.2).

Between 4000 and 5000 tons of mercury per year most probably enters the oceans as a result of the release of man-utilized compounds to the rivers and atmosphere. As yet, the bulk of this input has not been sought or identified in the marine environment. If the oceanic organisms behave as do their fresh-water counterparts, sea fish accumulate both organic and inorganic mercury compounds directly from solution with high concentration factors. On the other hand, mercury-containing substances are not taken up to an appreciable degree by plants.

Since much of the concern about marine pollution is homocentric, catastrophes can heighten our interest in the chemistries wrought by and upon the human ecosystem. A textbook example of scientific detective work isolated the cause of Minimata disease,

263 Chemical Invasion of Ocean by Man

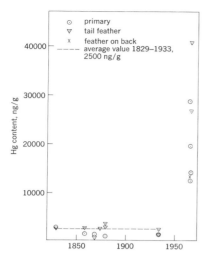

Figure 17.1 Environmental increases in mercury levels as recorded in the feathers of the eagle owl (*Bubo bubo*) collected from Sweden over the past 130 years

The increase in mercury in the 1940s and 1950s corresponds to increased usages of mercury alkyl compounds as seed dressings to combat fungal growths. The seeds were presumably ingested by the birds. Recent legislation in Sweden limiting the usages of such compounds has reduced the bird-feather concentrations of mercury to the levels obtained prior to the 1940s. (W. Berg et al., Mercury content in feathers of Swedish birds from the past 100 years, *Oikos*, *17*: 71–82, 1966.)

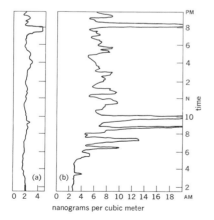

Figure 17.2 Mercury concentrations in air collected in San Francisco area
(a) Levels on day with fresh southwest winds. (b) Readings on smoggy day.
(S. H. Williston, Mercury in the Atmosphere, *Journal of Geophysical Research*, *73*: 7051–7055, 1968.)

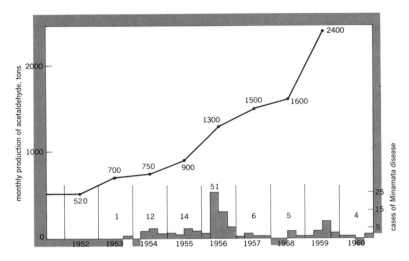

Figure 17.3 Production of acetaldehyde at Minamata factory and the number of cases of Minimata disease between 1952 and 1960
(K. Irukayama, The pollution of Minamata Bay and Minimata disease, *Advan. Water Pollution Research, 3:* 153–180, 1967.)

an illness that reached near-epidemic proportions in southwestern Kyushu, Japan, to mercury poisoning. By the end of 1960, 111 cases had been reported, and 41 deaths had occurred as of August, 1965 (Figure 17.3). Cats and rats living in the region also succumbed to the affliction. All victims had partaken of either fish or shellfish from Minimata Bay. The poisoning was due to methyl mercury chloride which somehow formed in the waste sludges of a factory that used mercuric oxide in sulfuric acid for the production of acetaldehyde. These solids were discharged to Minimata Bay, where the organomercury compound was accumulated by both fish and shellfish. Methyl mercury chloride, administered to cats, caused the various symptoms associated with the disease—unsteady movements, fits, tremor, and blindness. A second outbreak of Minimata disease occurred in Niigata City far removed from Minimata Bay. Here, 26 cases and 5 deaths occurred. Methyl mercury compounds in spent catalysts of an acetaldehyde factory discharged into the Agana River were the cause of the trouble.

Man appears to be responsible for an input of lead into the oceans equal to that of natural processes. In the Northern Hemisphere about 350,000 metric tons of lead, as the antiknock agent tetraethyllead, is burned in automobile internal combustion engines and subsequently introduced into the atmosphere. This airborne lead can be removed either by an inelastic collision with a surface, such as provided by vegetation, or by structures, or by rain. About 250,000 metric tons of lead is annually washed out over the oceans and about 100,000 metric tons over the continents in correspondence to their relative areas. Some of the land fallout eventually reaches the oceans as river runoff. This compares with an annual input of lead into the oceans through natural weathering processes of 150,000 metric tons. This impingement by man has raised the average lead content in surface waters of oceans in the Northern Hemisphere from about 0.01–0.02 to 0.07 µg/kg of sea water in the 45 years since the introduction of lead as an antiknock chemical (Figure 17.4). The effect of such an increase in lead content upon marine organisms is unknown.

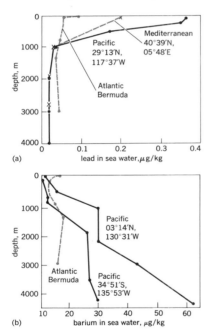

Figure 17.4 Barium and lead concentrations in sea waters as a function of depth
(a) Barium typifies many metals with increased contents at depth, most probably as a result of a transfer by plants and animals. (b) The source of the high lead in surface waters is attributed to the combustion of gasoline, containing tetraethyllead, in automobile engines. (T. J. Chow and C. C. Patterson, Concentration profiles of barium and lead in Atlantic waters off Bermuda, *Earth Planet. Sci. Lett, 1*: 397–400, 1966.)

Petroleum

The loss of petroleum to the sea as a result of man's utilization of this fossil fuel already has widespread repercussions. A continuous problem involves the transport of 10^9 tons of petroleum across the world ocean annually, an amount that each year increases about 4%. This is about one-half of the world's production. The estimated losses to the marine environment, through spillage and leakage, is about 0.1% of the total transport, or about 10^6 tons per year.

The primary production of organic matter in the sea is approximately 10^{10} tons per year; about a hundredth of a percent of this amount, or 10^6 tons, is in the form of hydrocarbons, materials similar to the components of petroleum. Thus the continuous input of petroleum hydrocarbons by man roughly corresponds to the natural hydrocarbon production.

Disasters, such as the breakup of the tanker *Torrey Canyon* and the oil releases in Santa Barbara, Calif., have injected about 100,000 tons of material into the marine environment for each event. The visible consequences of oil-covered beaches and dead birds result from the localized natures of these large inputs in both time and space.

Since petroleum components such as the hydrocarbons associate with the lipid-containing materials of the sea, the many reports of fish tasting of petroleum are not surprising.

Pesticide Residues

Perhaps the most abundant of the synthetic pollutants in the marine environment is *p,p'*-DDE, a degradation product of *p,p'*-DDT, the principal component of industrial DDT. These two substances, along with other chlorinated hydrocarbon pesticides, such as dieldrin and endrin, have been transported by the atmosphere and rivers from the continents to enter all levels of the marine ecosystem. DDT is the most widely used, as well as studied, of the chlorinated hydrocarbons. Thus DDT and DDE can illustrate the impact of these pesticides upon the oceanic plants and animals in whose fatty tissues they are concentrated.

At the base of the marine food chain are the single-celled algae whose photosynthetic activity is decreased by only a few parts per billion of DDT in sea water. The highest contents of these chlorinated hydrocarbons are found in marine birds. Skuas from the Antarctic, shearwaters from both the Atlantic and Pacific, black petrels nesting in the Gulf of California, and the Bermuda petrel of the northern Atlantic carry DDT residues at the multi-part-per-million level.

Widespread population declines and potential extinctions of some of these birds have been linked to their high body burdens of halogenated hydrocarbons. These pesticide residues have the property of inducing liver epoxidase enzymes which can degrade sex hormones and other steroids. One of the sex hormones, estrogen, governs various phases of calcium metabolism, such as its deposition in bone marrow. The calcium in the marrow is subsequently transported to the oviduct, where it becomes a part of the egg shell. A significant decrease in egg shell weights and thicknesses, accompanied by egg breakage and egg eating by parent birds, abandonment of nests, and abnormal breeding behavior, has been related to the population falloff of the peregrine falcon and to their high levels of pesticide residues.

Industrial Chemicals

Industrial chemicals, released as volatiles to the atmosphere or as discharges to rivers, can find their way to the marine environment. The recent observations of acetone, butyraldehyde, and methyl ethyl ketone in surface waters of the Florida Straits, the eastern Mediterranean, and the Amazon estuary may represent such injections. These aldehydes and ketones have several characteristics in common: low molecular weights, relatively low boiling points, high solubilities in water, and high utilizations and high productions in the chemical industry.

The fates of other volatile chemicals used in our society demand consideration. There is a high evaporation loss of dry-cleaning solvents, which are quite stable halogenated hydrocarbons, at a rate of about 350,000 tons per year—a value roughly equal to their production. About 2.5% of the total production of gasoline is lost by evaporation through transfer processes from production site to storage tanks and vehicles and through vaporization from automobile gas tanks and carburetors. This amounts to over 10^6 tons per year in the United States.

A group of halogen-containing organic materials, the polychlorinated biphenyls (PCB), are widely dispersed in the marine biosphere and may be the second most important synthetic chemical there. The weight ratio of the pesticide DDT to PCB in seabirds of the Pacific appears to be between 5 and 10 with individuals from industrial areas having higher PCB levels than those from more remote regions. Used in the plastics, paint, and rubber industries, PCBs have effects similar to the previously mentioned pesticide residues as inducers of steroid hydroxylate enzymes in birds. The mechanism of dispersal of PCB from industrial sites to the oceans is not clear, but the PCBs presumably enter the atmosphere as vapors.

Fossil Fuels

The burning of fossil fuels injects into the atmosphere a variety of substances which may enter the oceans (see Table 17.1). A National Academy of Sciences committee indicated that about 60% of these substances results from transportation, about 19% from manufacturing processes, about 13% from power plants, about 6% from space heating, and the remainder from the incineration of wastes. The United States, in comparison with the rest of the world, appears to be responsible for about one-third to one-half of such waste products.

Much of this material will be oxidized in the atmosphere and will be taken up by the oceans as carbon dioxide, sulfates, and nitrates, which are forms already present in sea water. It is estimated that the carbon dioxide level of surface ocean waters has increased several percent since the industrial revolution. The rate of injection of this gas is approaching the rate of fixation of carbon per year by the photosynthesizing plants of the sea. While society's production of carbon dioxide increases with time, the natural production of organic matter by plants is presumed to remain more or less constant. The effects of such substances upon

Table 17.1 Estimated rates of injection of materials into the atmosphere in grams per year

Material	United States	World
Carbon monoxide	7×10^{13}	2×10^{14}
Sulfur oxides	3×10^{13}	8×10^{13}
Hydrocarbons	2×10^{13}	8×10^{13}
Nitrogen oxides	8×10^{12}	5×10^{13}
Carbon dioxide	—	9×10^{15}
Smoke particles	1×10^{13}	2×10^{13}

life processes and geological phenomena in the oceans are as yet unknown.

The solid smoke particles that pass through the marine environment to the sediment may alter the overall chemical composition of the oceans by their acting as reaction sites for precipitation or oxidation-reduction processes involving the dissolved species in sea water. Practically no thought has been given to these possibilities as yet.

Radioactive Species

Radioactive species, released from the recent nuclear-device testing programs of the United States, the Soviet Union, Britain, China, and France or from discharges from nuclear reactors, may be found in all oceans and in all members of the marine biosphere. These activities have introduced such novel nuclides as radioactive Sr^{90} and Cs^{137} into the world's seas, as well as substantially increasing the amounts of cosmic-ray-produced nuclear species, such as C-14 and H-3, in the surface layers of the ocean and in the atmosphere. With these radioactivities, man has left his imprint in all parts of the world's oceans, which covers two-thirds of the Earth's surface to an average depth of 4 km. Up to the present time, there is no evidence that these pollutants have had any widespread adverse effect upon marine communities. On the other hand, their very presence has stimulated much support of research at national and international levels which has markedly broadened our oceanographic knowledge.

Domestic Wastes

There is a per capita discharge of about 15 kg of solid wastes to domestic sewers annually in the United States. With 30,000,000 persons living in the vicinity of the sea coast, around 500,000 tons per year of such substances enters the marine environment. If it is assumed that the inhabitants of the United States are responsible for about 20% of the world's domestic coastal effluents, several million tons of solid sewage are yearly contributed to the world ocean. Although this amount is small compared to the 10^{16} tons of solid phases resulting from the weathering of the Earth's crust pouring into the oceans by way of the rivers each year, the impacts of domestic discharges are quite evident at sewer outfalls through the alterations of natural communities and upon their introduction of sludges to the sediments.

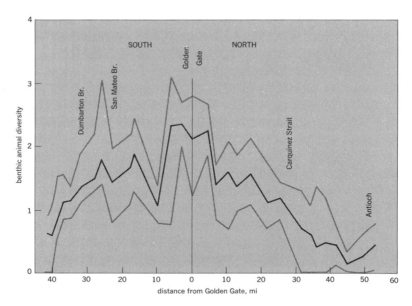

Figure 17.5 Variation of the diversity of benthic organisms as a function of distance from the Golden Gate Bridge
The points of high sewage input are at the extreme ends of the figure. (E. A. Pearson, P. N. Storrs, and R. E. Selleck, Some physical parameters and their significance in marine waste disposal, in T. A. Olson and F. J. Burgess, eds., *Pollution and Marine Ecology, Interscience,* 1967.)

In general, the number of species of organisms in the near vicinity of a sewer outfall decreases, although the total number of individuals may remain the same as or be higher than that in adjacent, unaffected areas. Such a situation is illustrated in the species diversity in San Francisco Bay, where waste discharges enter at areas north and south of the Golden Gate Bridge. The areas of input of such materials have a smaller number of species of benthic organisms (Figure 17.5).

Natural Processes
As a result of phenomena occurring at or near the Earth's surface, the components of the oceanic system change in form, concentration, reactivity, and distribution. The implications of such changes to an understanding of man's impingements upon the marine environment are considered in the following section.

Marine Chemistries

The characteristics of natural processes that have been studied within the oceanic system provide an important background for predicting the fates of materials introduced by man. Many marine chemistries can be described in the following ways: (1) The reactants are in extremely low concentrations, that is, micromolar and below. (2) The periods needed to produce discernible amounts of materials are measured in geologic units, hundreds to millions of years, in comparison to the time constants usually involved in laboratory experiments, ranging from fractions of a second to days. (3) The reactions take place at phase discontinuities, such as atmosphere-ocean, atmosphere-sediment, and atmosphere-biosphere.

Iron and Manganese Oxides

The formation of the ferromanganese minerals, accumulations of iron and manganese oxides, which occurs in all oceans and at all depths, illustrates the above three characteristics. The reactants in the sedimentary process, iron and manganese, have sea-water concentrations of 2×10^{-7} and 4×10^{-8} M, respectively, yielding the minerals in which their contents achieve levels of 15–20% by weight on the average. In some areas of the oceans they may cover wide expanses of the sea floor. Their patchy distributions preclude an average distribution figure, but it is of interest to note that in the southwestern Pacific a region with 26–46% of the bottom paved with manganese nodules has been found, while an area in the northern Pacific, centered at a latitude of 18–20°N, possesses a bottom coverage of the minerals between 10 and 100%.

The chemical reactions involved in the formation of the minerals involve, most probably, the oxidation of dissolved manganese in sea water. A simple and direct representation of the process is given by the equation below. This reaction probably occurs

$$Mn^{++} + 2OH^- + \tfrac{1}{2}O_2 = MnO_2 + H_2O$$

heterogeneously with a solid surface which is a good adsorber for oxygen, converting molecular oxygen to atomic oxygen. MnO_2 or Fe_2O_3 may provide the sites for the reaction since these oxides contain the metals in lower states of oxidation. Thus they can, in principle, absorb oxygen, forming metal ions with high oxidation states. Such substances then provide a source of atomic oxygen.

The ferromanganese minerals accrete at rates of millimeters per million years, or, perhaps in more dramatic units, at rates of atomic layers per year. This growth rate, measured by several independent radioactive techniques, reflects one of the slowest chemical reaction rates ever determined.

Concentration by Organisms

Members of the marine biosphere accumulate metal ions which exist in submicromolar concentrations. Certain tunicates concentrate vanadium, whose marine concentration is 2×10^{-8} M, and others its vertical periodic table neighbor, niobium, with a concentration of 10^{-10} M; still other tunicates concentrate neither element, but none can accumulate both elements. This observation emphasizes the most unusual specificities that have been encountered in recent investigations. Perhaps the comment of some Woods Hole Oceanographic Institution investigators that "for any given chemical element there will be found at least one planktonic species capable of spectacularly concentrating it" will direct attention to the range of possibilities for the uptake by organisms of unusual chemicals introduced by our advancing civilization.

Reactivities

The relative reactivities of elements in the marine environment can be measured by the lengths of time they spend in the ocean before precipitation to the sea floor. This period, the so-called residence time, can be calculated under the assumption that the world's oceans are a single, well-mixed reservoir of water at steady state. The input of substances from the rivers is complemented by an equal amount of material accommodated in the sediments; thus the composition of the oceans is taken to be invariant with time. This schematization of the oceans yields residence times for the alkali metals and alkaline earths of the order of millions to hundreds of millions of years, emphasizing their lack of reactivity in solution. Other elements, such as aluminum, iron, the rare earths, and thorium, have much shorter residence times, hundreds to thousands of years, which are periods of the order of, or less than, that of mixing times for oceanic water masses. These elements, in part, enter or exist in the oceans as particulate phases and rapidly settle to the sedimentary deposits. Some are the reactants in the chemical reactions for marine precipitates such as the ferromanganese minerals, and zeolites. Their entry into the oceans as solids or their high chemical reactivity or both can account for their low residence times.

Thus the chemical invasion of the oceans by man involves time periods quite different from those encountered in terrestrial water systems. Whereas a river renews itself annually and a lake in decades or centuries, the oceanic environment maintains its components for times orders of magnitude greater. Present-day introductions of materials will be measurable in the ocean for many thousands of years in the future.

Transport Paths

The distributions of materials introduced from the continents to the oceans are influenced by the transporting agency: rivers, winds, or glaciers. Man has imposed two additional paths: discharges from ships and sewage outfalls. Of consequence are the long distances of transport that take place. Pesticides applied to agricultural areas in Africa have been transported 6000 km across the Atlantic by the Northeast Trades to be found in dust collected in Caribbean islands. Pesticides introduced in Africa have also been observed falling into the Bay of Bengal, thousands of kilometers to the east.

Radioactive debris from the explosion of a Chinese nuclear device on May 14, 1965, was detected in its travels around the world at sampling sites in Tokyo (140°E, 36°N) and Fayetteville, Ark. (94°W, 36°N). The site of the nuclear detonation was at Lop Nor (90°E, 40°N). Sr^{89} and Sr^{90} were sampled from rains and were detected in June and July, 1965, indicating that a part of the materials had circled the world twice. The average velocity of the wind transport was about 16 m/sec. The movement of material took place in the tropospheric jet streams.

Clearly, the industrial or agricultural emissions from one country can adversely affect the marine resources of another. Pesticides or radioactivity in food fish can result from human activities many thousands of miles away. Problems of marine pollution are international, not provincial, and will require the dedicated effort of many nations toward solution.

Conclusions

An understanding of the consequences of man's chemical invasion of the oceans is of crucial importance. Can this intrusion affect the composition of the marine communities by altering the mortality rates of one or more species? Can certain areas be made uninhabitable for creatures of the sea as a result of the introduction of injurious substances? What is the probability of any undesirable substances injected by man being returned to him in his food products recovered from the marine environment?

A substantial portion of the scholarly activity of our society has been dedicated to an understanding of its surroundings. In part these studies have been motivated by curiosity, in part by the need to protect man's institutions against the ravages of nature, such as winds, rains, and floodwaters. The ability of mankind to significantly alter the characteristics of the Earth's surface emphasizes a third need for environmental studies. Man-imposed changes in the oceans can be detrimental to life and perhaps can result in the loss or restricted use of a valuable resource. The ability to predict undesirable results along these lines may lead to necessary and protective policies concerning such encroachments.

18 Chlorinated Hydrocarbons in the Marine Environment

SCEP Task Force

Introduction

It is most difficult to predict "things to come" for our society and for our environment. For society the rapidity of technological advances and concomitant changes in living patterns limit our descriptions of the future to a fraction of a decade at best. For the environment, alterations to our surroundings on a global basis take place much more slowly, with changes that often are imperceptible during such a period. The ability of man to manipulate his environment, the climate, and the various assemblages of living organisms is the hallmark of an advanced technology. The consequences are yet to be understood.

A few scientists have foretold the environmental catastrophes of the past. In the case of DDT, introduced to agricultural practices during World War II, there were warnings of its deleterious impact upon the environment several years after its first application as a pesticide. In 1946 Clarence Cottam and Elmer Higgins of the Fish and Wildlife Service wrote:

> From the beginning of its wartime use as an insecticide the potency of DDT has been the cause of both enthusiasm and grave concern. Some have come to consider it a cure-all for insect pests; others are alarmed because of its potential harm. The experienced control worker realizes that DDT, like every other effective insecticide or rodenticide, is really a two-edged sword; the more potent the poison, the more damage it is capable of doing. Most organic and mineral poisons are specific to a degree; they do not strike the innumerable animal and plant species with equal effectiveness; if these poisons did, the advantage of control of undesirable species would be more than offset by the detriment to desirable and beneficial forms. DDT is no exception to this rule. Certainly such an effective poison will destroy beneficial insects, fishes, and wildlife.
>
> Although many more investigations are needed in all these fields, it seems that the most pressing requirement is a study to determine the effects of DDT as applied to agricultural crops on the wildlife and game dependent upon an agricultural environment. About 80 percent of our game birds, as well as a very high

---

Prepared during SCEP by the following participants and members of a Workshop of the National Academy of Sciences Committee on Oceanography (NASCO) who were SCEP consultants: E. D. Goldberg, Chairman*; P. Butler; G..Ewing*; M. Ingham; D. Jenkins; P. Kearny; B. Ketchum; P. Meier; D. Menzel*; J. Reid*; R. Risebrough; L. Stickel*; G. Woodwell; and S. Burbank, Rapporteur.

* Indicates NASCO Workshop members. E. D. Goldberg was a full-time SCEP participant and also Chairman of the NASCO group.

percentage of our nongame and insectivorous birds, and mammals are largely dependent upon an agricultural environment. In such places application of DDT will probably be heavy and widespread; therefore it is not improbable that the greatest damage to wildlife will occur there.

Because of the sensitivity of fishes and crabs to DDT, avoid as far as possible direct application to streams, lakes, and coastal bays.

Wherever DDT is used, make careful before and after observations of mammals, birds, and fishes, and other wildlife.

The Task Force confirms the observations and predictions of Cottam and Higgins. But our knowledge of the fate and behavior of DDT in the marine environment is inadequate. Estuarine fish-eating birds are building up such high body burdens of DDT and its degradation products that they are suffering extensive failures in the reproductive process. Yet, we still do not really know the paths of DDT and of its residues from the continents, to the marine environment, to the fish.

This report is not intended to be an exhaustive survey of the literature. It has been prepared for the community of marine scientists to alert them to one of the more serious problems arising from the disposition of materials by man about his surroundings. The Task Force has attempted to present a sense of what DDT and its degradation products are doing to some of the ecosystems of the oceans. Emphasis has been placed upon this group of chemicals inasmuch as they are the most studied to date. Further, it has attempted to point out lacunae in our knowledge and to suggest remedial actions. Finally, it has extended and amplified upon the predictions of Cottam and Higgins, made over twenty-five years ago. We hope that a similar report will not be necessary twenty-five years hence.

Summary

The oceans are an ultimate accumulation site of the persistent chlorinated hydrocarbons. As much as 25 percent of the DDT compounds produced to date may have been transferred to the sea, but it is likely that the oceanic burden of polychlorinated biphenyls represents only a small fraction of total production. The amount of DDT compounds in the marine biota is estimated to be in the order of less than 0.1 percent of total production and has produced a demonstrable impact upon the marine environment.

Populations of fish-eating birds have experienced reproductive failures and population declines, and with continued accumulations of persistent chlorinated hydrocarbons in the marine ecosystem additional species will be threatened.

The decline in productivity of marine food fish and the accumulation of unacceptable levels of persistent chlorinated hydrocarbons in their tissues can only be accelerated by the continued releases of these pollutants to the environment.

Certain risks in the utilization of chlorinated hydrocarbons are especially hard to quantify, but they require most serious consideration. The rate at which such substances degrade to harmless products in the marine system is unknown. For some of the more persistent materials their half-lives are certainly of the order of years, but perhaps even of decades or centuries. If most of the remaining persistent chlorinated hydrocarbons are presently in reservoirs that will in time transfer their contents to the sea, we may expect, quite independent of future manufacturing practices, an increased level of these substances in marine organisms. And if, in fact, these compounds degrade with half-lives of decades or longer, there will be no opportunity to redress the consequences.

The more the problems are studied, the more unexpected effects are identified. In view of the findings of the past decade, our prediction of the hazards may be vastly underestimated.

Recommendations

A massive national effort should be made immediately to effect a drastic reduction of the escape of persistent toxicants into the environment with a view to achieving a virtual curtailment in the shortest possible time.

Programs should be designed to determine both the rates of entry of each pollutant into the marine environment and to make base-line determinations of the distribution of the pollutants among the components of that environment. These should precede a monitoring program that will follow long-term trends to record progress or document disaster.

The laws relating to the registration of chemical substances

Table 18.1 Production of Chlorinated Hydrocarbons in Units of 10^9 Grams/Year (United States only)

Year	DDT*	Aldrin-Toxaphene Group*	PCB†
1968	63.4	52.7	5
1967	47.0	54.6	
1966	64.2	59.3	
1965	64.0	54.0	
1964	56.2	47.9	
1963	81.3	48.2	
1962	75.9	48.3	
1961	77.9	47.2	
1960	74.6	41.2	
1959	71.2	39.5	
1958	66.0	44.7	
1957	56.6	34.3	
1956	62.6	39.4	
1955	59.0	35.0	
1954	44.2	20.5	
1953	38.4		
1952	45.4		
1951	48.2		
1950	35.5		
1949	17.2		
1948	9.2		
1947	22.5		
1946	20.7		
1845	15.1		
1944	4.4		
Total	1220	670	

Source: * Stanford Research Institute, 1969.
† Estimate based upon consumer information.

and to the release of production figures by government need to be examined and perhaps revised in the light of present evidence of environmental deterioration caused by some of these substances.

U.S. and World Production of Chlorinated Hydrocarbons

The U.S. production figures for DDT, for the Aldrin-Toxaphene* Group and for polychlorinated biphenyls (PCBs) are presented in Table 18.1. The United States utilizes perhaps 70 to 80 percent of its production of the Aldrin-Toxaphene Group and about 30 percent of the DDT manufactured. The remaining materials are exported (Stickel, 1968).

* Includes aldrin, chlordane, dieldrin, endrin, heptachlor, terpene, polychlorinates, and toxaphene.

279 Chlorinated Hydrocarbons in the Marine Environment

The collation of world production data for these materials is beset with difficulties due to the inadequate information available. Although figures may be found for a given country for a year or a series of years, a comprehensive set of the major producers has not yet been compiled. This hampers mass balance calculations for their dispersion about the earth and for an understanding of their behavior in nature. An urgent need for such numbers is clear. Even more pressing is the need for the production figures for the polychlorinated biphenyls, which have been in use since the early 1930s.

An inspection of the production and utilization figures that were available to us suggests that the total world production of DDT and the Aldrin-Toxaphene Group is probably not in excess of one and a half times that of the United States (Food and Agriculture Organization, 1968). Thus, as a first approximation the integrated world productions (that is, the total amount since production began) of DDT is 2.0×10 grams and the production of the Aldrin-Toxaphene Group is 1.0×10^{12} grams. Yearly production rates for DDT and for the Aldrin-Toxaphene Group of 10^{11} grams/year are used in calculations.

Transport of DDT Residues and PCBs to the Marine Environment

There are two principal routes for transport of DDT residues[*] from places of application on land to the ocean: rivers and atmosphere.

Surface Runoff

Total annual surface runoff of water from all continents has been estimated at 3.7×10^{19} cc. The maximum concentration of DDT residues in water reported in a survey of rivers of the western United States was about 100 parts per trillion (Manigold and Schulze, 1969; Bailey and Hannum, 1967). If all rivers of the world contain this maximum, 3.7×10^9 grams of DDT residues would be transported annually to the sea. This total would be approximately 3 percent of the annual production of the world. It is, however, almost certainly high, perhaps by a factor of 10. DDT residues transported by the Mississippi River into the Gulf

---

[*] DDT residues is a term defining DDT, DDE, and DDD. DDD and DDE are metabolites of DDT, and DDD is also a pesticide in its own right.

of Mexico have been estimated at 10^7 grams annually (Risebrough and Coulten, 1968). The total river output from the continental United States might be twice that of the Mississippi, approaching 2×10^7 grams. Domestic use of DDT between 1961 and 1968 has ranged between 5 and 8×10^{10} grams/year, thousands of times more than can be carried in the rivers. Therefore, it seems unlikely that much more than 1/1,000 of the annual production of DDT (10^8 grams) could reach the oceans by surface runoff.

Atmospheric Transport

DDT residues enter the atmosphere by several routes including aerial drift during application by rapid vaporization from water surfaces (Acree, Beroza, and Bowman, 1963) and by vaporization from plants and soils (Nash and Beal, 1970). Once in the atmosphere it may travel great distances, entering the sea in precipitation or in dry fallout. There are few data for estimating these rates of transfer. The most extensive sampling of DDT residues in precipitation has been in Great Britain, where total accumulation was measured at seven stations between August 1, 1966, and July 1967. The mean concentration for those rainwater samples was 80 parts per trillion (Tarrant and Tatton, 1968), about twice that reported for meltwaters of recent antarctic snow (Peterle, 1969). The DDT residues in South Florida precipitation averaged 1,000 parts per trillion in eighteen samples taken at four sites from June 1968 to May 1969 (Yates, Holswade, and Higer, 1970).

Total annual precipitation of water over the oceans has been estimated as 3.0×10^{20} cc (Sverdrup, Johnson, and Fleming, 1942). If this contained an average of 80 parts per trillion, a total of 2.4×10^{10} grams of DDT residues would be transported annually to the oceans, about one-quarter of the estimated total annual production of DDT. It is at least plausible that the atmosphere is the major route for transfer of DDT residues into the oceans.

Other chlorinated hydrocarbons with similar physical and chemical characteristics can be expected to have similar dispersal mechanisms. Dieldrin, which readily volatilizes into the atmosphere (Lichenstein et al., 1968), has been found in airborne particulates over the ocean (Risebrough et al., 1968) and in rainwater (Tarrant and Tatton, 1968). Polychlorinated biphenyls have also been detected in rainwater (Tarrant and Tatton, 1968). PCBs have not as yet been measured in river waters, but there is an indication that it is more abundant than the DDT components in biota from both the Pacific and the Atlantic (Risebrough, unpublished observations). The relative importance of river transport, aerial fallout, and dust discharge of PCB into the sea remains therefore to be determined.

Distribution of Chlorinated Hydrocarbons in the Marine Environment

Few data are available to document the concentration of chlorinated hydrocarbons (including PCBs) in the open ocean environment. Yet from a few observations it appears likely that DDT and its residues are distributed throughout the marine biosphere. Gray whales contain up to 0.4 ppm DDT residues in their blubber and sperm whales up to 6 ppm (Wolman and Wilson, 1970). The former feed largely on benthic organisms in the Chukchi and Bering Seas, and the latter on larger pelagic organisms. Seabirds (petrels and shearwaters), which feed on planktonic organisms far from land, have concentrations of DDT residues as high as 10 ppm (Risebrough, unpublished data). Migratory fish (tuna) carry up to 2 ppm of these same compounds in their gonads, and other marine mammals up to 800 ppm in their fat (Butler, 1969). In the latter two cases, it is not known whether these concentrations resulted from localized contact in coastal waters or were accumulated during the life of the organisms in the open ocean. In the coastal environment DDT and its residues range from undetectable levels to 0 to 5.4 ppm in oysters (Butler, 1969). Concentrations between these limits are highly variable locally, and even within the same estuary.

In spite of the paucity of useful data, some assumptions can be made for the marine environment, excluding estuaries, by assigning likely values for DDT residues to the biota of the open ocean and extending the calculations to a global basis. Such computations are valuable in that they identify potential sinks and provide order of magnitude estimates of most probable distributions. The following assumptions are made:

1. The standing crop of plankton (plant and animal) is 3×10^{15} g (Menzel, unpublished data).

2. The standing crop of fish is 6×10^{14} g, equal to 10 times the present annual fish harvest (Ryther, 1969).
3. The concentration of DDT residues in plankton averages 0.01 ppm.
4. The concentration of DDT residues in fish average 1.0 ppm.
5. A homogeneous distribution of DDT in the mixed layer (the upper 100 meters or so of seawater that is homogenized by the action of winds) has resulted from atmospheric transport from the continents.

By assigning the values in assumptions (3) and (4) to the calculated standing crop of organisms, (1) and (2), it is estimated that the total plankton now contain 3.0×10^7 g and fish 6×10^8 g of DDT residues, both insignificant fractions of the total annual input of these residues to the environment (10^{11} g). We propose these estimates as upper limits to the size of the pool.

With a saturation level of DDT* in water of 1 ppb (Bowman, Acree, and Corbett, 1960) and the volume of the mixed layer (upper 100 m of the ocean) as 0.025×10^{24} ml, the surface waters of the ocean are capable of accommodating a load of 7.5×10^{13} g of DDT or aproximately 38 times the total production to date. There is no indication, however, that DDT introduced into the marine environment is uniformly distributed in the mixed layer. Enrichment is likely to occur in the sea's surface film, which contains fatty acids and alcohols. If this is the case, predictions indicate (1) that DDT may be stripped from this film by bacteria and/or phytoplankton and thus enter the food chain; (2) that it adsorbs to airborne particles that sink through the water column (in this case the compound is probably ingested by grazing organisms that do not discriminate between living and inert particles); or (3) that it codistills with water or is injected back into the atmosphere as aerosols and is redistributed, leading to neither a net increase nor decrease in concentration at the surface.

Lacking any data for concentrations in the waters of the open sea, it is impossible to estimate directly how much is present which is not incorporated into living organisms. However, estimates of fallout in rain suggest that one-quarter of the world's production of DDT may have entered the ocean. Its areal distribution is prob-

---

* Saturation levels of DDD and DDE have not been measured as yet.

ably uneven, dependent upon weather patterns and proximity to major sources of input.

If only 0.01 percent (10^8 g) can be accounted for in pelagic marine organisms, 0.50×10^{12} g (one quarter of total production) should be present in solution and in the bottom sediments. In order to balance input with accountable fractions, the surface mixed layer volume (0.025×10^{24} ml) should contain concentrations of approximately 5×10^{-12} g/ml, given a residence time of five years and an annual input of 0.25×10^{11} g of DDT/year.

The Ecological Impact
The acute and chronic toxicity of chlorinated hydrocarbons has been identified by observing their effects under controlled laboratory conditions. The exposure of test populations of marine fauna to serial dilutions of these pollutants in flowing seawater has shown that they affect growth, reproduction, and mortality at concentrations currently existing in the coastal environments. These laboratory effects and their field counterparts may be summarized as follows.

Plankton
The addition of chlorinated hydrocarbons to laboratory cultures of molluscan larvae and the phytoplankton on which they feed causes, with increasing concentrations, decreased growth rates, developmental failures, and increased mortality (Ukeles, 1962; Davis, 1961). By extrapolation, toxaphene levels observed, for example, in one estuary of the southeastern United States in 1967 were high enough to have caused the death of a majority of the phytoplankton suitable as food for mulluscan larvae (Butler, 1969).

In the open ocean phytoplankton are at the base of the food chain and may act as primary concentrators of chlorinated hydrocarbons from the water. Laboratory evidence is available demonstrating inhibition of photosynthesis by DDT, dieldrin, and endrin in single-celled marine plants (Wurster, 1968; Menzel, Anderson, and Randtke, 1970a). It is doubtful, however, that these results are ecologically meaningful. The concentrations necessary to induce significant inhibition far exceed expected concentrations in the open ocean and, in the case of DDT, exceed by ten times its solubility (1 ppb) in water. One species tested was insensitive to concentrations of all three pesticides up to 1 ppm. Therefore, toxicity may vary interspecifically and, if not universally toxic, may exert some control on species succession in the near-shore environment.

If chlorinated hydrocarbons are concentrated in surface oil films, it is not improbable that concentrations there may reach levels sufficient to cause acute toxicity to plants. Considering that this layer may extend to 1 mm in depth, its effect on total production within the euphotic zone would be about 10^{-6} (100 m depth). Plants and/or bacteria, however, may provide an effective means of extracting these hydrocarbons from the surface film. Partition coefficients have not been experimentally determined from marine species but have been established as approximately 1,000 in the case of dieldrin for a freshwater alga. These organisms are available as food to grazing and surface-skimming feeders. The important fact here is not the effect of chlorinated hydrocarbons on primary production but rather that plants may be the vehicle for transferring these compounds from the water to higher trophic levels.

Experimental evidence from the calanoid copepod *pseudodiaptomns cornatus* has shown that the development of adults from nauplii is completely blocked when hatched from egg-bearing females maintained in seawater containing 10 parts per trillion DDT (Menzel, Anderson, and Randtke, 1970b). Significant mortality was observed at 5 parts per trillion. These concentrations of DDT are lower than expected in rainwater falling on the sea surface (80 parts per trillion). Data are not now available to extrapolate from these observations to the open oceans since, as noted earlier, no measurements of DDT concentrations are available from these waters.

Crustaceans

Bioassay tests show that laboratory populations of commercial species of shrimp and crabs as well as zooplankton are killed by exposure to chlorinated hydrocarbons such as DDT and PCB in the parts per billion range (Butler, 1964; Duke, Lowe, and Wilson, 1970). Continuous exposure of shrimp to DDT concentrations of less than 0.2 ppb caused 100 percent mortality in less than 20 days (Nimmo, Wilson, and Blackman, in press). Concentrations of this magnitude have been detected in Texas river waters flowing into commercially important shrimp nursery areas (Manigold and Schulze, 1969). We can be certain that significant mortalities of juvenile crustaceans are increasing in such contaminated areas. In California the declining production of Dungeness crabs is associated with observed DDT residues in the developing larvae. PCBs (Monsanto's Arochlor 1254) caused a 50 percent mortality in fifteen days to test populations of commercial shrimp at a concentration of 1 ppb (Nimmo, personal communications).

Mollusks

The chlorinated pesticides and PCBs characteristically interfere with the growth of oysters. One ppb of the PCB Arochlor 1254, for example, causes a 20 percent decrease in shell growth (Duke, Lowe, and Wilson, 1970). Many pesticides interfere with oyster growth at levels as low as 0.1 ppb (Butler, 1966b) in the ambient water. Mollusks generally concentrate these chemicals and thus serve as indicators of pollution levels in marine waters. Coastal monitoring samples have demonstrated that the magnitude of chlorinated hydrocarbon residues in mollusks is directly correlated with the application rates of these agricultural chemicals in adjacent river basins (Butler, 1967).

Fish

Marine fish are almost universally contaminated with chlorinated hydrocarbon residues. There is an expected concentration of such residues in lipid tissues such as the ovary. In the speckled sea trout on the south Texas coast, DDT residues in the ripe eggs are about 8 ppm. This level may be compared with the residue of 5 ppm in freshwater trout which causes 100 percent failure in the development of sac fry or young fish. There is presumptive evidence for similar reproductive failure in the sea trout. Sea trout inventories in the Laguna Madre in Texas have shown the following progressive decline: 30 fish per acre in 1964; 25 in 1965; 12 in 1966; 2.7 in 1968; and 0.2 in 1969. No data exist for 1967, as hurricanes destroyed all of the fishing gear (P. A. Butler, data from U.S. Bureau of Commercial Fisheries and Texas Parks and Wildlife Department). It is significant that no juvenile fish have been observed there in recent years, although in less contaminated estuaries 100 miles away there is a normal distribution of sea trout year classes (Butler, 1969). Declines in the productivity of fish in California coastal waters have not yet been correlated with pesticide residues. However, the sale of California mackerel has been halted, because DDT residues exceed permissible tolerance levels, even in the processed product.

Laboratory experiments have also established the concentration of several chlorinated hydrocarbons, including DDE, that damage reproductive success of birds, fish, and marine invertebrates. Only preliminary work on the effects upon marine organisms of PCB has been reported. Concentrations of one or more chlorinated hydrocarbons in species from the marine environment exceed those found to have deleterious effects in the laboratory and have been correlated with population decreases or reproductive failures of a number of marine species. Signs of incipient damage that can be expected to develop with continuing accumulation have also been reported.

Birds

In experimental studies, DDE in the diet resulted in thin eggshells and reduced hatching success of mallard ducks, p, p′-DDT produced the same effects, but to a lesser degree (Heath, Spann, and Kreitzen, 1969). Both DDE alone and a combination of p,p′-DDT and dieldrin in the diet resulted in thin eggshells; the test of the combination was carried out long enough to also show reduced hatching success (Porter and Wiemeyer, 1969; Wiemeyer and Porter, 1970).

Black duck egg samples at Atlantic coastal sites showed highest residues in states where duck reproduction is poorest. Nationwide sampling of wing tissues showed highest residues in the same areas (Reichel and Addy, 1968; Heath, 1969).

In three states from which bald eagle eggs were analyzed, highest residues of DDT residues and dieldrin occurred in the eggs from the state where reproduction is poorest and has declined to nearly zero (Krantz et al., 1970).

Eggshell thinning has occurred since the mid-1940s in a wide range of species of fish-eating birds and birds of prey, as shown by studies of museum series of eggs. Where shell thinning has occurred, the populations usually have declined also (Ratcliffe, 1967; Hickey and Anderson, 1968).

Sea eagle reproduction has failed, and outright mortality has occurred in the Baltic Sea in association with very high levels of DDT compounds and PCB in the tissues (Jensen et al., 1969).

Deaths of bald eagles (Mulhern et al., 1970), common loon (Butler, 1966), and peregrine falcons (Jefferies and Prestt, 1966) have been correlated with lethal amounts of chlorinated hydrocarbons in body tissues. Widespread mortality of many birds along the coast of the Netherlands and population decline of certain forms were traced to dieldrin contamination (Koeman et al., 1967 and 1968).

Samples of brown pelican eggs from twelve Atlantic and Gulf colonies showed greatest shell thinning in the Carolina colonies, where populations have declined precipitously. On the basis of museum series, shell thinning has occurred in all areas (Blus, 1970; Anderson and Hickey, 1970).

In the marine ecosystems of southern California, where concentrations of the DDT compounds in fish may exceed 10 parts per million (Risebrough et al., in press), two species, the bald eagle and the peregrine falcon, have disappeared from the wilderness environment of the Channel Islands (Herman, Kirven, and Risebrough, unpublished manuscript). In this area the brown pelicans and double-crested cormorants are no longer able to reproduce. Only one brown pelican was known to hatch in southern California in 1969 from about 1,200 nesting attempts (Risebrough, Sibley, and Kirven, forthcoming). In 1970, over 1,000 nesting attempts produced no young between the initial breeding attempts in March and July 1 (Gress, forthcoming). No double-crested cormorants were fledged in southern California in either 1969 or 1970 (Gress et al., unpublished). The cause in each case was a failure of the eggs to hatch because of breakage during incubation. Concentrations of DDE, PCB, and the other chlorinated hydrocarbons in the lipid fractions of the eggs were correlated inversely with shell thickness in brown pelican eggs from the Isles Coronados, Mexico, the Isles San Martin and San Benitos, Baja California, Mexico, the Gulf of California, four Florida localities, as well as Jamaica, Venezuela, Panama, and Peru (Risebrough and Anderson, forthcoming; Schreiber, forthcoming; Jehl, forthcoming).

Pelican eggs collected on Anacapa in 1962 showed a critical amount of thinning (Anderson and Hickey, 1970), but young pelicans were still produced on Anacapa in 1963 and 1964. Between 1962 and 1969, therefore, accumulations of chlorinated hydrocarbons in the southern California marine ecosystem passed a level critical to the brown pelican.

By comparing pelican eggs from California and Florida, where the relative amounts of DDE and PCB are very different, and by considering the eggs from the other localities, it was concluded that DDE is the major cause of shell thinning in the brown pelican (Risebrough and Anderson, forthcoming).

Shell thinning has also been detected in marine and coastal birds in northern California, including the common murre (Gress, Risebrough, and Sibley, unpublished manuscript), the ashy petrel (Risebrough and Coulter, unpublished), and American egret (Pratt and Risebrough, unpublished). Continued buildup of chlorinated hydrocarbons in this ecosystem and in other marine ecosystems around the world will cause reproductive failures in these and other marine species.

Biochemical Effects

Several physiological effects of chlorinated hydrocarbons account for shell thinning and for the abnormal behavior observed in contaminated populations. In affecting nerves, chlorinated hydrocarbons, including DDE, are believed to block the ion transport process by inhibiting ATPase in the nerve membrane (Matsumura and Patil, 1969; Koch, 1969, 1970), that causes the required energy to be made available.

Transport of ionic calcium across membranes such as those in the shell gland of birds is also an energy-requiring process dependent upon membrane ATPase. Inhibition of these enzymes by DDE could account for the concentration-effect curves obtained for shell thickness and DDE concentration in eggs of the brown pelican, double-crested cormorant (Anderson et al., 1969), and herring gull (Hickey and Anderson, 1968). DDE has also been found to inhibit the enzyme carbonic anhydrase (Peakall, 1970; Risebrough, Davis, and Anderson, 1970; Bitman, Cecil, and Fries, 1970), essential for the deposition of calcium carbonate in the eggshell and for the maintenance of pH gradients across membranes such as those in the shell gland. Inhibition of this enzyme by such drugs as sulfanilamide results in the production of thin-shelled eggs.

The chlorinated hydrocarbons, including DDE and PCB, induce mixed-function oxidase enzymes in the livers of birds and mammals which hydroxylate and render water soluble foreign, lipid-soluble compounds (Conney, 1967; Risebrough et al., 1968). Induction is usually a temporary phenomenon, ending when the inducing materials are themselves metabolized. DDE and the more heavily chlorinated PCB molecules are comparatively resistant to degradation by the induced molecules so that they may persist as inducers for some time. The induced enzymes may therefore become constitutive.

The steroid hormones such as estrogen and testosterone (Conney, 1967; Peakall, 1970) and thyroxine (Schwartz et al., 1969) are metabolized at higher rates when these enzymes are induced. Lower estrogen concentrations are present in pigeons fed p,p'-DDT (Peakall, 1970). Birds may also show symptoms of hyperthyroidism when fed a chlorinated hydrocarbon (Jefferies, 1969; Jefferies and French, 1969). An increasing number of instances of abnormal behavior are being reported in contaminated populations, including herring gulls and the brown pelicans (Gress, forthcoming). Our present state of knowledge indicates that these abnormalities also affect reproductive success and most likely result from hormone imbalance caused by the activity of the nonspecific enzymes induced by the chlorinated hydrocarbons. However, there are extensive researchers in this field with rapid changes in current theories.

Long-Term Effects upon Community Structure
These well-documented changes in the earth's living systems are part of a larger pattern of changes in the structure of the natural communities of estuaries, coastal regions, and the oceans. The pattern is familiar: it is the pattern associated with accelerated eutrophication and pollution of water bodies; it is analogous to the changes in structure of forests caused by a variety of disturbances including ionizing radiation (Woodwell, 1970). The reduction of structure leads progressively to shorter food chains in which hardy, broad-niched species with rapid reproductive rates predominate. On land, these are the simplified, erruptive insect and rodent communities of highly disturbed areas; in water, they are the equally simplified communities of eutrophic lakes and estuaries, in which harvestable fish populations are often depressed

and bird populations are dominated by scavengers such as the herring gull. The problem in water, however, is worse, in that the reduction of consumer populations is accompanied by a shift in plant species to hardy algae that are not consumed by grazers; their production instead accumulates, decaying anaerobically and further reducing the potential of the site for support of man.

While such changes are caused by many factors, the accumulations of persistent chlorinated hydrocarbons in estuaries and in coastal waters have become a significant factor accelerating this pattern of change.

Recommendation: A Chlorinated Hydrocarbon Base-Line Program for the Marine Environment

The paucity of analyses of chlorinated hydrocarbons in materials from the marine environment and from those parts of the atmosphere and from the continental hydrosphere which provide them limits our ability to make judgments on their present and future impacts upon ecosystems. Any predictions about the rates of buildup in a given reservoir are somewhat speculative.

Yet we cannot initiate an effective monitoring program until the present dissemination of these materials at the earth's surface is detailed. An entry to a reasonable monitoring program can be found in a base-line study where the concentrations of the chlorinated hydrocarbons in geological and biological components of the marine environment as well as in their transporting agencies are determined. Such an investigation can conceivably be carried out in a year.

To obtain a critical amount of data with the minimum expenditure of funds, a carefully planned investigation is necessary. By utilizing a single laboratory to manage the program, difficulties in standardization and in sample preparation and handling can be minimized. Further, in the detailed planning and sample collection the inputs from a variety of disciplines are necessary not only at the beginning of the work but also on a continuing basis. A large-scale program entering such a broad area will have diversions from the initial schedule of analyses as unexpected results come in.

The following outline indicates the types of samples to be

obtained. We envisage perhaps a thousand or so analyses during the first year's base-line program. Temporal, geographical, and spatial sampling procedures will be formulated for each of the groups of substances.

I. Samples of transport paths.
A. Major wind systems (sample in dust and as vapors).
 1. Prevailing westerlies—the jets.
 2. Trades.
 3. Regional winds such as monsoon and harmattans.
 4. Rain, snow, and dry fallout. Glacial samples provide fallout over past years (Windom, 1969).
B. The major rivers of the world (dissolved, suspended, and bed-load materials).
C. Sewer outfalls draining directly into marine environment from major population centers of the world.
D. Material released either by intent (dumping of wastes) or inadvertently (PCBs from antifouling paints on ships) during vessel transport.
II. Samples from reservoirs in the marine environment.
A. Seawater (particulate and dissolved loads).
B. Isolates from air-water interface. Surface active agents. Slicks.
C. Sediments.
D. The major fisheries food chains including plankton, the harvestable material, as well as predators of the harvestable material.
E. Selected organisms from estuaries, such as oysters.

The frequency of samples, either as a function of location, time, or type, will depend upon the resources available. Reasonable emphasis should be placed on historical records of inputs to the environment through the use of glacial (permanent snowfield) samples that record the atmospheric transport of material removed by precipitation and dry fallout. Museum specimens may provide a most useful index of the past body burdens of organisms.

A base-line program for chlorinated hydrocarbons will provide samples to investigate other materials being disseminated about the world ocean and which pose threats to marine organisms.

The identification of a material that may endanger an ecosystem often occurs after substantial disseminations have been made to the environment. A bank of samples, collected in sufficient quantities, will provide a historical record of invaluable use in such instances.

Examples of other materials that might be sought are such metals as mercury, cadmium, lead, and arsenic; such industrial chemicals as methyl chloride and vinyl chloride; such widely used chemicals as dry cleaning fluids (perchlorethylene and trichlorethylene) and freon; and such dust components as asbestos and fly ash.

Recommendation: Removal of Obstacles to Public Access to Chemical Production Data*

Among the causes contributing to the lack of available data on the chlorinated hydrocarbons† is a legal structure that allows manufacturers of a given material, when there are no more than two producers, the right to hold their production figures as privileged information.

The Task Force recognizes the economic rationale that deters the release of production figures by such manufacturers. Further, it understands that our government is charged by law with the protection of that proprietary interest. Indeed, it approves in general the principle that governmental action should not artificially affect competition.

However, the majority of us feel that there are times when it is not in the public interest for government to maintain as privileged such data as are necessary for research into, and an assessment of, the state of our environment. In that regard, we also recognize the possibility that it is not always competitive concerns alone that determine the less than candid posture assumed by industry concerning production figures.

We recommend that the laws relating to the registration of chemical substances and to the release of production figures by the Department of Commerce and/or the Bureau of the Census be reexamined and revised in the light of present evidence of environmental deterioration. With particular reference to the polychlorinated hydrocarbons, the protection afforded manufacturers

---

* This recommendation is not necessarily supported by all of those members of the SCEP Task Force who were not also members of the NASCO Workshop.
† For example, the Monsanto Chemical Company has refused to release its production figures for PCBs, although requested to do so by many scientists and government officials.

by government is itself an artificial obstacle to effective environmental management. In view of other impediments, technological, methodological, and financial, such protection is clearly inappropriate, given widespread public concern about environmental affairs.

References

Acree, F., Jr., Beroza, M., and Bowman, M. C., 1963. Codistillation of DDT with water, *Journal of Agricultural and Food Chemistry, 11:* 278–280.

Anderson, D. W., and Hickey, J. J., 1970. Oological data on egg and breeding characteristics of brown pelicans, *Wilson Library Bulletin, 82:* 14–28.

Anderson, D. W., Hickey, J. J., Risebrough, R. W., Hughes, D. F., and Christensen, R. E., 1969. Significance of chlorinated hydrocarbon residues to breeding pelicans and cormorants, *Canadian Field Naturalist, 83:* 92–112.

Bailey, T. E., and Hannum, J. R., 1967. Distribution of pesticides in California. *Journal of the Sanitary Engineering Division, Proceedings of the American Society of Civil Engineers, 93:* 27.

Bitman, J., Cecil, H. C., and Fries, G. F., 1970. DDT—induced inhibition of avian shell gland carbonic anhydrase: a mechanism for thin eggshells, *Science, 168:* 594–596.

Blus, L. J., 1970. Measurements of brown pelican eggshells from Florida and South Carolina, *Bioscience, 20:* 867–869.

Bowman, M. C., Acree, F., and Corbett, J., 1960. Insecticide Solubility, Solubility of carbon 14 DDT in water, *Journal of Agriculture and Food Chemistry, 8.*

Butler, Philip A., 1964. Commercial fishing investigations and effects of pesticides on fish and wildlife, *Fish and Wildlife Service Circular, 226:* 65–77.

Butler, Philip A., 1966a. Fixation of DDT in estuaries, *Transactions 31st North American Wildlife and Natural Resources Conference:* 184–189.

Butler, Philip A., 1966b. Pesticides in the marine environment, *Journal of Applied Ecology, 3* (Supplement): 253–259.

Butler, Philip A., 1967. Pesticide residues in estuarine mollusks, presented at National Symposium on Estuarine Pollution, Stanford University, August 1967, pp. 107–121.

Butler, Philip A., 1969. Monitoring pesticide pollution, *Bioscience, 19:* 889–891.

Conney, A. H., 1967. Pharmacological implications of microsomal enzyme induction, *Pharmacological Reviews, 19:* 317–366.

Cottam, C., and Higgins, E., 1946. DDT: its effect on fish and wildlife, *U.S. Department of Interior, Fish and Wildlife Service Circular, 11.*

Davis, Harry C., 1961. Effects of some pesticides on eggs and larvae of oysters (*Crassostrea virginica*) and clams (*Venus mercenaria*), *Commercial Fisheries Review, 23,* No. 12: 8–23.

Duke, T. W., Lowe, J. I., and Wilson, A. J., Jr., 1970. Polychlorinated biphenyl (Acrolor 1454[R]) in the water, sediment, and biota of Escambia Bay, Florida, *Bulletin of Environmental Contamination and Toxicology, 5,* No. 2: 171–180.

Food and Agriculture Organization (FAO), 1968. *Production Yearbook* (Rome: FAO).

Gress, F., 1970. Reproductive Success of the Brown Pelicans on Anacapa Island in 1970, *Transactions of the San Diego Society of National History, 1970* (forthcoming).

Gress, F., Risebrough, R. W., Jehl, J., and Kiff, L., 1970. Reproductive failures of the double crested cormorants resulting from shell thinning (unpublished).

Heath, R. G., 1969. Nationwide residues of organochlorine pesticides in wings of mallards and black ducks, *Pesticides Monitoring Journal, 3:* 115–123.

Heath, R. G., Spann, J. W., and Kreitzen, J. F., 1969. Marked DDE impairment of mallard reproduction in controlled studies, *Nature, 224:* 47–48.

Herman, S. G., Kirven, M. N., and Risebrough, R. W., 1970. Pollutants and raptor populations in California (unpublished).

Hickey, J. J., and Anderson, D. W., 1968. Chlorinated hydrocarbons and eggshell changes in raptorial and fish-eating birds, *Science, 162:* 271–273.

Jefferies, D. J., 1969. Induction of apparent hyperthyroidism in birds fed DDT, *Nature, 222:* 578–579.

Jefferies, D. J., and Prestt, I., 1966. Post-mortems of peregrines and lannens with particular reference to organochlorine residues, *British Birds, 59:* 49–64.

Jefferies, D. J., and French, M. C., 1969. Avian thyroid: effect of p,p'-DDT on size and activity, *Science, 166:* 1278–1280.

Jehl, J., 1970. Shell thinning in eggs of the brown pelicans of western Baja California, *Transactions of the San Diego Society of Natural History, 1970* (forthcoming).

Jensen, S., Johnels, A. G., Olsson, M., and Otterlind, G., 1969. DDT and PCB in marine animals from Swedish waters, *Nature, 224:* 247–250.

Koeman, J. H., Oskamp, A. A. G., Veer, J., Brouwer, E., Rooth, J., Zwart, P., v.d.Brock, E., and van Genderer, H., 1967. Insecticides as a factor in the mortality of the sandwich tern (*Sterna sandvicensis*): a preliminary communication. *Medelinger Rijksfaculteit Landbouwweter schappen Gent, 32:* 841–853.

Koeman, J. H., Veer, J., Brouwer, E., Huisman-deBrouwer, L., and Koolen, J. L., 1968. Residues of chlorinated hydrocarbon insecticides in the North Sea environment, *Helgoländer wissenschaftliche Meeresuntert suchungen, 17:* 375–380.

Koch, R. B., 1969–1970. Inhibition of animal tissue ATPase activities by chlorinated hydrocarbon pesticides, *Chemical-Biological Interactions, 1:* 199–209.

Krantz, W. C., Mulhern, B. M., Bagley, G. E., Sprunt, A., IV., Ligas, F. J., and Robertson, W. B., Jr., 1970. Organochlorine and heavy metal residues in bald eagle eggs, *Pesticides Monitoring Journal* (forthcoming).

Lichenstein, E. P., Anderson, J. P., Fuhremann, T. W., and Schultz, K. R., 1968. *Science, 159:* 1110.

Manigold, D. B., and Schulze, J. A., 1969. Pesticides in selected western streams—a progress report, *Pesticides Monitoring Journal, 3,* No. 2: 124–135.

Matsumura, Fumio, and Patil, K. C., 1969. Adenosine triphosphatase sensitive to DDT in synapses of rat brain, *Science, 166:* 121–122.

Menzel, D. W., Anderson, J., and Randtke, A., 1970a. Marine phytoplankton vary in their response to chlorinated hydrocarbons, *Science, 167:* 1724–1726.

Menzel, D. W., Anderson, J., and Randtke, A., 1970b. The susceptability of two species of zooplankton to DDT (unpublished).

Mulhern, B. M., Reichel, W. L., Locke, L. N., Lamont, T. G., Belisle, A., Cromartie, E., Bagley, G. E., and Prouty, R. M., 1970. Organochlorine residues in bald eagles 1966–1968 (unpublished).

Nash, R. G., and Beal, M. L., Jr., 1970. Chlorinated hydrocarbon insecticides: root uptake versus vapor contamination of soybean foliage, *Science, 168:* 1109–1111.

Nimmo, D. R., Wilson, A. J., Jr., and Blackman, R. R., 1970. Localization of DDT in the body organs of pink and white shrimp (unpublished).

Peakall, D. B., 1970. p,p'-DDT: effect on calcium metabolism and concentration of estrodol in the blood, *Science, 168:* 592–594.

Peterle, T. J., 1969. DDT in antartic snow, *Nature, 224:* 620.

Porter, R. D., and Wiemeyer, S. N., 1969. Dieldrin and DDT effects on sparrow hawk eggshells and reproduction, *Science, 165:* 199–200.

Ratcliffe, D. A., 1967. Decrease in eggshell weight in certain birds of prey, *Nature, 215:* 208–210.

Reichel, W. L., and Addy, C. E., 1968. A survey of chlorinated pesticide residues in black duck eggs, *Bulletin of Environmental Contamination and Toxicology, 3:* 174–179.

Risebrough, R. W., and Anderson, D. W., 1970. Pollutants and shell thinning in the brown pelicans: *Transactions of the San Diego Natural History Society, 1970* (in press).

Risebrough, R. W., and Coulter, M. C., 1968. Chlorinated hydrocarbons and shell thinning in the ashy petrel (unpublished).

Risebrough, R. W., Davis, J. D., and Anderson, D. W., 1970. *Effects of various chlorinated hydrocarbons on birds* (unpublished).

Risebrough, R. W., Huggett, R. J., Griffin, J. J., and Goldberg, E. D., 1968a. Pesticides: transatlantic movement in the Northeast Trades, *Science, 159:* 1233–1236.

Risebrough, R. W., Menzel, P. B., Martin, D. J., and Olcott, H. S., 1970. DDT residues in Pacific marine fish. *Pesticides Monitoring Journal* (in press).

Risebrough, R. W., Reiche, P., Peakall, D. B., Herman, S., and Kirven, M., 1968b. Polychlorinated biphenyls in the global exosystem, *Nature, 220:* 1098–1102.

Risebrough, R. W., Sibley, F. C., and Kirven, M. N., 1970. Reproductive success of the brown pelicans on Anacapa Island in 1969. *Transactions of the San Diego Society of Natural History, 1970* (in press).

Ryther, John, 1969. Photosynthesis and fish production in the sea, *Science, 166:* 72–76.

Schreiber, R. W., 1970. Pollutants and shell-thinning of eggs of brown pelicans in Florida, *Transactions of the San Diego Natural History Society* (in press).

Schwartz, H. L., Kosyreff, U., Surks, M., and Oppenheimer, J. H., 1969. Increased deiodination of L-thyroxine and L-trilodothyronine by liver microsomes from rats treated with phenobarbitol, *Nature, 221:* 126–163.

Stanford Research Institute (SRI), 1969. *Chemical Economics Handbook* (Menlo Park, California: SRI).

Stickel, L. F., 1968. Fish and Wildlife Service Special Scientific Report—Wildlife No. 119, U.S. Department of Interior.

Sverdrup, H. V., Johnson, M. W., and Fleming, R. H., 1946. *The Oceans* (New York: Prentice-Hall, Inc.).

Tarrant, K. B., and Tatton, J., 1968. Organo-pesticides in rainwater in the British Isles, *Nature, 219:* 725–727.

Ukeles, Ravenna, 1962. Growth of pure cultures of marine phytoplankton in the presence of toxicants, *Applied Microbiology, 10,* No. 6: 532–537.

Wiemeyer, S. N., and Porter, R. D., 1970. DDE thins eggshells of captive American kestrels, *Nature* (forthcoming).

Windom, H. L., 1969. Atmosphere dust records in permanent snowfields: implications to marine sedimentation, *Bulletin of the Geological Society of America, 80:* 761–782.

Wolman, A. A., and Wilson, A. J., Jr., 1970. Occurrence of pesticides in whale, *Pesticide Monitoring Journal, 4*.

Woodwell, G. M., 1970. Effects of pollution on the structure and physiology of ecosystems, *Science, 168:* 429–433.

Wurster, C. F., 1968. DDT reduces photosynthesis by marine phytoplankton, *Science, 159:* 1474–1475.

Wust, G., Oberflächen Salzgehalt, 1936. Verdunstung Und Nederschlag Auf Dem Weltmeere. Lan Derkunoliche Forschung: Festschrift Norbert Krebs. Edited by H. Louis and W. Panzen, Stuttgart, pp. 345–359.

Yates, M. L., Holswade, W., and Higer, Aaron L., 1970. Pesticide residues in hydrobiological environments, Water, Air and Waste Chemistry Section of the *American Chemical Society Abstract, 1970.*

19
Ocean Pollution by Petroleum Hydrocarbons

Roger Revelle,
Edward Wenk,
Bostwick H. Ketchum, and
Edward R. Corino

Nature of the Problem

At the present time, the most conspicuously detrimental effects of oil pollution of the ocean are localized in extent and are caused by accidental spills in near-shore areas. These loci of concern, however, potentially include the coastal zones of every continent and every inhabited island so that the problem of accidental spills is of worldwide significance. Projections of future growth in ocean transport and offshore production of petroleum indicate that both the frequency and the damaging effects of local accidents are likely to increase.

Although accidental oil spills cause the most evident damage to ocean resources, they make up a small percentage of the total amount of oil entering the marine environment. At least 90 percent of this amount originates in the normal operations of oil-carrying tankers, other ships, refineries, petrochemical plants, and submarine oil wells; from disposal of spent lubricants and other industrial and automotive oils; and by fallout of airborne hydrocarbons emitted by motor vehicles and industry. The extent and character of the damage to the living resources of the sea from this "base load" of oil pollution is little known or understood. In the long run it could be more serious, because more widespread, than the localized damage from accidental spills.

The magnitude of oceanic oil pollution is likely to increase with the worldwide growth of petroleum production, transportation, and consumption. World crude oil production reached 2 billion tons per year in 1969, and production of 3 billion and 4.4 billion tons per year is predicted for 1975 and 1980, respectively.

Sources of Petroleum Hydrocarbons in the Sea

Petroleum hydrocarbons enter the sea:
1. Directly
   a. in accidental spills from ships, shore facilities, offshore oil wells, and underwater pipe lines;

Prepared during SCEP.

b. from tankers flushing oil tanks at sea;
   c. from dry cargo ships cleaning fuel tanks and bilges;
   d. from leakage during normal operation of offshore oil wells;
   e. from operation of refineries and petrochemical plants;
   f. in rivers and sewage outfalls carrying industrial and automotive wastes; and
2. As "fallout" from the atmosphere, probably as particles or in rain.

We shall consider all these sources except accidental spills as constituting the base load of oil pollution in the sea.

Accidental Oil Spills

At present, the average annual influx to the ocean from accidental oil spills throughout the world is probably about 200,000 tons. Most of these spills are relatively small. Out of 714 recorded accidental spills in U.S. waters in 1968, approximately half were from ships and barges, most of which were docked at the time of the accident. About 300 spills occurred from shore facilities of various types, and a few resulted from ships dragging anchor across submarine pipelines in bays.

Even under carefully controlled conditions accidental oil spills in port are negligible. Milford Haven, a relatively new British oil port, is adjacent to a national park, and great efforts have been made to control and prevent oil pollution. In 1966 the annual turnover at Milford Haven was 30 million tons with losses amounting to 2,900 tons or 0.01 percent of the total amount handled (Blumer, 1969a).

Accidental oil spills resulting from stranding or collision of large tankers and from accidents to offshore drilling or producing wells deservedly attract much public attention because of the extensive damage done to beaches, recreational areas, and harbors. The wreck of the *Torrey Canyon,* which discharged 118,000 tons of crude oil in the sea, is the best known example although somewhat smaller tanker wrecks have occurred elsewhere, such as off Nova Scotia and Puerto Rico. All large accidental spills to date have occurred fairly near shore, and the spreading sheet of oil has drifted or has been blown by winds onto beaches and into shallow water areas. Present efforts to contain and to dispose of the oil before it does extensive damage have been singularly ineffective. Agents such as talc, clay, and carbonized sand have been used to sink the oil. Various dispersing agents have been developed which break up the oil into minute droplets that are subsequently dispersed throughout the water. Earlier versions of these chemical dispersants were more toxic than the oil, but a number of essentially nontoxic dispersants are now available. Even with a nontoxic dispersant, dispersed oil is more toxic to marine life than an oil slick on the surface, primarily because of its increased availability to the organisms. With all our vast inventory of chemical agents, the best and safest means of disposal is apparently still absorption on chopped straw, if conditions permit.

The danger of large-scale accidents is increasing with the increasing size of tankers. Four 327,000-ton ships are already in operation; vessels of 500,000 dead weight tons will soon be constructed, and 800,000-ton vessels have been projected within the next few years. These monster ships have so much draft and inertia and are so difficult to handle that a stranding or collision is more likely to result in a destructive wreck than with smaller ships. A loss of one of the new large tankers under conditions where it would be impossible to off-load the oil would add around 20 percent to the amount of petroleum entering the oceans in a single year.

Although handling difficulties increase with size, the increase is not directly proportional to size. Moreover, larger ships means fewer ships, and, therefore, traffic can be considerably reduced. Fewer ships also means crews can be limited to highly qualified personnel, and they can be better trained. The larger tankers could also afford to install highly sophisticated navigation gear which might be prohibitively expensive for the many smaller ships.

Spectacular "blowouts" from offshore oil well drilling and production make up a surprisingly small fraction of the total influx of oil to the ocean environment. For example, the widely publicized Santa Barbara blowout has so far produced only between 3,000 and 11,000 tons of oil. Similarly, the accident to a producing well off the Louisiana coast, which began on February 10, 1970, and lasted until the end of March, released only about 4,300 tons of oil. These figures emphasize the enormous amount of damage that can be done by a relatively small amount of oil concentrated over a relatively small, previously uncontaminated area. With present drilling and production technology, accidents of this kind are nearly inexcusable. Preventing them depends on institutional changes, not technical ones.

Sources of the Base Load of Oil Pollution in the Sea

Most oil production occurs at some distance from processing and marketing areas and consequently much crude oil is transported in oceangoing tankers. In 1969, 1.3 billion tons, or about 65 percent of total oil production, was carried in tankers. Projections by the U.S. Department of Transportation indicate that the amount of oil moved by tankers will increase to 2.8 billion tons by 1980.

Normal tanker operations (ballasting, tank cleaning) were estimated to have introduced 530,000 tons of oil to the sea in 1969. Eighty percent of the world fleet used control measures ("Load on Top" or LOT). If LOT were practiced faithfully, these ships would contribute only 30,000 tons of the total losses compared to 500,000 tons from the 20 percent not using such measures. If LOT were used on all tankers, only 56,000 tons would be expected to be lost to the ocean through normal operations in 1975 and 75,000 tons in 1980. If 20 percent of the fleet continued to operate in the present fashion, total losses in 1975 and 1980 would be 800,000 and 1.06 million tons, respectively.

Nontankers, dry cargo ships of greater than 100 gross registered tons, are estimated to have discharged 500,000 tons to the ocean in 1969, primarily from pumping bilges and cleaning operations. This estimate is of low reliability because available data are very limited. The total amount, however, is comparable to that generated by the tanker fleet.

Offshore oil production is estimated to discharge during normal operations about 100,000 tons per year. At present, offshore production accounts for about 16 percent of total crude production. This percentage is expected to increase in the future, as new underwater fields are discovered and new technology permits extension of drilling and production into deeper water. Estimates of losses for 1975 are 160,000 to 320,000 tons and for 1980 are 230,000 to 460,000 tons. The smaller figures assume that offshore production will continue to represent 16 percent of world production, and the larger figures assume 32 percent. In both cases the assumption is made that no improvement in pollution abatement will occur. Many of the new wells will be drilled off the coast of nations that do not have the technological capabilities to enforce good drilling and production procedures or to deal with massive spills.

About 300,000 tons of oil are lost to the sea each year through normal operations of refineries and petrochemical plants. This estimate is based on extensive data from the American Petroleum Institute and private surveys by refineries and industry organizations. With present pollution control measures this figure could grow to 450,000 tons in 1975 and 650,000 tons in 1980. If some improvements in pollution control are made, as predicted by the U.S. Federal Water Quality Administration, oil lost to the sea from refineries and petrochemical plants could drop to 200,000 tons in 1975 and 440,000 tons in 1980.

Industrial and automotive waste oils and greases contitute a significant source of oil pollution in the marine environment. These include all petroleum products, except fuel, used and discarded in the operation of motor vehicles and industrial production, for example, spent lubricants, cutting and hydraulic oils, coolants, and solvents. Much of the disposal of these wastes occurs by dumping on land. An estimate of the quantity eventually finding its way into the ocean can be made from measurements of the hydrocarbon concentrations in river waters, multiplied by the total river discharge, plus the amounts contributed by sewage treatment plants which discharge directly to the oceans. Data assembled by E. R. Corino (1970) suggest that the average concentration of hydrocarbons in U.S. rivers entering the sea is 85 parts per billion. Multiplying this figure by the estimated annual river discharge of 1,750 billion tons (Revelle, 1963) gives approximately 150,000 tons of hydrocarbons annually carried to the oceans from the United States or about 450,000 tons for the entire earth. Perhaps as much as 150,000 tons of oil and grease are discharged to the ocean in municipal sewage effluents from U.S. cities and towns (Study of Critical Environmental Problems, 1970). A large fraction of oils and greases in sewage do not originate from petroleum. If we assume that one-third of sewage oils and greases are petroleum hydrocarbons and multiply by three to give the world total we arrive at 100,000 tons per year from this source. Thus all industrial and automotive petroleum wastes entering the ocean may be about 550,000 tons. This amount should increase at about the same rate as total oil production, namely, to about 825,000 tons by 1975 and 1.2 million tons by 1980.

All the preceding estimated direct losses to the marine environment made up approximately 2.2 million tons per year in 1969:

Accidental spills	0.2 million tons
Tanker operations	0.5 million tons
Other ships	0.5 million tons
Offshore production	0.1 million tons
Refinery operations	0.3 million tons
Industrial and automotive wastes	0.6 million tons
Total	2.2 million tons

The total is expected to increase to between 3.3 and 4.8 million tons by 1980, as indicated in Table 19.1. Petroleum hydrocarbons entering the sea from all the above sources are about 0.1 percent of world oil production. If the possible fallout of airborne hydrocarbons on the sea surface is added, the total amount of oil and oil products contaminating the ocean may be as much as 0.5 percent of world production.

Table 19.1 Estimated Direct Petroleum Hydrocarbon Losses to the Marine Environment (not including airborne hydrocarbons deposited on the sea surface)

	Millions of Tons				
	1969	1975		1980	
		Min.	Max.	Min.	Max.
Tankers	.530	.056	.805	.075	1.062
Other Ships	.500	.705	.705	.940	.940
Offshore Production	.100	.160	.320	.230	.460
Refinery Operations	.300	.200	.450	.440	.650
Oil Wastes	.550	.825	.825	1.200	1.200
Accidental Spills	.200	.300	.300	.440	.440
Total	2.180	2.246	3.405	3.325	4.752

Sources: See text of this paper and also Corino (1970).

To give these figures perspective, we can make two historical comparisons.

Oil pollution of the marine environment existed long before the first oil well was drilled. This pollution came from natural seeps on the sea floor. There have never been any measurements of the quantity of oil entering the ocean from such natural seepage areas, but two lines of evidence indicate that it must be quite small, compared to the present amounts of oil entering the ocean because of human activities. First, if much oil had continually seeped into the ocean, all of the petroleum reserves would have long since disappeared. For example, if 100,000 tons of oil per year entered the ocean from natural seepages, within a few million years this would exceed the total estimated oil reserves of the entire earth. Second, we know from the Santa Barbara and Louisiana well accidents that any natural oil seepage producing even a few thousand tons of oil per year would have resulted in very conspicuous slicks of oil spreading over large areas of the sea surface. No such large natural slicks have ever been observed. Typically, natural seeps produce quite small quantities of oil which occasionally bubble up to the surface and produce small slicks. We estimate, therefore, that oil coming into the marine environment before the human use of petroleum began must have been considerably less than 100,000 tons per year, less than 5 percent of the present 2.2 million tons a year injected directly from land and marine sources.

Another point of comparison with today's annual influx of oil comes from the sinking of tankers and ships in World War II. The number of U.S.-controlled tankers sunk or seriously damaged during that war was ninety-eight, with an estimated average cargo capacity of 10,000 tons. Thus, up to 1 million tons could have been released to the ocean from sinkings of U.S. ships. Perhaps another 3 million tons of oil were lost through the sinkings of Japanese oil-carrying ships by American submarines and of British and other allied ships by German submarines. The total quantity of oil lost in the ocean during the six years of World War II thus may have been about twice the annual direct influx to the ocean at the present time. As far as we know, no permanent damage was done to the ocean ecosystem by these rather large releases, perhaps in part because most of them occurred far from land in relatively deep water, and in part because much of the oil may have escaped into the sea very slowly, as the sunken tanks corroded away (ZoBell, 1962).

A great variety of hydrocarbons is produced by marine plants. ZoBell (1962) estimates the average concentration of solid and liquid hydrocarbons in these plants as at least 10 parts per million. Combining this with Menzel's (1970) estimate of 3×10^9 tons for the standing crop of marine plankton gives a minimum of 30,000 tons of hydrocarbons present in marine plant tissues at any given time. This is only about 1.5 percent of the input of petroleum into the sea. However, the total organic production of marine plankton is about 100 times the standing crop. If hydrocarbons are produced in proportion, about 3 million tons of hydrocarbons enter the ocean from organic activity each year.

The direct influx of petroleum hydrocarbons to the ocean is small compared to the emission of petroleum products and chemically produced hydrocarbons to the atmosphere through evaporation and incomplete combustion. The emission of petroleum hydrocarbons to the air each year is about 90 million tons (Department of Health, Education, and Welfare, 1970; Goldberg, 1970), roughly forty times the amounts of these substances entering the ocean directly from ships, shore installations, rivers, and the sea floor. Most of the hydrocarbons emitted to the atmosphere may be oxidized to harmless substances within a relatively short time. It is known that others are combined with nitrogen oxides and ozone to produce substances that are highly toxic to land plants. A fraction of the petroleum hydrocarbons emitted to the atmosphere exists as, or is absorbed on, very small particles, or becomes caught in rain, just as happens to DDT and other chlorinated hydrocarbon pesticides. Much of this fraction may settle out on the surface of the ocean. If 10 percent of the petroleum hydrocarbons emitted into the atmosphere eventually find their way to the sea surface in this way, the total hydrocarbon contamination of the ocean would be about five times the direct influx from ships and land sources. This quantity should be expected to increase about as rapidly as the total petroleum production, which means more than doubling by 1980.

Physical Concentration and Distribution of Oil Pollution

Neither the base load of hydrocarbons nor the concentrated accidental sources can be expected to be distributed uniformly throughout the ocean. Obviously the intensity will be greatest near the sources and unloading points and the most heavily affected areas will be near the coasts.

It is likely that most of the oil entering the sea from ships, rivers, and the sea floor ends up in a narrow zone near shore at most only a few kilometers in width. Some of this oil will become absorbed on clay, silt, sand grains, and other particles and will settle to the bottom. The oil remaining in the water will evaporate or become oxidized. Biodegradation of the bottom-deposited oil will also gradually occur, but fractions of the bottom-deposited oil will continue to disperse into shallow overlying waters for months or years. This inshore zone is the most sensitive to severe damage to the living resuorces of the sea from direct pollution by oil.

Submarine reservoirs of petroleum are likely to be found on the continental shelves of almost every continent, and the incidence of local contamination from underwater drilling and production on the continental margins will ultimately be widespread.

Sources from ships as a result of tank cleaning, bilge pumping, and accidents will be expected to follow the pattern of tanker and other cargo routes, with the highest concentrations near ports and harbors and in semienclosed seas such as the Mediterranean, the Black, and North seas, the Persian Gulf, and the Gulf of Mexico. The total area of these water bodies is slightly over 2 percent of the area of the ocean, but perhaps one-fourth of the total oil pollution from ships and land sources may occur in them. The future development of oil production in the Alaskan North Slope and the Canadian Northern Archipelago may produce serious contamination in the Arctic Ocean. Regional international agreements may be the most effective way to deal with the concentration of pollution in such semienclosed seas.

On the high seas, winds and ocean currents will bring about a convergence and retention of concentrations of hydrocarbons in the subarctic and equatorial convergence zones such as the Sargasso

Sea. Workers from the Woods Hole Oceanographic Institution have found that oil gobules and tar balls are more abundant in the Sargasso Sea than the Sargassum weed for which the sea is named.

Probably most of the hydrocarbon fallout from the air onto the sea surface occurs in the mid-latitudes of the Northern Hemisphere. These latitudes contain the trajectories of the winds blowing from the industrialized countries. If hydrocarbons deposited from the air formed a surface film over most of the North Atlantic, its thickness might be about 1,000 angstroms. Such a film should be detectable by suitable optical methods and might have physical as well as biological effects. It is more likely than most if the oil is in small particles, droplets, or tarry lumps, of the kind described by Horn, Teal, and Backus (1970), and that much of it settles quickly below the surface. As we shall see, oil films and droplets near the surface and DDT and other oil-soluble chlorinated hydrocarbons may have combined effects on the high seas which may do serious damage to open ocean ecosystems.

Modes of Hydrocarbon Removal from the Oceans

Hydrocarbons in the sea are diluted and dispersed by natural mixing and eventually disappear by microbial or physical oxidation, evaporation, and burial in the bottom sediments.

Hydrocarbons dissolved or suspended in the water column are eventually destroyed by bacteria, fungi, and other microorganisms. Some workers have found that the most toxic compounds are also the most refractory to microbial destruction, though the evidence is somewhat conflicting on this point (ZoBell, 1962).

No single microbial species will degrade any whole crude oil. Bacteria are highly selective and complete degradation requires numerous different bacterial species. Bacterial oxidation of hydrocarbons produces many intermediates which may be more toxic than the hydrocarbons; therefore, organisms are also required that will further attack hydrocarbon decomposition products.

The oxygen requirement in marine bacterial oil degradation is served. Complete oxidation of one gallon of crude oil requires all of the dissolved oxygen in 400,000 gallons of air-saturated seawater at 60°F (ZoBell, 1962). (This is equivalent to a layer of water one foot deep covering 1.2 acres.) Oxidation may be inhibited in areas where the oxygen content has been lowered by previous pollution, and the bacterial degradation may cause additional damage through oxygen depletion.

The rate of oxidation is strongly affected by the temperature of the water, being at least ten times slower at 40° F than at 80°, and much slower still when the water is near freezing temperature. Estimates by ZoBell (1962) indicate that oil dispersed in continuously oxygenated seawater, containing an abundance of oil-consuming bacteria at 80°F, may be oxidized at a rate of one gallon of oil per 400,000 gallons of water in about 2.5 days. At such a rate, 145 gallons of oil would disappear in a year. For this characteristic temperature of tropical near-surface ocean waters, the rate of oxygen uptake in the water and its nitrate and phosphate content will limit the rate of oxidation of the oil rather than temperature. Nitrate and phosphate, which are essential for the growth of oil-consuming bacteria, are present in very low concentration in the upper 100 meters of most tropical, open-ocean waters. Oxygen is continually replenished from the atmosphere in waters near the surface, but below about 10 meters this replenishment is usually extremely slow.

In the water of high latitudes, where the temperature is below 40°F, at least a month would be required to oxidize one gallon of oil in 400,000 gallons of seawater, even if the oxygen content were continuously replenished. In bays, entuaries, and shallow coastal waters most of the oil would settle to the bottom during this period and any further decomposition would be greatly slowed by the lack of oxygen in the bottom sediments.

Consequences of Oil Pollution
Depending upon their location, character, and concentration, petroleum hydrocarbon pollutants in the ocean can produce the following unwanted consequences:
1. Poisoning of marine life filter feeders such as clams, oysters, scallops and mussels; other invertebrates; fish, and marine birds.
2. Disruption of the ecosystem so as to induce long-term devastation of marine life.
3. Degradation of the environment for human use by reducing economic and recreational values on either a short- or long-term basis and by changes of esthetics of the marine environment.

Crude oil and oil fractions poison marine organisms through different effects.

a. Direct kill through coating of surfaces. Hundreds of thousands of oceanic birds freeze to death in winter because their feathers become fouled with oil which displaces the insulating air layer next to the skin. Many air-fouled birds are unable to fly; some lose their buoyancy and sink, others drift helplessly ashore (ZoBell, 1962).

b. Direct kill through contact poisoning.

c. Direct kill through exposure to the dissolved or colloidal toxic components of oil at some distance in space and time from the source.

d. Incorporation of sublethal amounts of oil and oil products into organisms, resulting in reduced resistance to infection and other stresses (one of the causes of death of birds surviving the immediate exposure to oil).

Disruption of the ecosystem may occur through

a. Destruction of the generally more sensitive juvenile forms of organisms.

b. Destruction of the food sources of higher species.

c. Possibly through interference with the communications systems of organisms.

Environmental degradation occurs on both sandy and rocky beaches and in bays and estuaries. From twenty pounds to a ton of oil and tar per mile of beach, in separate globs and in coatings on sand grains has been observed on many beaches. Even a ton of oil per mile along U.S. beaches would represent a small fraction of the oil entering the sea near the United States each year. In bays and estuaries, boats, fisherman's nets, piers, quays, wharfs, mooring lines, buoys, and lobster pots become smeared and fouled with oil. Severe fire hazards are created near docks and piers by oil on the water surface and by oil-soaked debris.

Toxicity of Oil

All crude oils and many refinery products at sufficient concentrations are poisonous to marine organisms. Responsible for this immediate toxicity are principally three complex fractions (Blumer, 1969a). The *low-boiling saturated hydrocarbons* have until quite recently been considered harmless to the marine environment. It has now been found that this fraction, which is rather readily soluble in seawater, produces at low concentration anesthesia and narcosis and at greater concentrations cell damage and death in a wide variety of lower animals; it may be especially damaging to the larval forms of marine life.

The *low-boiling aromatic hydrocarbons* are the most dangerous toxic fractions. Benzene, toluene, and xylene are acute poisons for man as well as other organisms. Naphthalene and phynanthrene are even more toxic to fish than benzene, toluene, and xylene.

The *higher-boiling crude oil fractions* are rich in multiring aromatic compounds. It was at one time thought that only a few of these compounds, for example, 3, 4-benzopyrene, were capable of inducing cancer, but carcinogenic fractions containing 1, 2-benzopyrene and alcobenzedrene have been isolated by Carruthers, Stewart, and Watkins (quoted in Blumer, 1969a). Biological tests have shown that extracts obtained from high-boiling fractions of the Kuwait oil are carcinogenic.

These tests were made by topical application of concentrated materials. While there is ample reason to be concerned about the possible effects of the presence of carcinogenic agents in the marine environment, a considerable amount of research needs to be done to assess the extent of the actual danger. This includes answers to questions of biodegradability, possible accumulation in the food chain, and the fate of naturally occurring carcinogens which are known to be present in the oceans.

We know that a far wider range of polynuclear aromatic compounds than benzopyrene and benzanthene are potent tumor initiators. Four and five ring aromatic hydrocarbons show carcinogenic activity, and it is relevant that substituted polycyclic compounds predominate in the aromatic fractions from crude oil.

Some data on biodegradability of carcinogenic hydrocarbons exists. Sisler and ZoBell (1947) state that polycyclic hydrocarbons were found to be from 11 percent to 68 percent oxidized to carbon dioxide and water by mixed cultures of bacteria within four days at 90°F. Degradation of polycyclic hydrocarbons is also discussed by Beerstecher (1954) and Davis (1967). Their data indicate that polycyclic hydrocarbons are subject to biodegradation to varying degrees depending on conditions. Fuller investigation is needed to answer the question of possible accumulation or concentration of these fractions by marine organisms.

Observed Effects of Accidental Oil Spills

There is considerable uncertainty concerning the biological effects of accidental oil spills. Oil from the *Torrey Canyon* disaster killed many marine birds, but most of the damage to other animals and plants living in the intertidal zone and in shallow water is said to have been caused by the highly toxic detergents used to disperse the oil (Holme, 1969). Except for areas heavily sprayed with detergent, no effects were noted on fisheries or plankton. Similarly, though thousands of marine birds were killed by the Santa Barbara oil slick and a number of dead seals, porpoises, and sea elephants were found, no effects were noted on open-ocean plankton or fishes (Holmes, 1969). On the beaches and in the intertidal zone, entire plant and animal communities were killed by a layer of encrusting oil which was often one or two centimeters thick. Elsewhere, barnacles, eel grass, and a species of red algae were killed, but other intertidal organisms were not severely damaged, at least during the first few months after the beginning of the spill.

Both the *Torrey Canyon* and the Santa Barbara disasters occurred offshore in relatively deep water, and most of the toxic fractions of the oil may have evaporated or been dispersed to harmlessly low levels in large volumes of water before the oil drifted into the shallow waters near shore. In contrast to these deep-water spills, the effects of two relatively small oil spills in shallow water have been carefully observed, and here extensive damage was demonstrated.

The wreck of the *Tampico* in Baja California, Mexico, in 1957 almost totally destroyed the marine fauna over an area of about a square kilometer. Among the dead species were lobsters, abalone, sea urchins, starfish, mussels, clams, and hosts of smaller forms (North, 1961). The kelp plants growing near the wreck were undamaged, though some kinds of oil are known to be toxic to kelp.

On September 16, 1969, a wrecked barge released 240 to 280 tons of No. 2 fuel in Buzzards Bay off the shores of West Falmouth, Massachusetts. Scientists at the Woods Hole Oceanographic Institution have followed the effects of this event carefully. There was an immediate massive kill of animals of all kinds—lobsters, fish, marine worms, and mollusks (Hampson and Sanders, 1969).

Dredge hauls in three meters of water collected fish, worms, crustaceans and other invertebrates. Ninety-five percent were dead, and the balance were moribund. Follow-up studies showed that the oil persisted for over ten months in organisms and in bottom sediments to a depth of twelve meters—the greatest depth in the area (Blumer, 1969b; Blumer, Souze, and Sass, 1970). The accumulation and persistence of oil in the bottom sediments was unexpected since it had previously been assumed that light oil would remain on the surface and evaporate rather quickly.

Both the *Tampico* and the Buzzards Bay spills involved refined petroleum products, heavy diesel and No. 2 fuel oil, respectively, which are considerably more toxic than the crude oil that constitutes the vast majority of accidental spills. The data from other large spills mentioned earlier which did not show massive persistent damage may reflect this difference in toxicity as well as differences in location of the spills. Clearly, additional research is needed to assess the severity and duration of damage to the environment resulting from spills of petroleum and of oil products refined from petroleum.

Effects of the Base Load of Oil Pollution

We have only fragmentary information about the biological effects of the base load of oil pollution in estuaries and coastal waters or on the high seas. One of the difficulties in assessing oil damage in coastal waters is that many other pollutants are also present in this zone and it is hard to separate their different effects. Indeed, the effects may not be separable, but instead additive or mutually reinforcing.

The so-called "tainting" of oysters and other shellfish by oil which makes the flesh unpalatable has been recognized for many years; up to one milligram of polycyclic aromatic hydrocarbons has been detected per kilogram of shucked oysters harvested from moderately oil polluted waters (ZoBell, 1962). Mackin (1970) found that tainted oysters purged themselves of oil within two to four weeks when removed from the source of pollution. However, recently, Blumer, Souze, and Sass (1970) have reported evidence that, under some circumstances at least, all the oil does not "pass through the gut without harm" but a fraction may penetrate the wall of the gut and become incorporated and stabilized in the lipid pool of the organism. They suggest that within the body lipids, even relatively unstable hydrocarbons could be preserved and transferred along the food chain. Such transfer has not been observed. The evidence is too scanty to permit conclusions as to whether the incorporation and stabilization of petroleum fractions in body lipids occurs generally with massive exposures or at all with chronic exposure. Spooner (1968; Spooner and Spooner, 1968) has observed oil passing through the gut of a number of organisms including chitons and mussels.

The threshold for lethal damage by direct poisonous effects is not now quantitatively known, and data from different sources are not in agreement. A large body of data on toxicity of oil, nontoxic dispersants, and mixtures of the two (Spooner, 1968; Mackin, 1970; Oda, 1968; State of Michigan, 1969; Lane, 1969) to a variety of marine organisms including shellfish, fish, and various larvae indicate that acute toxicity levels may be of the order of 1,000 ppm for dispersed crude oil. One researcher (Mackin, 1970) found that concentrations of dispersed oil as high as 8,000 ppm were not toxic to oysters after eight days. Other limited data (Tarzwell, private communication) indicate that concentrations as low as 25 ppm of oil and dispersant may be toxic to filter feeding mollusks (clams, oysters, scallops, mussels). There are also considerable data on undispersed oil (Smith, 1968; Spooner and Spooner, 1969; Spooner, 1969; Mackin, 1970; ZoBell, 1962; Galtsoff et al., 1936; Mackin, 1950) to indicate that limited exposures to oil concentrations in excess of 1,000 ppm are not acutely toxic to shellfish. The toxicity may increase when the oil is chemically dispersed even with nontoxic dispersants. Whether this is a chemical synergism or simply results from the increased availability of oil is not known.

If all the oil entering the coastal waters of the United States from rivers, municipal sewage, refinery operations, and ship discharges were to remain in the waters of a zone one kilometer wide and 10 meters deep, a concentration of oil of about 25 parts per million parts of water would be reached throughout a strip 3,000 kilometers long. This is a major fraction of the length of the coastal zone of the United States. The fact that the populations of clams, scallops, and oysters have not yet been destroyed by pollution in U.S. coastal waters, except in certain estuaries and other restricted areas, suggests that oxidation, evaporation, diffusion, and incorporation in bottom sediments have occurred sufficiently rapidly to maintain the concentrations of dissolved and suspended petroleum hydrocarbons below toxic levels along most of the coastal zone.

Oil pollution, even at very low levels, may be responsible for long-term damage through effects on the communication and information-gathering system of marine animals. Many biological processes are mediated by extremely low concentrations of chemical messengers in the seawater. The introduction of hydrocarbon fractions into the environment may cause destructive changes in the behavior of lobsters and other invertebrates. For example, it has long been known that lobsters are attracted to crude oil distillate fractions, especially kerosene.

One possibly quite serious effect of oil dispersed over wide ocean areas in small particles, droplets, and thin slicks will be that the oil will tend to concentrate chlorinated hydrocarbons, including DDT, in a thin layer near the sea surface. The result will be much higher concentrations of chlorinated hydrocarbon pesticides in this thin surface layer than would exist if the pesticides were dispersed throughout the mixing zone that is the top several hundred meters of the ocean. A major part of the airborne chlorinated hydrocarbons carried to the ocean will be deposited on 20 percent of the ocean's surface, between 50 degrees and 15 degrees north latitude, where the winds blowing off the industrialized countries and the countries with large agricultural areas travel over the ocean, carrying their load of chlorinated hydrocarbons. Most of the petroleum hydrocarbons near the surface of the open sea will also be present in this same area.

Measurements by Seba and Cochrane (1969) of the effects of a natural slick in Biscayne Bay, Florida, showed that the concentration of a single chlorinated hydrocarbon (dieldrin) in the top 1 millimeter of water containing the slick was more than 10,000 times higher than in the underlying water. Since the water was about 10 meters deep, about half of the total volume of pesticide in the water column was dissolved in the slick. Assuming that the thickness of the slick was 1,000 angstroms, its mass would be 10^{-5} g/cm^2 and the concentration of the measured pesticide in the slick material was about 1 part in 10,000 compared with 1 part in a trillion in the underlying water.

Cochrane and his coworkers estimate (private communication) that the concentration of all chlorinated hydrocarbons taken together, including DDT and its breakdown products, aldrin, dieldrin, BHC, lindane, and PCB, in the upper hundred meters of the waters of the Caribbean and the Florida current is close to 10 parts per trillion. If this concentration were uniform over the entire ocean area between 15 and 50 degrees north latitude, the total mass of these substances present in the top 100 meters would be 70,000 tons, perhaps a third of the world's annual production. This is roughly consistent with the estimate made elsewhere in the SCEP Report that about one-fourth of the annual production of chlorinated hydrocarbons is transported through the atmosphere to the ocean each year, and that the residence time in the top few hundred meters is about five years. With the partition coefficient between water and oily materials found by Seba and Cochrane (1969), the concentration of chlorinated hydrocarbons in the top 1 millimeter of water in areas where slicks and globules of petroleum hydrocarbon are present at the surface would be 0.1 part per million. This concentration is toxic to many marine animals.

Bacteria and other organisms will remove chlorinated hydrocarbons from the surface layer by various biological processes, but because of the high oil solubility of these substances, globules of oil and oil slicks will combine with the convective and turbulent motions of the water to act as a pump or blotter that "pulls" chlorinated hydrocarbons back up to the surface and retains them there. We know that the small larval stages of fishes and both the plant and animal plankton in the food chain tend to spend part of the night hours quite near the surface, and it is very likely that they will extract, and concentrate still further, the chlorinated hydrocarbons present in the surface layer. This could have seriously detrimental effects on these organisms and their predators, but it will not significantly reduce the amount of chlorinated hydrocarbons in the surface waters, because the total mass of living creatures in the sea is relatively so small.

Conclusions for Action
1. One distinguishing characteristic of the problem of oil pollution is that while most of the evident damage occurs in waters

and coastal areas under national jurisdiction, international agreements and control measures are needed to reduce or eliminate many of the causes of the pollution. Present national and international arrangements to control oil pollution are inadequate because they lack effective means of enforcement. International agreements should include provisions for easily and uniformly applied sanctions which would ensure that the costs to polluters are either greater than the benefits they receive or at least sufficient to clean up the pollution and repair the damage.

2. Perhaps a quarter of total oil pollution from ships and land sources occurs in semienclosed seas such as the Mediterranean, the Black, and North seas, the Persian Gulf, and the Gulf of Mexico, which have a total area slightly over 2 percent of the area of the ocean. Because of the very slow rate of oxidation of oil in cold ocean waters, the future development of oil production in the Alaskan North Slope and the Canadian Northern Archipelago may produce serious contamination in the Arctic Ocean if adequate safeguards and pollution control facilities are not employed. Regional international agreements may be the most effective way to deal with the concentration of pollution in such semienclosed seas.

3. Much more knowledge is needed about the biological effects of oil pollution. Research should be undertaken on the following specific lines:

   a. To determine the levels of concentration of different petroleum fractions in dissolved or colloidal form which are toxic to sensitive marine organisms such as filter feeding clams, oysters, and scallops, and to the larval stages of other invertebrates and fishes. The physiological mechanisms of damage also require further study.

   b. To elucidate the possible effects of different petroleum fractions on the communication and information gathering systems and on the reproductive, feeding, and defensive behavior of marine invertebrates and fishes.

   c. To find the extent of concentration of persistent pesticides and other chlorinated hydrocarbons in oil films, slicks, globules, and masses on the surface in the open ocean, and the effects on birds, fish, and plankton.

4. More information is also needed on the distributions of pe-

troleum and other hydrocarbons at the surface of the open sea and in the water column and the bottom sediments of estuaries and near-shore areas. This will require not only monitoring of these distributions but also studies of the rates and mechanisms of evaporation, solution, dispersion, and bacterial or physical oxidation of different petroleum fractions, and the rates of fallout of hydrocarbons in particulate form from the atmosphere on to the sea's surface.

Particular attention should be given to the fate of oil in the ocean and to its persistence and biological degradation under different conditions of bacterial population, temperature, and oxygen.

5. Several lines of technical development designed to reduce or remove direct oil pollution in the sea should be pursued. These include:

a. Better methods of removing oil from large accidental spills such as the *Torrey Canyon* and Santa Barbara incidents. Containment of the oil where it can do least harm, followed by physical removal, is most satisfactory, but better methods to accomplish this are needed. Burning large quantities of oil at the sea surface near shore can be hazardous and may produce severe local air pollution. Use of chemical dispersants can remove oil from the surface where, in some cases, its presence may be extremely hazardous or harmful, but it does not remove the oil from the ocean since it disperses it in the water. In the dispersed state, the oil is more subject to biodegradation but also is more available to marine organisms. At some concentrations, oil dispersed even with nontoxic chemicals can poison organisms, especially filter feeders. The question whether even the limited use of nontoxic dispersants can be justified where other methods are not possible cannot be conclusively answered at this time because data on the fate and effects of oil in the sea are so incomplete.

b. More effective and economical antipollution measures for refinery and petrochemical operations.

c. More economical and easily used techniques, which do not involve the discharge of oil residues in the sea, for cleaning the bilges and fuel tanks of merchant ships.

6. Regulations and other institutional changes are needed to reduce or eliminate the discharge or leakage of waste oils and

greases into rivers, lakes, estuaries, and coastal areas. Such regulations should be based on quantitative definitions of required water quality for different uses.

7. More effective international control measures for oil-carrying tankers should be developed:

a. To ensure that all tankers use "load on top" or other antipollution procedures.

b. To prevent strandings, and collisions: this may require monitoring and control of tanker tracks at all times, just as transoceanic aircraft are now regulated, and also provisions for tankers to load and unload only at certain specified safe locations.

8. For all ships, measures requiring that dirty ballast and washing be retained aboard and discharged to shore facilities for treatment should be investigated. Permissible limits of oil content in waters discharged to the sea should be specified.

9. Because many future offshore oil drilling and production operations can be expected off the coasts of countries that are not technically equipped to enforce effective antipollution regulations, international standards for safe procedures and an international mechanism for their enforcement should be developed.

References

Beerstecher, Ernest, 1954. *Petroleum Microbiology: An Introduction to Microbiological Petroleum Engineering* (Houston: Elsevier Press).

Blumer, M., 1969a. Oil pollution of the ocean, *Oceanus, 15* (2): 2–7.

Blumer, M., 1969b. Oil pollution of the ocean, *Oil on the Sea*, edited by D. P. Hoult (New York: Plenum Press).

Blumer, M., Souze, G., and Sass, J., 1970. Hydrocarbon pollution of edible shellfish by an oil spill, Reference No. 70–1, Woods Hole Oceanographic Institution (unpublished manuscript), January.

Corino, Edward R., 1970. Industrial wastes—oils of petroleum origin (Paper prepared for SCEP).

Davis, J. B., 1967. *Petroleum Microbiology* (New York: Elsevier Publishing Co.).

Galtsoff, P. S., Prytherich, H. F., Smith, R. O., and Koehring, V., 1936. Effects of crude oil pollution on oysters in Louisiana waters, *Bulletin of the U.S. Bureau of Fisheries, 18:* 143–210.

Goldberg, E. D., 1970. Atmospheric transport (manuscript prepared for SCEP).

Hampson, G. R., and Sanders, H. L., 1969. Local oil spill, *Oceanus, 15* (2): 8–10.

Holme, Robert A., 1969. Effects of *Torrey Canyon* pollution on marine life, *Oil on the Sea*, edited by D. P. Hoult (New York: Plenum Press).

Holmes, Robert W., 1969. The Santa Barbara oil spill, *Oil on the Sea*, edited by D. P. Hoult (New York: Plenum Press).

Horn, M., Teal, J. M., and Backus, R., 1970. Petroleum lumps on the surface of the sea, *Science, 168:* 245–246.

Lane, C. E., 1969. (Unpublished report, Institute of Marine Sciences, University of Miami).

Mackin, J. G., 1970. Effects of crude oil and bleedwater on oysters and aquatic plants, Progress Report (College Station: Texas A&M Research Foundation).

Michigan, Department of Natural Resources, 1969. A biological evaluation of six chemicals used to disperse oil spills (Lansing).

North, W. J., 1961. Successive biological changes observed in a marine cove exposed to large oil spillage, University of California, Institute of Marine Resource, Ref. 61–6, pp. 1–33.

Oda, A., 1968. A report on laboratory evaluations of five chemical additives used for the removal of oil slicks on water, Ontario Water Resources Commission, Toronto, Canada.

Revelle, R., 1963. Water, *Scientific American,* September.

Seba, D., and Cochrane, E., 1969. Surface slicks as a concentrator of pesticides, *Pesticide Monitoring Journal,* December.

Study of Critical Environmental Problems (SCEP), 1970. *Man's Impact on the Global Environment* (Cambridge, Massachusetts: The M.I.T. Press).

Sisler, F. D., and ZoBell, C., 1947. Microbial utilization of carcinogenic hydrocarbons, *Science, 106.*

Smith, J. E., 1968. *Torrey Canyon: Pollution and Marine Life* (New York: Cambridge University Press).

Spooner, M. F., 1968. Preliminary work on the comparative toxicities of some oil spill dispersants and a few tests with oils and COREXIT, Marine Biological Association of U.K., Plymouth, England (unpublished manuscript).

Spooner, M. F., and Spooner, G. M., 1968. The problems of oil spills at sea, Marine Biological Association of U.K., Plymouth, England (unpublished manuscript).

U.S. Department of Health, Education, and Welfare, 1970. Summary of emissions in the United States, 1970 Edition, May.

ZoBell, C. E., 1962. The occurrence, effects and fate of oil polluting the sea, *Proceedings of the International Conference on Water Pollution Research,* Section 3, No. 48 (London), September.

20
Phosphorus and Eutrophication

Excerpt from SCEP Work Group on Ecological Effects

Of the major nutrients—nitrogen (N), phosphorus (P), and potassium (K)—which man introduces into the environment, phosphorus can cause the most serious pollution problems. Eutrophication of lakes is already a major local problem; eutrophication of estuaries is a potential global problem.

Some phosphorus production data are given in the report of Work Group 5. Verduin has found that most phosphorus enrichment of surface waters comes from sewage treatment plants and only secondarily from agricultural fertilizers (Verduin, 1966). Table 20.1 shows the relative amounts of phosphorus entering rivers from municipal raw sewage and urban and rural runoff. Primary sewage treatment removes about 10 percent of the phosphate and secondary treatment about 30 percent (American Chemical Society [ACS], 1969). Although much of the phosphate may settle during some phase of treatment, it is released back into the effluent during subsequent sludge digestion. In view of the small nutrient mass in the sludges, their separation from the main stream of the sewage makes little difference to the eventual nutrient enrichment of ocean areas.

Table 20.1 1968 U.S. Mean Phosphorus Content of Runoff

Source	Percentage of Total U.S. Discharge[a]	Magnitude (metric tons)[b]
Municipal raw sewage	60	262,000
Urban runoff	23	100,000
Rural runoff	17	74,000

[a] Municipal raw sewage and urban runoff percentages were computed using the 17 percent figure for rural runoff given in Federal Water Pollution Control Administration (FWPCA), 1968, and the following data from American Chemical Society (ACS), 1969: of the total U.S. urban drainage, raw sewage constitutes 20,000 lb/sq mile/yr, and all other sources contribute 7,800 lb/sq mile/yr.
[b] These magnitudes were computed using the percentages in the table and the following data from ACS, 1969: in U.S. surface waters, there are 280 million lb/year of phosphates from household detergents and 680 million lb/year from other sources for a total of 960 million lb/year or approximately 436,000 metric tons.

Reprinted from Study of Critical Environmental Problems (SCEP), 1970. *Man's Impact on the Global Environment* (Cambridge, Massachusetts: The M.I.T. Press), pp. 144–149.

Nutrient runoff from agricultural land is not yet a major global problem. Most fertilizers are applied in developed countries, and consequently the estuaries in these countries are the principal sinks for agricultural nutrients. Phosphorus from farmland enters streams primarily through runoff from feedlots and erosion of topsoil. Soil particles have a strong affinity for phosphate, thus reducing the amount of phosphorus in runoff.

Eutrophication of Lakes

The eutrophication of lakes is a serious problem in practically all countries and may serve as a model for what is likely to happen to estuaries and possibly the coastal ocean if present trends continue. Many lakes throughout the world are overnourished (eutrophic) or are becoming overfed by effluents in which the key ingredient is phosphorus from domestic sewage, eroded soil, and farm manure.

Lakes that have been clear and clean for thousands of years have become repulsive and odorous within ten years after manmade effluents are introduced (Hasler, 1969). Excess nutrients change the algal community from one of great diversity of species to one of a few species which cause nuisances. Costs of purifying water for human use rise, property values drop, and recreational features deteriorate. When eutrophication progresses to its extreme stages, the water becomes deficient in dissolved oxygen, and all useful species (including fish) die. Eutrophication in lakes can, however, be alleviated by stopping waste discharge (for example, the cases of Madison, Wisconsin, lakes, and Seattle's Lake Washington described in Hasler, 1969).

Although eutrophication is caused by a mix of nutrients, potassium is usually present in excess and tends to remain in the water, while nitrogen levels can be supplemented through the biological fixation of atmospheric nitrogen. Phosphorus tends to be precipitated in sediments, and phosphorus levels cannot be supplemented. Thus, phosphorus is the most likely substance limiting algal growth, and it serves as the best general indicator of nutrient enrichment (Rohlich, 1969; ACS, 1969; Vollenweider, 1969).

One of the most graphic examples of lake eutrophication in the United States is the case of Lake Erie. The Federal Water Pollution Control Administration (FWPCA) estimated rural runoff phosphorus at 0.1 metric ton per square mile per year and urban runoff at 0.25 metric ton per square mile in the Lake Erie drainage basin (FWPCA, 1968). Municipal waste accounted for 70 percent of the phosphorus contributed to Lake Erie in 1967 and is projected to contribute 80 percent in 1990.

Estuaries and Coastal Ocean Areas

Estuaries are semienclosed basins where water flowing from continents mixes with surface and subsurface ocean waters. Water circulation in an estuary is a two-way process. Surface layers of less saline water flow generally seaward over subsurface layers of more saline water flowing generally landward. The subsurface flow replaces water and salt entrained in surface waters, while extensive mixing occurs across the pycnocline (the layer of strong density difference separating the surface and subsurface layers). Estuarine-type circulation is not restricted to estuaries but is typical of most coastal ocean areas where freshwater inputs from rivers and precipitation exceed water loss through evaporation.

This circulation pattern produces several important effects. It results in estuaries and coastal ocean areas accumulating riverborne sediments and wastes, as well as wastes dumped by adjacent urban areas and sediments (commonly sands) moving along the coast. (Harbor dredging and shipping facility construction improve an estuary's sediment-trapping efficiency and also present the problem of disposing of the dredged wastes.) More important, as a result of this circulation pattern, estuaries trap nutrients. This results in the well-known productivity of coastal waters. Surface organisms feed on the nutrients (especially nitrogen and phosphorus), die, sink to the bottom, and decompose. When productivity is excessively large, this may exhaust the dissolved oxygen in the bottom waters, killing all normal benthic organisms except bacteria. Landward-moving bottom waters return the released nutrients to the surface, supporting the growth of more phytoplankton. Thus, nutrients tend to be retained and recycled in the estuary (or coastal ocean water) rather than move seaward to mix with the open ocean.

Estuaries receive riverborne wastes from upriver and adjacent communities and industries as well as the natural discharge of nutrients and solids from river and rain. Table 20.2 illustrates the

Table 20.2 Example of Possible Nutrient Discharges to Estuaries from Various Sources (Metric tons per year)

	Nitrogen (N)	Phosphorus (P)
Sewage:[a] population of 10 million, per capita generation 400 liters/day	6,000	15,000
River water:[b] discharge 500 m³/sec (approx. 15,000 cfs)	300	7
Subsurface seawater:[c] 20 volumes of seawater mixing with each volume of river water	6,000	900
Storm-water runoff:[a] 75 cm (30 inches) per year area	6,000	800

Sources:
[a] Weibel, 1969.
[b] Bowen, 1966.
[c] Ketchum, 1969.

relative size of each contribution for a hypothetical community of 10 million people covering an area about 12,000 square miles, on a river discharging about 500 cubic meters per second (15,000 cubic feet per second). Each volume of river water mixes with 20 volumes of nutrient-carrying subsurface water. (Note: No allowance has been made for the industrial wastes commonly discharged directly into the estuary.) This table illustrates the local origin of most estuarine problems involving excessive nutrient discharge and accompanying ecological imbalance. The dominant source of phosphate is sewage (primarily from detergents containing phosphate), while nitrogen in various forms is supplied in nearly equal volumes by sewage treatment plants, storm-water runoff, and subsurface seawater (Verduin, 1966).

Most of the world's large urban centers (in both developed and less-developed countries) are located on estuaries and discharge their wastes into the coastal waters (Cronin, 1967). Since estuaries are the permanent residence, passage zone, or nursery area for about 90 percent of commercially important fish, pollution here could have far-reaching effects (McHugh, 1967; FWPCA, 1970). For example, over 20 percent of the world's minimal protein needs could come from the sea by the year 2000, provided estuaries are still healthy (Ricker, 1969). Moreover, national pollution can have international effects. For example, the pollution of Alaskan and Canadian estuaries, the principal nursery of Pacific salmon, would affect not only the United States and Canada but also Japan and the USSR, who fish the salmon on the high seas.

The concentration of industrial nations around the North Atlantic basin and the common fate of estuaries in industrialized nations lead us to recommend careful investigation and consideration of international monitoring. Yet, because of the usual local origin of estuarine problems, effective control is most likely to be accomplished unilaterally.

Recommendations

1. We recommend that nutrients in areas of high concentrations, such as sewage treatment plants and feedlots, be reclaimed and recycled.
2. We recommend that the dumping of industrial wastes into sewage systems be restricted, so that toxic wastes do not interfere with nutrient recovery and recycling.
3. We recommend that nutrients not be used in materials that are discharged in large quantities into water or air. We recommend, for example, that phosphates in detergents be replaced with new materials, being certain that the substitute does not itself create a new problem.
4. We recommend that the institutional structures responsible for defining, monitoring, and maintaining water quality standards over large areas be improved. The multiplicity of authorities involved in river basins, estuaries, and coastal oceans makes effective control nearly impossible.

References

American Chemical Society (ACS), 1969. *Cleaning Our Environment: The Chemical Basis for Action* (Washington, D.C.: ACS).

Bowen, H. J. M., 1966. *Trace Elements in Biochemistry* (London and New York: Academic Press).

Cronin, L. E., 1967. The role of man in estuarine processes, *Estuaries*, edited by G. H. Lauff (Washington, D.C.: American Association for the Advancement of Science).

Federal Water Pollution Control Association (FWPCA), 1968. *Cost of Clean Water*, vol. III (Washington, D.C.: U.S. Government Printing Office).

Hasler, A. D., 1969. Cultural eutrophication is reversible, *Bioscience, 19*.

Ketchum, B. H., 1969. Eutrophication of estuaries, *Eutrophication: Causes, Consequences, Correctives*, edited by G. Rohlich (Washington, D.C.: National Academy of Sciences).

McHugh, J. L., 1967. Estuarine nekton, *Estuaries,* edited by G. H. Lauff (Washington, D.C.: American Association for the Advancement of Science).

Ricker, W. E., 1969. Food from the sea, *Resources and Man* (San Francisco: W. H. Freeman and Co.).

Rohlich, G., ed., 1969. *Eutrophication: Causes, Consequences, Correctives* (Washington, D.C.: National Academy of Sciences).

Verduin, J., 1966. Eutrophication and Agriculture, paper presented at American Association for the Advancement of Science Symposium, Washington, D.C.

Vollenweider, 1969. *Eutrophication* (Paris: Organization for Economic Cooperation and Development).

Weibel, S. R., 1969. Urban drainage as a factor in eutrophication, *Eutrophication: Causes, Consequences, Correctives,* edited by G. Rohlich (Washington, D.C.: National Academy of Sciences).

Part V   Measurements and
                                Monitoring

One of the major difficulties which faced the SCEP participants in their attempt to assess man's impact on the global environment was the poor state of the data base with respect to important ecological processes and to the rates, routes, reservoirs, and effects of pollutants. The development of a more complete understanding of these areas will require extensive research, measurements, and monitoring.

During SCEP, one of the Work Groups focused on the various needs and methods available for monitoring critical environmental problems. Several portions of their report which relate to terrestrial and oceanic ecosystems are reprinted here. A major section of the report dealing with an ocean base-line sampling program has been deleted because this subject is treated in much more detail in the two other papers in this series. The chairman for the Work Group on Monitoring was G. D. Robinson, Center for the Environment and Man, Inc., and the participants were A. P. Altshuller, National Air Pollution Control Administration; Richard D. Cadle, National Center for Atmospheric Research; Robert Citron, The Smithsonian Institution; Seymour Edelberg, Massachusetts Institute of Technology; Gifford Ewing, Woods Hole Oceanographic Institution; Dale W. Jenkins, The Smithsonian Institution; Jules Lehman, National Aeronautics and Space Administration; Henry Reichle, National Aeronautics and Space Administration; Morris Tepper, National Aeronautics and Space Administration. Jonathan Marks of Harvard Law School was the rapporteur.

An active participant in the SCEP Work Groups on Ecological Effects and on Monitoring was Dr. Dale W. Jenkins, Director of Ecology of the Smithsonian Institution. For some time, he has been involved in various efforts to design biological monitoring systems, including activities in the Smithsonian and the National Academy of Sciences. Following the Study, Dr. Jenkins prepared the paper included here. It is the product of his wide experience and exposure to the pros and cons of various interest groups in this very complex area.

Participants in SCEP concluded that there is simply not enough presently known to design a monitoring system for the oceans. In order to answer critical questions about the dispersion and effects of pollutants so that priorities for future research and monitoring can be established, several members of the Study undertook the first steps of designing a program of base-line measurements. The two papers by Drs. Goldberg and Gross and Drs. Goldberg, Arnason, Gross, Lowman, and Reid are only initial formulations of such a program—their inadequacies emphasize the need for additional inputs. One such input was developed in detail at the Seminar on Methods of Detection, Measurement, and Monitoring of Pollutants in the Marine Environment, sponsored by the U.N. Food and Agriculture Organization in December 1970 and chaired by Dr. Edward D. Goldberg.

The final paper in this series contains the results of an extensive survey conducted by Mr. Robert Citron of major international environmental monitoring programs that utilize intergovernmental or international nongovernmental organizational machinery to coordinate their activities. These data sheets provide a capsule view of operational, planned, and proposed programs and indicate where more information can be obtained about each.

21
Work Group on Monitoring   SCEP Report
(Abridged)

The Concept of Monitoring

When the miner's canary died, it was time to get out of the mine. The canary "monitored" the mine air and gave an indication of potential disaster due to odorless, invisible methane. The immediate action necessary was clear; long-term solutions could be considered later.

But when we are concerned with a global environmental problem, this type of monitoring is insufficient. Because we cannot escape from the earth, we must have more than a sentinel to sound an alarm if a critical threshold is passed; we must know what it is that kills our "canary," where it comes from, and how to turn it off at the source.

Accordingly, we think "monitoring" is best conceived of as systematic observations of parameters related to a specific problem, designed to provide information on the characteristics of the problem and their changes with time. The parameters and problems with which we have been concerned are those of the global environment. And though any monitoring program will provide information useful to dealing with local and regional problems, our concern has been with identifying existing and potential monitoring systems capable of securing the information necessary to deal with the critical global problems identified by the Study of Critical Environmental Problems (SCEP).

For every one of the global problems that have been identified, we find we have insufficient knowledge of either the workings or the present state of the environmental system (see reports of the Work Groups on Climatic Effects and Ecological Effects). This hinders us as we attempt to design monitoring that will not only warn us of change but also provide information upon which we can base rational and efficient remedial action. In most instances we can suggest a likely analogue of the canary, but we do not know what action would be best once our bird shows up sick. Further, we are persuaded by our colleagues that global systems both physical and biological are so complex that the ultimate

Reprinted from Study of Critical Environmental Problems (SCEP), 1970. *Man's Impact on the Global Environment* (Cambridge, Massachusetts: The M.I.T. Press), pp. 167–176, 186–191, 212–222.

consequences of any disturbance cannot at present be predicted with confidence.

For these reasons our report is concerned not only with monitoring in its sense of providing warning of critical changes but also with measurements of the present state of the system (the "base line") and with measurements in support of research into the workings of the system. We mention the need for this research where it is apparent to us; we have not attempted to provide a complete assessment of research needs. In general, however, we have agreed that research is most needed in providing a closer specification of the present state of the planet and in developing a more complete understanding of the mechanisms of interaction between atmosphere, ocean, and ecosystem.

Monitoring Techniques and Systems

Before we turn to an analysis of the monitoring aspects of the critical global environmental problems identified by SCEP, we feel it necessary to look more generally at the current state of monitoring techniques and systems, both to provide a framework within which the later recommendations can be seen and to point out areas on which emphasis should be placed as the effort to obtain information for environmental management continues.

Economic and Statistical Monitoring

If we are concerned with predicting the accumulation of a pollutant in the environment, its rate of input must be known. For example, to evaluate the global CO_2 problem, we require a long-term prediction of the total atmospheric content, which, in turn, depends in part on statistical projections of fuel production and consumption. To evaluate the contribution of SO_2 to the global particulate problem, we require projections of natural, industrial, and energy-conversion emissions of SO_2 and SO_2-precursors which take into account possible control and abatement measures.

If we are concerned with evaluating the effects of alternative control technologies on pollutant levels, we need quantitative information about the flow of materials which will be altered by control technology to include inputs, wastes, and end products at each stage of the process (Ayres and Kneese, 1968).

Both these kinds of activities seem to us to be essential forms of monitoring. To some degree they are already being carried out, for example, in industry as a part of the management process, and in government as a part of economic policy making and of already existing regulatory activities.

Yet, the focus of the gathering and synthesis of data concerning industrial, agricultural, domestic, and energy-producing activities has not commonly included an "environmental effects" component. Moreover, research into the natural pathways and degradation of pollutants once they are deposited in the environment has as yet yielded little quantitative data. Nor is our knowledge of the functioning of particular ecosystems sufficient to allow us to quantify effects of a particular pollutant when it, for example, eradicates one or a group of species within the ecosystem. This has meant that we lack information on which to base projections and models for decision making. It has also meant that the organization of data on which to base an analysis of the global environmental problems considered by this conference has been a tedious, often approximative, and sometimes impossible task.

New methods for gathering and organizing economic and statistical data must be developed if we are to have the "handle" we need to deal with environmental problems. New centralized collection and collation points are needed. Federal regulatory bodies in the United States, such as the National Air Pollution Control Administration (NAPCA), have begun this task for their own areas of responsibility. But there exists no effective organization summing data across traditional areas of environmental responsibility, such as air and water pollution. Nor do we have any comparable international organization or, indeed, any effective standards to ensure, for example, that the industrial production data collection going on across the world will be of comparable precision and focus. All these tasks must be accomplished if we are to use effectively our economic and statistical monitoring potential.

RECOMMENDATIONS

1. We recommend the development of new methods for gathering and compiling global economic and statistical information, which organize data across traditional areas of environmental responsibility, such as air and water pollution.
2. We recommend the propagation of uniform data-collection standards to ensure, for example, that industrial production data

collection being carried out across the world will be of comparable precision and focus.

Physical and Chemical Monitoring

Physical and chemical monitoring methods are used to determine the amount of a contaminant in a sample of soil, water, air, or organism. Physical methods are also used to determine a property of an environmental system as a whole, such as the refractive index or the albedo of the atmosphere.

There are numerous examples of this type of monitoring. NAPCA's network of sampling stations monitors the quality of urban air. Weather satellites monitor the formation of hurricanes. The essence of good monitoring of this type is to measure what is needed, and no more, with the precision that is needed, and no more, and to maintain standards indefinitely.

Traditionally, monitoring of this type is carried out in networks of fixed stations. The entire operation may be completed at these stations or a sample may be taken to a central laboratory for examination or analysis. In either case, central coordination of methods and central standardization is necessary. Monitoring is now extended to measurements on ships, aircraft, and satellites. Moreover, it has become an international activity, and international coordination of standards is necessary.

Measurements of solar radiation at the ground form a good if little-known example. They are made in numerous countries by national meteorological services, universities, agricultural research stations, and so forth. They are also collected, edited, and published at a Soviet observatory under arrangements guided by the World Meteorological Organization. These measurements are not of uniform quality, but the best which can be identified have been standardized by instruments compared internationally on an ad hoc basis. The last major comparison was arranged bilaterally between an American manufacturer and a Soviet university. The necessary nominal institutional anchor for these comparisons is the Radiation Commission of the International Association of Meteorology and Atmospheric Physics within the framework of the International Council of Scientific Unions. To give another example, our knowledge of the CO_2 content of the atmosphere is due to the interest, skill, and cooperation of small groups of scientists in the United States and Sweden (Pales and Keeling, 1965; Keeling, 1970).

RECOMMENDATION

We recommend the development and expansion of current physical and chemical monitoring systems and techniques. Specific recommendations are included in a later section.

Biological Monitoring

Even though our interest in environmental pollution stems from our concern about its effects on living organisms, the concept of using such organisms, either individually or as a population or species, as tools to monitor the state of the environment is still a relatively untested one. Moreover, although the study of natural ecosystems has long been an important scientific activity, the observation and evaluation of changes in these finely tuned systems have not yet been systematized to yield warnings about harmful contaminants.

Yet living organisms can serve as excellent quantitative as well as qualitative indices of the pollution of the environment. Plants and animals are continally exposed and can act as long-term monitors that integrate all environmental effects to reflect the total state of their environmental milieu. They can show the pathways and points of accumulation of pollutants and toxicants in ecological systems. Their use can remove the extremely difficult task of relating physical and chemical measurements to biological effects.

There are numerous examples of the use of biological organisms as monitors. The miner's canary has already been mentioned. Rats are being used today as air pollution monitors, for CO. There is currently a widespread and effective pesticide monitoring network in the United States based on the analysis of biological material (Murray et al., 1970).

Yet further planning and coordination is necessary if biological monitoring is to play its optimum role in environmental information gathering. The pesticide monitoring network, already noted, could have by now made a greater contribution to our understanding of this critical problem had it been initially combined with air and soil monitoring into a comprehensive program for dealing with the problem of pesticide pollution instead of having been implemented as a valuable but inadequate response to the discovery of *symptoms* of pesticide poisoning in animals.

Planning for biological monitoring of global environmental problems should consider three basic methods: (1) international networks of biological surveys; (2) international networks of terrestrial and aquatic ecological base-line monitoring stations; and (3) the use of biological organisms as sentinels, detectors, and indicators and for biological assay of man-produced contaminants and changes.

1. Biological surveys are broad geographical evaluations of the population size, reproductive success, and health of species of living organisms. Such surveys can be accomplished by the census of organisms during such critical periods in the life cycle as breeding, nesting, overwintering, migration, and flowering. Commercial and other harvest data on organisms such as fish and game birds are already available examples of information that could make up part of such a survey. Production figures from plant-food crops and lumber can also be used, provided information about the growing cycle and treatment with fertilizers and pesticides is known so that the data can be analyzed.

2. A global network of ecological base-line monitoring stations would show general changes of major significance to the flora and fauna of the world. A complete system would require establishing stations in each of the following biomes: deciduous forests, conifer forests, tropical forests, savannas, grasslands, deserts, tundras, estuaries, and various ocean areas. The terrestrial base-line stations would enclose permanent, protected natural areas with a long record of ecological and meteorological study. The site would best be at least 4 kilometers square with a large surrounding area relatively free from human intervention and direct contamination. A meteorological station and a contaminant sampling station would be located near the base-line station to permit correlation of biological results with physical and chemical measurements. Comparable samplings of temperature, rainfall, solar radiation, and the composition of the atmosphere would be made near aquatic stations.

Within the area of the base-line station detailed ecological studies would be combined with biological surveys of the kind just described. The results obtained in these stations would be compared with those obtained at impact stations located to study the direct biological effects of man, especially those caused by modern agricultural practices, cities, and industry.

3. Many species of biological organisms have great sensitivity to pollutants. Many animals and plants can be used as "early warning" sentinels for particular pollutants. Others can be used to provide graphic records of pollutants; by looking at certain plants, for example, it is possible not only to identify the presence of certain pollutants in a given area but also to gain information on this concentration and its variation in time. Other organisms, simply by their presence in a particular place, signal that a particular pollutant is also present; the proliferation of algae in lakes, for example, indicates an excess of nutrients like phosphorus.

RECOMMENDATIONS

1. We recommend early implementation of a set of ecological base-line stations in remote areas that would provide both specific monitoring of the effects of known problems and warnings of unsuspected effects.

2. We recommend central coordination and, where necessary, modification of national and regional surveys of critical populations of fish, birds, and mammals from commercial catches, harvests, and surveys. This would provide an early warning system by monitoring highly sensitive and vulnerable species.

Modern Technology and Monitoring

A major problem in those monitoring systems that require precise measurement and rigorous attention to detail is to combine the continuity and security of institutional control with the continued devotion of the scientists and skilled technicians who are often required to work in isolated regions. Further, the extreme care that is necessary in sample handling makes the standardization and interpretation problem extremely difficult when measurements must be made independently at each one of many stations within a global network (*in situ* monitoring) (Pales and Keeling, 1965).

For these and other reasons we must continue to investigate

the possibility of monitoring remotely, by automatically reporting or man-operated instruments carried in earth satellites or airplanes.

The unmanned earth satellite, especially, has many virtues as a vehicle for monitoring equipment. It can provide global coverage in very short periods of time. It can carry equipment without subjecting it to environmental stress. If large amounts of information from widely distributed points are required or if a particular monitoring system can be "piggybacked" on a satellite with other funtions, there may be cost advantages. *In situ* techniques in general yield data whose precision has not yet been equaled by remote techniques for measuring the same parameter. The techniques now available for use in satellites must be improved if they are to provide the high-precision data that are required.

It must be pointed out that interesting and useful data have already been obtained from satellites for use in the atmospheric and oceanographic sciences. This, combined with preliminary information about satellite monitoring experiments planned in the next few years, indicates that satellite solutions to many environmental data-gathering problems could be available in the not-too-distant future.

What is clearly needed is a series of evaluations of appropriate satellite techniques, including both scientific feasibility studies and cost-benefit analyses. Once scientific feasibility is established, the aim of such an evaluation would be to determine the optimum system for providing the required information, be it a satellite, ground, or mixed system.

In addition to this comparative evaluation of remote and *in situ* monitoring, there is need for continuing research into monitoring per se. In recent years new and powerful techniques have been developed in radiometry, radar, and spectroscopy which are operated in the microwave, infrared, and visible regions of the electromagnetic spectrum. Furthermore, digital data-processing systems have been developed which permit rapid handling and analysis of large data flows. There is a need for continuing research into these systems, as well as into the integration of diverse requirements to ensure optimum use of resources, and into scientific and administrative coordination.

RECOMMENDATIONS

1. We recommend generally a series of evaluations of appropriate satellite measurement and monitoring techniques, including both scientific feasibility studies and cost-benefit analyses, aimed at determining the role of satellites in an optimum monitoring system for the problems dealt with by SCEP.
2. We recommend continuing research into the application to environmental monitoring of radiometry, radar, spectroscopy, digital data processing and other newly available techniques.

Monitoring Chlorinated Hydrocarbons and Toxic Heavy Metals

We here group together two very different classes of materials used by man in very different activities. But the substances are closely allied from the aspect of monitoring. Their threat to the environment—the poisoning of living things other than those which man, mistakenly or not, wants to poison—is the same. Their routes and reservoirs in the environment are the same. It happens also that the required monitoring technique, quantitative chemical analysis of trace substances, is similar in the two cases. Hence, at least in general, we can and should monitor the substances in the same places and with the same degree of detail.

We are dealing here with persistent poisons, which man has found cause to use either because they kill pests or are important in industrial processes. The chlorinated hydrocarbons not only kill target insect pests but also many beneficial biological control organisms. They are accumulated in harmful amounts in the higher levels of various food chains.

Lead, mercury, and certain other heavy metals can be harmful to men and animals if inhaled. They may be harmful to plants and animals in streams, lakes, and oceans. When they reach the earth's surface, by the scrubbing action of rain, for example, they may have harmful effects on plants. The heavy metals are introduced into the air as particles by automobiles, many industries, and by power plants. They are a part of the waste effluents dumped into rivers by many industrial processes.

The chlorinated hydrocarbons, since they include such a wide variety of products, are put into the environment by several different routes. In the application of pesticides to agricultural lands, a significant proportion is lost directly to the air, from which it later falls in rain; another portion is washed out of the soil into watercourses; another portion remains in the soil and is degraded there. Some chlorinated hydrocarbons ultimately accumulate in the oceans.

The knowledge from which the above descriptions of routes and reservoirs is taken is primarily qualitative; what quantitative data we have is generally rough, with the exception of some world production and use figures for the chlorinated hydrocarbons, United States emission figures for the heavy metals, and isolated data on the accumulation of such substances in organisms. There is a need for base-line surveys and monitoring in all these areas.

We must discover what has happened to the quantities of chlorinated hydrocarbons and heavy metals we have produced over the past decades. We must map the current flows of such materials in the environment, identifying, for example, how much persistent insecticide is transported in the atmosphere, how much of this is deposited as rain, and where it is deposited.

Our monitoring effort, then, must include the atmosphere, soil, oceans, rivers, and biological material. (For amplification of the foregoing discussion, see reports of Work Groups 2, 5, and 6.)

COMPILATION OF SOURCE, USE, AND DISPERSAL DATA

In order to evaluate the importance and size of the problems involved, it is necessary to have the world production and use data, including especially data on dispersal, emission, and direct pollution into the air, water, and soil. We should know the geographical area of dissemination into the environment as well as the volume and dates of dispersal.

At present, statistics on production and use are available for most of the world for both the chlorinated hydrocarbons and heavy metals (the Soviet Bloc data are lacking). Few accurate U.S. and world data are available on the escape of heavy metals to the environment.

MONITORING FOR CHLORINATED HYDROCARBONS
AND TOXIC HEAVY METALS

SAMPLING AND ANALYSIS

The techniques for collecting air, soil, and water samples for chlorinated hydrocarbon and heavy metal analysis are well developed and are routinely carried on in a number of laboratories. (Feltz,

1969, National Air Pollution Control Administration [NAPCA], 1969; Cadle, 1970; Carver, 1970; Murray, 1970; Sand, 1970.)

Air

Most of the organic chlorides of concern have low but significant volatility and tend to be associated with airborne particles at the relatively high concentrations at which measurements are generally made. Since the particles have a short residence time in the air (probably a few days), the organic chlorides probably have a short residence time as well. Accordingly, a fairly large number of collection points seem necessary to assure adequate sampling.

The analytical methods employed assume for the most part that the chlorinated hydrocarbons are associated with airborne particles and start by the collection of such particles on a filter. However, when the chlorinated hydrocarbon concentrations are very low, as in the ambient atmosphere, this assumption may lead to gross errors since the chlorinated hydrocarbons may be largely in the gas phase. Thus, it is recommended that at least at first the atmosphere should be analyzed for both the gas and particulate phases of these substances. The usual analytical technique involves the collection of particles on a filter, extraction of the filter, and a chromatographic analysis of the extract. Special techniques may have to be developed for gas-phase analysis.

Since the heavy metals are associated with atmospheric particles, their residence time in the atmosphere is relatively short, probably a half-life of a few days. Thus, a large number of sampling points is advisable, of the same order as for the organic hydrocarbons. The U.S. Public Health Service already monitors a large number of metals routinely through its National Air Sampling Network. Many of these 185 urban and 51 nonurban stations could provide the needed samples (NAPCA, 1969).

The particles containing the metals can be collected by filtration and analyzed in a central laboratory. Probably the same collection system as described for sulfates later in this report will suffice. Probably 10 to 15 metals would be monitored, certainly including lead, mercury, copper, zinc, arsenic, cadmium, and vanadium. Determination of the concentrations of this number of metals is relatively easy once samples have been obtained and prepared for analysis.

Water and Soil Samples

Samples can be collected in prepared containers from lakes and rivers with a depth-integrating sampler. Bottom sediments from rivers can readily be obtained with either a bed material sampler or a piston core sampler containing a Teflon liner.

Collections of samples in estuaries and open ocean locations can be done by use of collectors presently employed in oceanographic activities. Water samples of about 200 liters can be obtained by large volume samplers.

Pollution in sediments accumulated from water runoff from the land usually is deposited on the bottom surface and may be a good index of general pollution levels in the local hydrologic environment. Both the piston-type bed material sampler and commercial core samplers are adequate for collecting sediment samples.

Soil samples should consist of cores 2 inches in diameter and 3 inches deep.

Required sensitivity for detecting trace amounts of chlorinated hydrocarbons in water, suspended particles, bottom sediments, atmospheric dryfall, and biota is 5 ppt. The electron capture gas chromatograph is needed for minimum identification. Microcoulometry, coupled with gas chromatography, may be used for confirmation when concentrations permit. Analysis is performed with extracts injected into two columns having different retention properties. Positive identification can usually be obtained by corroboration of results using at least two types of gas chromatography columns. Additional confirmation when required can be made using specific detectors such as chloride, sulfur, or nitrogen microcoulometry and gas chromatography phosphorus detectors. Analysis by a third column with different retention times can be helpful, and, if infrared spectroscopy is available, special microtechniques applied to separated materials can aid in positive identification.

SAMPLING NETWORK

In order to obtain systematic sampling of all significant air, water, land, and biota systems, a network of some 200 integrated sites strategically located around the world seems minimally necessary (or a larger number if some are specialized, for example, sampling only the atmosphere). To set exact numbers and locations, more information of the kind that would be obtained through the proposed ocean base-line sampling program is needed.

In general, however, each land-monitoring site and the sites selected for snow and ice fields would be required to sample dryfall, rainfall, ground surface cores, and land plants on a periodic basis, perhaps monthly. Dryfall and rainfall sampling would require multiple samplers to insure that individual specimens are obtained for chlorinated hydrocarbon and heavy metal analysis. The lake, river, estuary, continental shelf, and deep ocean sites would be required to sample water surface film, suspended particulates, the water itself, and marine biota, perhaps monthly. All sites but the deep ocean station would obtain bottom sediment samples somewhat less frequently. Sample frequencies and priorities would obviously change as such a program developed and as base-line information on pollution levels and variability was obtained.

Monitoring of chlorinated hydrocarbons and heavy metals is presently under way in many countries. In trying to set up a program of sampling to provide integrated and systematic information on a global scale, these programs as well as facilities being employed for other monitoring and surveying tasks should be used if at all possible. In fact, a first step toward developing the kind of comprehensive network just described would be the preparation of a survey summarizing and evaluating existing programs and facilities worldwide (for an example of such a survey, see Citron, 1970).

Without this kind of initial survey, it is difficult to estimate costs of a comprehensive monitoring program. In general, capital and operating costs would include the costs of sample collection, sample analysis, operating a central standards laboratory, data processing, and logistical support.

RECOMMENDATIONS

1. We recommend that chlorinated hydrocarbons and toxic heavy metals be monitored with a view to answering two questions:

What has happened to the material that we have used in the environment or allowed to escape into it?

What will happen to material that we use or release, in specified circumstances, in the future?

To provide answers to these questions, we recommend that current monitoring efforts be systematized and extended to provide adequate records of

a. The concentrations of these materials in living things and their effect on life.

b. The production, use, and escape to the environment of these materials.

c. The concentrations of these materials in the soil and air, in precipitation, and in lakes, rivers, and oceans.

2. We recommend a system of about 200 stations that, according to their location, would monitor air content, soil content against depth, water content against depth, dry fallout; washout by rain, bottom sediment content, and content of selected key species.

Our knowledge of existing concentrations and effects is not sufficient to make firm specifications for the system. Hence the number and location of stations should be kept under continuous review.

Monitoring Oil

An extensive discussion of the problem of oil pollution of the world's waters is included in the report of Work Group 2.

TECHNIQUES FOR MONITORING

Oil pollution is found in the ocean in at least six forms that require specific monitoring methods:

1. Heavy contamination from initial spills.

2. Thin films derived from the spreading of thick petroleum films and from biological sources. These are from 100 to 1,000 angstroms thick and are too thin to detect by methods that depend on interference effects or on the action of the refractive properties on visible or infrared electromagnetic waves. However, there are numerous physical effects of such films on the water, of which the most familiar is the smoothing of capillary ripples (Adam, 1941). This smoothing is easily photographed, particularly in the sun's glitter. At vertical incidence the rippled areas are difficult to distinguish from smooth areas. In this case, and at night, infrared detection is feasible and convenient.

3. Dispersed oil where molecules are so thinly dispersed on the water as not to be in close molecular contact. This does not constitute a film, and such aggregates are virtually indetectible *in*

situ; samples must be skimmed off for chemical or physical manipulation.

4. Lumps of oil some tens of millimeters across are formed when a thick film is broken up chemically and physically on the sea. These lumps appear to have a residence time on the sea surface of several months. They are probably consumed by biological and chemical reactions and eventually sink to the bottom.

5. Oil is found in the volume of the sea or on the bottom, either in dissolved form or absorbed to particles or bubbles.

6. Oil is found ingested in marine flora or fauna.

Of this list, the thick films are all too conspicuous and can be detected or monitored to any desired degree. They are of local distribution and occur most frequently near loading or drilling operations. Ship patrols are well suited to recovery of samples for chemical assay, while aircraft with or without air-to-sea sampling capability are convenient for rapid reconnaissance over large areas. For satellite monitoring, photographic or return-beam vidicon systems aimed at the sun's glitter path are practical, and photographic systems are already proved detectors of oil films.

The most obvious monitoring problem posed by thin films lies in the difficulty of discriminating ubiquitous thin, perhaps innocuous films, from the potentially dangerous thick ones. Methods that monitor in immediate contact with the surface or from low flight altitudes can discriminate between the two by observing interference effects (Newton rings, for example). This is a major advantage of these methods.

It does not at present appear appropriate to suggest a specific monitoring program for oil lumps on the sea (Horn, Teal, and Backus, 1970).

In the foregoing, techniques have been outlined that locate but do not usually identify the source of oil films. The latter requires physical capture of a sample for spectrographic, chromatographic, and other chemical analyses.

The methods now in use or under development for oil film identification consist of stripping it from the sea by using its adhesion to a wire screen or to a rotary drum, or of separating water from oil by centrifuging, or by scavenging as by ferric hydroxide sol. Other methods can doubtless be devised (Garrett, 1962, 1964; Harvey, 1966).

RECOMMENDATIONS

We do not make specific recommendations for widespread monitoring of oil. However, we do recommend more intensive investigations of local spills and effects. We think that any provision for global monitoring should await the outcome of base-line studies, such as the one proposed previously in this report, and of further investigation into the relation between oil and the transport and concentration of chlorinated hydrocarbons.

Monitoring Surface Changes

POTENTIAL SURFACE CHANGES

As the human population of the earth increases, greater and greater changes in the surface features of the planet take place. Forests are cleared for use as farmland, farmland is buried under the concrete of cities and roads. All the fossil and nuclear energy converted by man is ultimately dissipated as heat into the atmosphere and surface waters. These and other changes to the physical geography might ultimately affect the climate of the planet. Examples of land use or surface changes that, on a large enough scale, could have a global effect are the following:

1. Change of vegetation type. Changes from tropical forest to agricultural land may have a significant effect on the ability of the biosphere to absorb part of the carbon dioxide released to the atmosphere as a result of the burning of fossil fuel. It is also possible that the forests play a significant role in the transfer of water from the surface to the atmosphere in the tropics. Large-scale changes in the amount of forestland in these regions might have a significant effect on the general circulation of the atmosphere (see report of Work Group 1).

2. The spreading of deserts. New deserts may develop as a result of the diversion of groundwater, or existing deserts may change size as a result of man's activities on or near the deserts. Changes in desert size might affect the climate in two possible ways: (1) through the change in reflectivity of the region, and (2) through the greatly increased amounts of dust introduced into the atmosphere by the winds.

3. The filling of bays and estuaries. The tidal marshes that border bays and estuaries are extremely important to the growth of many marine species. For example, species as diverse as the shrimp

or bluefish spend a critical part of their life cycles in these regions. As the marshes are filled to provide space for housing and other activities, the species that are dependent on these areas are lost, leading to the loss of important ocean fisheries. The trend should be monitored before a multitude of local encroachments has produced an irreversible global effect.

4. Release of stored energy. This adds to the solar energy that drives the atmospheric circulation. In certain circumstances, it might lead to global as well as local climatic change. Much of the heat might be injected into rivers or estuaries, with serious effects on the life-forms existing there.

MONITORING TECHNIQUES

Land-use statistics can be derived from a multitude of sources, but over much of the world the data are probably inadequate. No system exists at present for compiling these data in a central file, for standardizing the data, and for filling in gaps in the data where they are known to exist (see report of Work Group 2). The major problems here seem to be data accuracy and obtaining data from the remote regions of the world.

A second approach to the monitoring of land-use changes is to utilize an airplane or an earth-orbiting satellite which can observe all or part of the earth. Satellite-type sensors, designed for land-use studies, are under development and are scheduled for flight during the next few years. By using multispectral scanning devices, instrument outputs that are compatible with large, high-speed digital computers can be obtained. The outputs can be interpreted not only as to whether the scene is soil, cropland, or forest but also as to type of soil, crop, or forest. Ground resolutions better than those required for climatological monitoring can be achieved now (Goldberg, 1969). Since these climatological surveys do not need the short time resolution that is required of these scanners for agricultural uses, it is quite unlikely that clouds would pose a significant problem. Surveys of 1-year duration should probably be flown at 5- or 10-year intervals. For this category the technology exists and will be in use in the near future. Costs for the instruments are of the order of $1 million (since the development costs have already been absorbed for the most part). It is quite likely that a survey could be conducted during the next few years using existing sensors and satellites. For use on a more limited geographical scale—for example, estuary studies—aircraft are of great utility.

THERMAL POLLUTION

Thermal pollution might play a significant role in certain fisheries through its effect on the usefulness of the estuary as a spawning or nursery ground for fish that are found offshore in their mature state (Snider, 1968). Since thermal pollution is caused primarily by the discharge of heat from large generating plants, it can be monitored (at least in most countries) by monitoring the level of output and sources of electrical power. Estimates (possibly crude but quite meaningful) of the effect of a plant can be made from its power output and from knowledge of its method of heat dissipation. This information can often be ascertained simply from a knowledge of plant location. Monitoring thermal pollution, then, can be most easily done by a study of national statistics.

RECOMMENDATION

We recommend monitoring surface changes by attention to land use and economic statistics and by the study of the output of satellite observations of land use and heat balance.

Monitoring Nutrients in the World's Waters

It appears to us that the release of nutrients into rivers, lakes, and estuaries will contribute to a global problem if effects in coastal waters appreciably damage world fish populations (see report of Work Group 2). The problem is really comprised of a multitude of local problems, many of which are already being monitored locally. The techniques for such monitoring are sufficiently outlined elsewhere (National Academy of Sciences, 1969).

RECOMMENDATIONS

We recommend monitoring nutrients only on a detailed, local basis. Important quantitative information can be obtained by attention to production and use data. We recommend no global or coordinated actions. We emphasize that we recognize the current and potential importance of this problem and add our recommendation that its examination should be continued and expanded.

Considerations for Implementation

No unified action whose purpose is to change the course of man's impact upon the environment can be begun without some basis for an agreement that the change is necessary.

As we have reported on our ability to identify and measure global environmental problems, we have often referred to limitations that hinder an attempt to discover or present facts and conclusions that would provide that basis for agreement. We have said that sufficient base-line data do not exist; that existing mechanisms for keeping track of man's industrial activities do not include an "environmental effects" component; that we lack an adequate model of the workings of an environmental system; that we need better instrumentation; that more monitoring stations are needed.

Though we have said that such deficiencies should be remedied and have, in many cases, suggested specific steps toward a solution, we have generally not explored the organizational possibilities for overcoming the deficiencies. Nor have we attempted to survey the numerous organizations—governmental and nongovernmental, national and international—which are already engaged in the monitoring, measurement, and research activities we have discussed.

Yet, it is our opinion that global problems are most efficiently studied and monitored on a global basis by a combination of national and international effort. We have viewed our work as the first stage of a larger exercise in exploring the dimensions, both scientific and political, of that effort. In what follows we recommend a general direction of inquiry for that exercise.

As we have said before in this report, we think that the most effective method for gathering information would be an integrated global network of fixed and mobile stations, each equipped to sample and analyze various pollutants in various subsystems of the environment. We also think that satellites can be important elements of such a network.

But we do not think that sufficient knowledge exists to establish such a system. A large amount of preliminary work must be done before any such attempt is made. We need comprehensive base-line surveys of the environment of the kind previously outlined in this report. We need a comprehensive survey of current monitoring activities and facilities around the world. We face long discussions among all those involved in measurement and monitoring of the global environment on objectives, on what should be monitored, on standards for measurement and analysis of data, and on financial support. Assuming that any system which is set up will involve some combination of national and international activities, we also need agreements concerning divisions of responsibility and control within the system.

Such a list of requirements—one which is far from comprehensive—manifests the difficulties that stand in the way of any attempt at establishing a global network, difficulties that do not even begin to take into account the genuine differences in priority given to "the problem of the environment" by various nations.

Given the problems of negotiating, planning, and setting up such a network, it may be many years before it is providing decision makers with comprehensive, integrated information about the global environment. Because of this time lag, and because of the immediate need for new kinds of information about the problems which SCEP has identified, we stress that the setting up of this network should not impede continuous and necessary expansion of current measuring and monitoring activities.

Therefore, we recommend that an immediate study of global monitoring should be instituted to examine the scientific and political feasibility of integration and to set out steps for establishing an optimal system. One major component of that study should be a consideration of how existing and expanded monitoring activities, particularly those whose main value lies in continuity and homogeneity, can be merged into the integrated system with a minimum loss of data during the transition.

We believe that the proposed study might consider the following organization principles as it develops a structure for a monitoring system:

1. As many national and international organizations as possible should be participating members of the system. Participants should be willing to modify their own monitoring activities in relation to the integrated monitoring plan.
2. The participants should define the problems to be monitored and evaluated.
3. The global monitoring system should be supervised by an

agent of the participants—a truly international body, composed primarily of scientists and engineers, which would be responsible for determining how the measurements and monitoring should be carried out, but not with *what* problem to monitor.

a. The supervisory body should not be separated either from scientific research or from the political scene; it should be able to give and take advice in both areas; its tasks and outputs must be very clearly defined. It should be obligated to maintain communications with the "user" of the monitoring information and to respond to "user" needs.

b. The supervisory body should be initially charged with as great a responsibility for determining the process by which the monitoring system would be established as possible—the fewer constraints imposed on technical operations by the initial agreements setting up the system, the better.

c. The supervisory body should have a technically competent monitoring staff to fill gaps and provide instruction.

4. The global system should contain a center or centers concerned with the indefinite maintenance of physical, chemical, and biological standards and with the control of procedures.

5. The global system should contain a "real time" data analysis mechanism. This would assure, for example, proper maintenance of measuring standards by allowing for prompt feedback to monitoring units in terms of modification of measurement parameters, levels of accuracy, and frequency of observation.

6. The implementation of the global system by the supervisory body should proceed by

a. The preparation of feasibility studies clarifying the objectives for measurement, examining the applicability of instrumentation, and completing a cost-benefit analysis of alternative approaches for solving the given problems.

b. The preparation of comprehensive plans which would, among other things:

(1) Establish fiscal hardware, and man-power requirements.

(2) Clarify the nature of the problem, whether it was to monitor a well-defined environmental problem, assess the magnitude of a suspected problem, or provide information needed to understand the global environment.

(3) Define standards and procedures for sampling data processing and analysis.

(4) Define the output of the system.

RECOMMENDATIONS

1. We recommend an immediate study of global monitoring to examine the scientific and political feasibility of integration of existing and planned monitoring programs and to set out steps necessary to establish an optimal system.

2. We recommend the expansion of current measuring and monitoring activities in accordance with our recommendations in the rest of this report to satisfy the immediate need for new kinds of information about the problems SCEP has identified.

3. We recommend that, because some components of what might ultimately be an integrated global monitoring system are so obviously needed, study and implementation of them be attempted independently of the investigation of an optimal global monitoring system. We specifically recommend a study of the possibility of setting up international physical, chemical, and biological measurement standards, to be administered through a monitoring standards center with a "real time" data analysis capability, allowing for prompt feedback to monitoring units in terms of such things as measurement parameters, levels of accuracy, frequency of observations, and other factors.

References

Adam, N. K., 1941. *The Physics and Chemistry of Surfaces* (London: Oxford University Press).

Ayres, R. U., and Kneese, A. V., 1968. Environmental pollution, *Federal Programs for the Development of Human Resources,* Vol. 2, a report submitted to the Subcommittee on Economic Progress of the Joint Economic Committee, U.S. Congress (Washington, D.C.: U.S. Government Printing Office).

Bolin, B., and Bischof, W., 1970. Variations in the carbon dioxide content of the atmosphere, *Tellus,* forthcoming.

Cadle, R. D., 1966. *Particles in the Atmosphere and Space* (New York: Reinhold).

Cadle, R. D., 1970. Atmospheric chemistry and aerosols, background paper prepared for SCEP (unpublished).

Cadle, R. D., and Allen, E. R., 1970. Atmospheric photochemistry, *Science, 167*.

Cadle, R. D., Bleck, R., Shedlovsky, J. P., Blifford, I. H., Rosinski, J., and Lazrus, A. L., 1969. Trace constituents in the vicinity of jet streams, *Journal of Applied Meterology, 8*.

Cadle, R. D., Lazrus, A. L., Pollock, W. H., and Shedlovsky, J. P., 1970. The chemical composition of aerosol particles in the tropical stratosphere, *Proceedings of the American Meteorological Society Symposium on Tropical Meteorology* (unpublished).

Cadle, R. D., and Thuman, W. C., 1960. Filters from submicron-diameter organic fibers, *Industrial and Engineering Chemistry, 52*.

Callender, G. S., 1961. Temperature fluctuations and trends over the earth, *Quarterly Journal of the Royal Meteorological Society, 87*.

Carver, T. C., 1970. Estuarine monitoring program, *Report of the Subcommittee on Pesticides of the Cabinet Committee on the Environment* (Washington, D.C.: U.S. Government Printing Office).

Citron, R., 1970. *National and International Environmental Monitoring Programs* (Cambridge, Massachusetts: Smithsonian Institution), forthcoming.

Collis, R. T. H., 1966. Lidar: a new atmospheric probe, *Quarterly Journal of the Royal Meteorological Society, 92*.

Commission for Air Chemistry and Radioactivity of the International Association of Meteorology and Atmospheric Physics of the International Union of Geodesy and Geophysics, 1970. *Report to the World Meteorological Organization on Station Networks for Worldwide Pollutants*.

Feltz, H. R., 1969. *Monitoring Program for the Assessment of Pesticides in the Hydrologic Environment* (Washington, D.C.: Water Resources Division, U.S. Geological Survey).

Flowers, E. C., McCormick, R. A., and Kurfis, K. R., 1969. Atmospheric turbidity over the United States, 1961–66, *Journal of Applied Meteorology, 8*.

Garrett, W. D., 1962. Collection of slick-forming materials from the sea, Naval Research Laboratory Report 5761; also in *Limnology and Oceanography, 10*.

Garrett, W. D., 1964. The organic chemical composition of the ocean surface, Naval Research Laboratory Report 6201; also in *Deep Sea Research, 14*.

Goldberg, I., 1969. Design considerations for a multi-spectral scanner for ERTS, *Proceedings of the Purdue Centennial Year Symposium on Information Processing* (unpublished).

Harvey, G. W., 1966. Microlayer collection from the sea surface: a new method and initial results, *Limnology and Oceanography, 11*.

Hesstvedt, E., 1970. Vertical distribution of CO near the tropopause, *Nature, 225*.

Horn, M. H., Teal, J. M., and Backus, R., 1970. Petroleum lumps on the surface of the sea, *Science, 168*.

Intersociety Committee on Methods for Ambient Air Sampling and Analysis, 1969. Tentative method of analysis for nitrogen dioxide content of the atmosphere, *Health Laboratory Science, 6*.

Junge, C. E., 1963. *Air Chemistry and Radioactivity* (New York: Academic Press).

Keeling, C. D., 1970. Is carbon dioxide from fossil fuel changing man's environment?, *Proceedings of the American Philosophical Society, 114*.

McCormick, R. A., and Ludwig, J. H., 1967. Climate modification by atmospheric aerosols, *Science, 156*.

Machta, L., 1970. Stratospheric water vapor, background paper prepared for SCEP (unpublished).

Martell, E. A., 1970. Pollution of the upper atmosphere, background paper prepared for SCEP (unpublished).

Minzner, R. A., and Oberholtzer, J. D., 1970. Space applications instrumentation systems, National Aeronautics and Space Administration Technical Report C-136.

Mitchell, J. M., 1969. Climatic change—an inescapable consequence of our dynamic environment (paper delivered at the American Association for the Advancement of Science, Boston).

Murray, W. S., et al., 1970. National pesticide monitoring program, *Report of the Subcommittee on Pesticides of the Cabinet Committee on the Environment* (Washington, D.C.: U.S. Government Printing Office).

National Air Pollution Control Administration (NAPCA), 1969. *Air Quality Criteria for Particulate Matter* (Washington, D.C.: NAPCA).

Pales, J. C., and Keeling, C. D., 1965. The concentration of atmospheric carbon dioxide in Hawaii, *Journal of Geophysical Research, 70*.

Robinson, E., and Robbins, R. C., 1969. Sources, abundance, and rate of gaseous atmospheric pollutants (Menlo Park, California: Stanford Research Institute).

Robinson, E., and Robbins, R. C., 1970. Gaseous nitrogen compound pollutants from urban and natural sources, *Journal of the Air Pollution Control Association, 20*.

Robinson, G. D., 1962. Absorption of radiation by atmosphere aerosols, as revealed by measurements at the ground, *Archiv für Meteorologie, Geophysik, und Bioklimatologie, B.12*.

Rohlich, G., ed., 1969. *Eutrophication: Causes, Consequences, Correctives* (Washington, D.C.: National Academy of Sciences). See especially Chapter IV, Detection and measurement of eutrophication.

Sand, P. F., 1970. National pesticide monitoring program, *Report of the Subcommittee on Pesticides of the Cabinet Committee on the Environment* (Washington, D.C.: U.S. Government Printing Office).

Shapley, H., ed., 1953. *Climatic Change: Evidence, Causes, and Effects* (Cambridge, Massachusetts: Harvard University Press).

Snider, G. R., 1968. Nuclear power versus fisheries (paper presented at the annual meeting of the Isaac Walton League, Portland, Oregon).

Stevens, C. M., 1970. Natural and man-produced emissions of carbon monoxide (unpublished).

West, P. W., and Gaeke, G. C., 1956. Fixation of sulfur dioxide as disulfitomercurate (II) and subsequent colorimetric estimation, *Analytical Chemistry, 28*.

Woodwell, G. M., and Whittaker, R. H., 1968. Primary production in terrestrial ecosystems, *American Zoologist, 8*.

Global Biological Monitoring Dale W. Jenkins

Introduction

Global monitoring of the life and environment of the planet is becoming more urgent as man exerts and increases his powerful influences. The severe ecological problems presently existing and the inherent and potential dangers have been fully presented in various reports.

Global environmental monitoring has been highly biased toward abiotic factors in the past. Detailed knowledge of the compositions and changes in the soil, water, and atmosphere and their contaminants are relatively useless unless we know their effects on man and the plants and animals of the world. Biological monitoring is an essential part of environmental monitoring, which is of great importance in quantifying the effects of physical and chemical environmental factors.

Living organisms are an excellent index of the amount of pollution and contamination of the environment. Plants and animals are continually exposed and act as long-term monitors. Animals show a wide range of response to environmental changes and contaminants, from subtle changes of decreased growth or change in enzyme function to violent immediate death. Plants show responses from lowered growth rate, leaf marking, to sudden death.

Three different methods of study are required to establish base lines and to survey properly and monitor the conditions of the plants and animals of the world. This requires international networks (1) using biological organisms as sentinels, biological assay organisms, and as detectors and indicators of man-produced contaminants and changes, (2) biological censuses and surveys, and (3) ecological base-line monitoring stations.

Plants and animals can serve effectively as biological monitors for pollution and environmental changes, and a variety of methods and techniques have been developed. They serve as sensitive indicators, as recorders of environmental conditions, as bioaccumulators, and as bioassay organisms. Use of biological monitors offers great potential for detecting, in the early stages, changes and undesirable trends in the biology of an environment, and in determining the effectiveness of control measures put into operation.

Prepared for this volume.

Global biological censuses and surveys of selected plant and animal species are effective ways to evaluate man's impact on the ecology of the earth. Global surveys and censuses of the most critical species are required, particularly those high in the trophic or food chain levels such as fish, birds, and mammals including man. These may be wide ranging over several continents or seas. Change in abundance, reproductive rate, or geographic range of a species indicates environmental changes that may forebode more important dangers to other species including man.

In addition to use of sensitive biological monitors and surveys and censuses of selected species of plants and animals, it is also necessary to study natural ecological communities to establish base lines.

Ecological stations are required to measure background and variations of environment and biota to determine global base lines and trends. They should be located in permanent sites away from disturbance and exposure, in protected natural areas, and representative of various biomes. Data from these stations should be compared with data from impact stations that would measure direct pollution and effects of environmental changes usually as a result of man. Impact stations should be located in or near cities, industrial areas, and river mouths.

Biological surveillance of natural plant and animal communities is an effective way to evaluate man's impact on the environment. To observe whether or not change is taking place, the nature, abundance, distribution, diversity, and condition of plant and animal components of natural communities must be known.

To make the results of biological monitoring meaningful, it is necessary to measure also the physical and chemical factors of the environment that have biological significance. In addition to natural factors, man-caused factors must be evaluated.

Selection of environmental parameters is complicated by the fact that there are about 2.5 million known chemical compounds, and each year some 500 new chemical compounds that go into widespread usage in highly industrialized countries, with little attention to long-term biological effects. This is further complicated with regard to biological monitoring since several biological organisms are often needed for an adequate evaluation of the effects of a single chemical compound. Use of standardized reference species for bioassay is required to evaluate effects of environmental variables, especially biologically active chemical compounds.

The biologically active compounds with great biological significance include biocides, such as insecticides, herbicides, and related toxic substances. These compounds usually (but not always) have some degree of physiological and ecological specificity (selectively toxic); resistance to biochemical degradation; strong tendency to sequential concentration in living organisms (in food chain or web); and a capacity for delayed onset of toxication.

These substances come from chemical control of plant and animal agricultural pests, control of disease-bearing or pest invertebrates, and industrial production of chemical compounds and waste products. Chemical monitoring of these usually complex substances in small quantities in the environment or in the bodies of organisms is a difficult and formidable task. Biological effects are often more readily and easily monitored using bioassay organisms.

Biological Monitor Organisms

Biological organisms can be used very effectively in monitoring the environment and as reference points in determining direct or indirect effects on man. The major advantages of biological monitors are (1) they integrate all environmental effects and reflect total environment including pollutants and changes; (2) they remove the extremely difficult task of relating physical and chemical measurements to biological effects; (3) they reveal rates of change in the environment; (4) they reveal trends in the environment; and (5) they show the pathways and points of accumulation of pollutants and toxicants in ecological systems.

Many species of biological organisms have great sensitivity to pollutants and act as biological monitors or indicators, sensors, or detectors. These monitors can act as "early warning" systems or devices providing immediate warning of dangerous levels of pollutants, particularly of certain air contaminants. When signs of damage begin appearing in sensitive plant indicators, we know that concentrations of air pollutants are being approached which might be poisonous to animals and man.

Monitor organisms also provide a representation or record

of the environmental conditions that have prevailed and are graphic records of the time and nature of past pollutants. By looking at certain plants it is often possible to tell which poisonous materials were present and at what time and to some extent the approximate concentration of the pollutant by the degree of damage.

The effect of a pollutant on an individual organism may be death, pathology, physiological effects such as decreased growth and stunting, illness, behavioral or performance effects, decreased reproduction, mutation, and effects on progeny. The pollutant can also cause a change in susceptibility by adaptation or by sensitization and resulting allergic reaction.

In all types of biological monitoring the specific sensitivity and range of response of the organisms is used. The different uses and objectives determine the type of monitor. Sentinels and bioassay monitors are specific organisms selected and used as bioreagents to warn or make bioassay determinations. Specific organisms occurring naturally in the environment can be used as detectors or indicators, and accumulators for monitoring pollution or change.

Study of multiple species or communities of organisms is also of great value in biological monitoring of pollution, especially the species composition, diversity of species, and change in species. A special case of importance is the predation of accumulators by predators, often at the end of a long food chain. The death of predators may be of great importance in calling attention to the presence of pollutants not otherwise discovered.

The types of biological monitors proposed are the following:
1. *Sentinels*—Highly sensitive organisms introduced into the environment as early warning devices to give an alarm of an approaching or presently dangerous situation.
2. *Bioassay monitors*—Selected organisms used as bioreagents to detect or monitor the presence and/or concentration of pollutants. The organisms may be indigenous or introduced test species.
3. *Detectors*—Individual species occurring in the environment that show a measurable response to pollutants or environmental changes such as pathology, mutation, death, change in physiology, reproduction, and so on.
4. *Indicators*—Specific species of organisms whose presence in-

dicates probability of pollution or other dangerous organisms associated with pollution.

5. *Accumulators*—Organisms that absorb and accumulate a pollutant in measurable, usually large, quantities. These species can also be used as monitors to measure the concentration of a pollutant that occurred or is present in the environment.

The criteria for selecting biological monitoring organisms are these:
1. Cosmopolitan
2. Abundant
3. Sensitive to pollutant—fragile
4. Show well-defined response
    a. change or mutate
    b. die or decrease
    c. replacement
5. Nontarget species—not object of control
6. Changes visible by remote sensing

Sentinel and Bioassay Organisms

Sentinel organisms such as the "mine canary" have been used by man as a warning where death indicated a dangerous concentration of invisible, odorless, methane. To warn of an epizootic of yellow fever in jungle areas, "sentinel" captive monkeys are put in the forest, and when they become sick and die, man is warned of the danger of the virus.

Sentinel organisms are selected on the basis of being highly sensitive to specific pollutants and exhibiting definite responses such as pathology, identifiable physiological or behavioral changes, or death. It is essential that the sentinel organisms be grown and exposed under controlled conditions to allow adequate interpretation of the responses to contaminants. An understanding of environmental interactions can be used to produce a wide range of sensitivity in monitoring organisms. For example, in plants the stomata must be open to allow most air pollutants to enter the plant.

In air pollution warning studies many species of plants have been used as sentinels. A sensitive strain of tobacco plant has been used to detect ozone 0.05 ppm/2 h, and white pine 0.06 ppm/2 h. Only 0.03 ppm/2 h of ozone combined with 0.24 ppm/2 h of SO_2

caused severe injury in tobacco plants. Tobacco, pinto beans, geranium, petunia, and begonia were used for monitoring SO_2 (0.28 ppm) and ozone (0.11 ppm). Sensitive gladiolus plants were injured by 0.1 ppb. Radish and alfalfa were sensitive to 0.1 ppm/2h of chlorine. Ethylene has been shown to damage orchid flowers at 5.0 ppb. *Cattleya* orchid flowers were abnormal after 6 hours exposure to 0.05 ppm of ethylene. Beans, petunia, and tomato showed severe damage from peroxyacatyl nitrates at 0.05 ppm/8 h and at 0.5 ppm/1 h.

Bioassay monitors may be exposed to air, water, or soil in the field or test site, or samples of air or water may be taken to the laboratory and bioassayed.

Bioassay is used in a larger sense to measure the biological effects of physical forces or fields, chemical, or biological organisms and their products. Bioassay is used extensively, and the data are usually obtained with reference to humans, but the extrapolation from plant and animal test organisms to humans is difficult and frequently subjective.

In bioassays there is a requirement to standardize all of the test conditions including the dosage, route, and time, and above all the bioreagent or test species with regard to nutrition, disease, and environment. It is often necessary to establish human dosage data by extrapolation from a number of species of test organisms. The use of bioassay to determine the toxicity of a pollutant is the most effective and accurate method of assessing potential danger. It is necessary to standardize the dosage and route of entry of the material being tested. In animal studies the route of entry may be inhalation, the eye, oral, on skin, or injected subcutaneous, intradermal, intravenous, intraperitoneal, or intracranial. Dosages may be given as a high single dose or as low doses over a period of time. Aquatic organisms may be tested as described or exposed or grown in the water. In static tests the organisms are exposed to or take up just the amount of material in a container. In flow-through tests the test organisms are exposed to the same amount of material continuously. Terrestrial plants can be exposed to air, treatment of leaves, or absorption by roots, and in the case of aquatic plants by absorption by the entire plant.

Rats are being used as air pollution monitors especially for

CO in France. Carbon monoxide is highly toxic to animals at low levels but not to plants in the concentrations observed as air pollution. The plants listed earlier as sentinels are also used as bioassay organisms for monitoring air pollution.

In aquatic tests two methods are used: (1) static bioassay in which the test solution is not changed during the period of exposure and (2) the flow-through bioassay in which the test solution is renewed continually. Toxicity data for flow-through tests are reported as median tolerance limit (TL_m). Most criteria for toxic substances must be based on a bioassay made for each specific situation dictated by lack of information and the wide variation in situations, species, water quality, and nature of pollutant.

Alternative tests for water may be carried out using one species of diatom, one invertebrate, and two species of fish.

Determination of the toxicity of known or unknown effluents can be made by exposing endemic organisms and determining the response.

Test organisms should be selected on the basis of economic importance, sensitivity, and/or importance in the food web of economically important organisms. For example, shrimp are highly sensitive to pesticides and are good bioreagents. In aquatic studies, polluted water is tested by using local species of fish as bioreagents. Stocking of streams or lakes with fish is similar to use of sentinel organisms since the effect of contaminants is discernible in the introduced fish. In using living organisms as monitors of aquatic habitats, it is more usual to select certain species of organisms as standardized or reference indicators. The polluted or contaminated water is introduced in a container with the bioassay organisms and the toxicity or effect determined. The goldfish *Carassius auratus* and the entomostracon *Daphnia magna* were used as standard test animals under controlled conditions. The goldfish is relatively resistant to poisons, injury, and anoxemia, while the *Daphnia* is more sensitive even than trout, so that the two test animals provided an effective range of sensitivity. The diatom *Nitschia linearis* was used as a standard organism, also insects, snails, and fish. The organisms are acclimated to the laboratory conditions and exposed to contaminants under controlled conditions. Snails were exposed to pollutants and after failing to respond to stimulii were returned to uncontaminated rearing water and if not recovered in forty-eight hours were determined to be dead.

The global network on insecticide resistance in disease vector insects (*Anopholes* and *Culex* mosquitoes, houseflies, and so on) is an example of an existing global network.

One hundred and seventy-five stations monitor pesticides in shellfish (oysters, mussels, clams). There are fifty stations for monitoring freshwater fishes in the continental United States, and there are fifty-seven stations in Alaska.

The monitoring of "hard" pesticides in birds is done by analysis of living tissue. Three species are being studied, the mallard duck, black duck, and the introduced starling. This method for locating pesticide "hot spots" seems effective. The New England states show the highest counts, with California second. Shell thickness, and hence hatchability, is adversely affected by some persistent pesticides.

Pesticide resistance in the common mosquito fish (*Gambusia affinis*) could be a biological parameter for measuring on a global basis. This North American fish has been introduced nearly globally for the control of mosquitoes. Usefulness of this fish is indicated by the fact that populations no more than 180 miles apart demonstrate 2,000-fold differences in degree of resistance to persistent pesticides. These differences seem to be dependent on their history of exposure.

Presence of DDT and other pesticides is being monitored in humans. It is possible to determine with whole-body counters the presence of γ-emitting radionuclides in living human beings. Cesium-137, from fallout and following nuclear explosions, is now found in everyone; before 1950 this radioelement was not a "naturally" occurring radioelement in man. Thus it is the classic example of an environmental change that has occurred within our lifetime. All humans also contain measurable levels of K^{40} and low levels of Ra^{226} and its daughter products, for these elements are truly naturally occurring radioelements.

Our network station should contain a whole-body counter and should measure the γ-rays emitted by a group of about twenty normal (that is, not involved in any nuclear industry) individuals with a frequency of once a year.

Humans could be used as monitoring organisms. One proposal is to monitor human mutagenesis by screening blood samples for fifteen proteins. Blood samples would be collected at birth. About 300,000 samples would be screened each year. Chromosome breakage would be analyzed in about 10,000 samples.

Sentinel and bioassay organisms that have been used are listed in Table 22.1. Most of these can be used nearly globally; however, some would be more suitable for tropical areas and others for temperate areas.

Table 22.1 Bioassay and Sentinel Organisms

A. Water:	B. Air:
Nitzschia linearis (diatom)	*Musca domestica*
Lemna minor (duckweed)	*Drosophila melanogaster*
	Canary
Invertebrates	Guinea pig
	Mice
Anopheles spp (larvae)	Rats
Aedes aegypti (larvae)	Primates
Mayflies, stoneflies (larvae)	Man
Asellus aquaticus	
Daphnia magna	**Plants**
Artemia salina	
Balanus	Tobacco
Shrimp	Gladiolus
Snails	Orchids
Mytilus edulis	Geranium
Crassostrea gigas	Begonia
	Tomato
Fish	Petunia
	Pinto bean
Salmo salar (salmon)	Radish
Fundulus (killifish)	Peas
Betta splendeus (fighting fish)	
Lebistes reticulatus (guppy)	**C. Soil**
Salmo gairdnerii (salmon)	
Leuciscus rutilus	Soil microorganisms
Ameiutus nebulosus (bullhead)	
Euponotis sp (sunfish)	cellulose decomposers
Carassius auratus (goldfish)	nitrogen fixing bacteria
Gambusia affinis (top minnow)	total microorganisms (respiration)
	Soil mites
	Collembola
	Tenebrionid beetles
	Lumbricus terrestris (earthworm)
	Plants
	sensitive to soil conditions
	bioaccumulators

A homogeneous, highly susceptible strain of *Drosophila melanogaster* is used for bioassay of DDT in soil. Slightly less than 10 ppm of p,p' DDT can be detected in twenty-four hours.

Detector, Indicator, and Accumulator Organisms

In addition to using organisms as introduced sentinels or as bioassay monitoring organisms, observations of individual indicator or accumulator species, naturally occurring in the environment being studied, provide valuable information on pollution or environmental changes.

Detectors directly detect the presence of pollutants and other changes in environment. The specific types of pathology and other responses frequently identify the specific pollutant.

Indicators are specific species of organisms whose presence indicates probability of pollution or of other dangerous organisms associated with pollution. The most common example is *Eschericia coli,* which is measured to determine the sanitary quality of water and to determine probable presence of other fecal coliform bacteria that may be pathogenic. Different levels of *E. coli* count have been standardized with reference to water for drinking, swimming, and fishing.

Accumulators are organisms that take up and accumulate a pollutant in measurable, usually large, quantities. These species can be used as monitors to measure the concentration of a pollutant that occurred or is present in the environment.

If organisms have low sensitivity or response, they can accumulate pollutants or even concentrate them. Some organisms deposit environmental materials such as calcium or silicon for protective shells. Organisms can also detoxify, divert, complex, or remove contaminants. A common attribute of many microorganisms is to convert various contaminants, especially heavy metals, to other compounds.

Biological magnification is a chronic effect of toxic pollutants (such as heavy metals, pesticides, radionuclides, bacteria, and viruses). Many organisms have the ability to remove substances from the environment and store them in their tissues in nontoxic levels. In animals, if the toxic substances are released in their bloodstream, they will cause death. Sex products may be formed that contain sufficiently high levels of pollutant so that normal development of the young is impossible. Biological magnification and storage of toxic substances in herbivorous and carnivorous animals at lower trophic levels may gradually build up and kill higher trophic level predators.

DDD was applied to control gnats in a California lake at 0.2 mg/l. Sequential concentration of DDD was shown in a food web. In plankton DDD was found as high as a 265-fold increase; in visceral fat in frogs and carp, 2,000-fold; bluegills to 12,500-fold; bullhead fish to 135,000-fold; in grebes, 80,000-fold; in largemouth bass, 85,000-fold, and in whitefish, up to a 118,750-fold increase over the maximum DDD applied. DDD could not be detected in the lake water two weeks after application.

Monitoring should be undertaken of the types and quantities of biocides and other toxic chemicals transmitted by various means and reaching nontarget species. Bioaccumulators such as oysters and other mollusks and plants can be useful as indicators, and the levels of concentrations in predatory animal species are of special importance.

When a persisting chemical compound is made available to biological networks in ecosystems, its movements and biological effects may involve many species, locations, and degrees of effect. Among the benefits the world has received from the use of DDT, high value should be placed on the increased knowledge and appreciation that have been gained of the global unity of the biological network, and therefore, of the global effects of local chemical usages.

Selection of indigenous detector, indicator, and accumulator organisms for use in monitoring stations is difficult since the indigenous fauna and flora are different in various parts of the world. Abundant, cosmopolitan, and well-known species have been suggested in Table 22.2, based on common usage and background data on sensitivities or ability to accumulate compounds.

Biological Surveys and Censuses

Survey and census of the biota are important and very difficult since judicious selection is required in determining the kinds of organisms to survey. Biological populations change in response to climatic conditions, food supply, social interactions, disease, or unknown factors in addition to man-introduced contaminants. It is therefore essential to monitor the physical and chemical envi-

Table 22.2 Indigenous Detector, Indicator, and Accumulator Organisms

Eschericia coli	Carp
Phytoplankton diatoms	Shark
Phytoplankton desmids	Smelt
Algae—marine attached	Anchovies
Cladonia rangiferina	Tuna
Tree lichens	
Pine	Gull
Astragalus	Cormorant
Mosses	Ducks
Collembola	Grebe
Mayfly, stonefly, dragonfly	Hawks
Musca domestica	Owls
Aedes, Anopheles	Starling
Blatta, Pediculus	Pheasant
Daphnia magna	
Shrimp	Microtus
Crabs	Vole
Mussel	Rabbits
Mytilis edulis	Reindeer
Lumbricus terrestris	Carnivores
Pike	Seals
Trout	Whales
Salmon	Humans

ronment near the living organisms to be able to correlate the causes for population changes and responses.

Biological monitoring requires a broad global network of surveys and censuses of the most critical species, particularly those high in the trophic or food chain levels such as fish, birds, and mammals including man. These may be wide ranging over several continents or seas.

Biological surveys are broad geographic surveys of the population size, reproductive success, and of the health of species of living organisms. Surveys can often be accomplished best by census of the organisms at certain times when they are breeding, nesting, overwintering, or migrating. Commercial or other harvest data are of extreme value for shellfish, fish, game birds, and mammals.

Plant food crop and lumber yield production figures can be used if the conditions and harvest data are understood.

Predatory fish, birds, and mammals are particularly subject to being killed by pesticides since they are high in the trophic levels of food chains and webs. They are victims of biological magnification of pesticides which results from bioaccumulation of the pesticides at each trophic level until very large quantities are found concentrated at the highest levels. This makes fish, birds, and mammals excellent bioassay indicators of the amount of pesticide in the environment. These organisms should be carefully surveyed and regular censuses are required.

Frequently in nature, the presence of a contaminant is not discovered except by decreased numbers of predators dependent on a certain species. Also, the decrease of a species may be less conspicuous than the rapid increase of a more tolerant replacement species. Any rapid, major change in species populations should be a warning to investigate the cause.

The study of rare and endangered species is of great importance in world environmental monitoring since they are usually highly vulnerable to man's activities and pollutants.

Fish

Fish are adversely affected by pesticides, particularly DDT, the aldrin group of chlorinated hydrocarbons, heavy toxic metals such as mercury, and severe eutrophication. Loss of higher level predators results in large populations of undesirable forage fish. Predator fish accumulate high levels of pesticide and become unsafe as human food.

At the present time commercial catch records are available for shellfish and fish throughout much of the world, particularly oceanic species. Major decreases in catch or reports of major kills and mortality should be reported. Fish are excellent monitors of the estuaries and oceans, and populations and reproduction success should be carefully studied.

Fish catch records include anchovy, menhaden, hake, sardine, smelt, herring, tuna, mackerel, flounder, cod, haddock, perch, sturgeon, and many others. In addition to commercial catches in the ocean, estuaries, rivers and lakes, creel catch records from the sports fishing industry and various fish surveys of breeding areas can be used. Observations of fish schools using various techniques including infrared sensing from aircraft and ships are being used. Study of rare and endangered species of fish is an important area to consider in relation to environmental changes and pollution.

Commercial shellfish harvests are reported mainly from continental shelves and shorelines, estuaries, bays, and seas. These include oysters, clams, scallops and other mollusks, including squid and octopus. The edible crustacea such as crab, shrimp, and lobster are also harvested and catches reported. Shellfish are highly vulnerable to pollution by biocides and toxic heavy metals and are excellent biological indicators. The mollusks are also biological accumulators and can be used for bioassay.

Birds

Birds are perhaps the best biological indicators of changes in the environment. At present surveys are made of bird populations, especially game birds in North America, Europe, Japan, and other countries. Breeding or nesting surveys are made of songbirds or nongame birds as well as upland game birds and ducks, geese and aquatic game and nongame birds. Censuses are made of wintering bird populations, especially where they are concentrated in specific overwintering grounds. Surveys of migrating birds are also being made regularly. Censuses are taken of the hunting kills of game birds such as pheasant, quail, grouse, ducks, geese, and many other species.

It is necessary to compile all of these data, to standardize census methods, and to regularly report all data on bird populations, reproduction, harvests, and observed mortality and kills. Of special importance are censuses of carnivorous and raptorial birds such as hawks, eagles, and related birds.

Bird banding and migration studies also contribute to knowledge on abundance and activity of migratory and other birds. Special reports are published on bird kills due to oil, DDT, PCBs, and other toxic materials.

The monitoring of DDT and chlorinated hydrocarbon pesticides in birds is done by analysis of living tissue. Three species are being studied in the United States, the mallard duck, black duck, and the introduced starling. This method for locating pesticide "hot spots" seems effective.

Mammals

Mammals are very good biological indicators. Rodent populations are highly variable and subject to extreme population cycles. Surveys should be made of the rodent population including rabbit and squirrel harvests. Wild ungulates are of importance, and deer, antelope, caribou, wild sheep, wildebeest, and other ungulate surveys and game harvests should be compiled as well as reproductive success data.

Carnivorous predatory animals are of special importance since they accumulate contaminants at the end of food chains. These animals are usually well known in the cat family but less observed in the weasel and related carnivorous groups. Trapping and fur harvest data and fur industry records should be carefully analyzed.

Mammal surveys also include wintering and breeding population surveys and dry season surveys as at watering holes during migrations.

Harvest data are available for ungulates including deer, elk, antelope, caribou, wildebeest, and other African ungulates, and llama. Harvest of small game such as rabbits, squirrels, and related rodents are also reported. Data are also available for certain big game mammals, seals, and whales.

The fur catch of weasel, mink, fox, wolf, sable, beaver, bear, seal, and a large variety of other animals is reported. Additional data are available on collections for zoos, pets, and research animals. Banding of migratory mammals, especially bats, provides additional data.

The study of rare and endangered species of mammals is of great importance to global environmental monitoring.

Plants

Surveys of plants and vegetation should be carried out on a broad scale. Changes in growth rate, damage, and morphological changes such as leaf spotting, browning, or other evidences of contaminants as well as killing should be observed. Certain species of plants are known to be sensitive to specific contaminants and should be particularly observed such as orchids, white pine, tobacco, beans, and other wild and cultivated plants.

Yield and harvest data are of value if the growing and treatment conditions are fully known so that the data can be analyzed.

Various plants are known to be highly sensitive to low levels of different air pollutants. Over 200 species of wild and cultivated plants have been shown to be sensitive and damaged by various pollutants. These are listed and illustrated. Specific plants such as lichens, orchids, and certain grasses and legumes are known to be highly sensitive and cosmopolitan in distribution. These and other selected plants should receive special study.

Biological Monitoring of Ecological Base-line Stations
A global network of ecological base-line monitoring stations should be established that will show general changes of major significance to the flora and fauna of the world. These stations should be located in representative biome regions on land and in estuaries and the ocean. In addition to biological monitoring, they should be designed to monitor general changes in climate including possible man-made changes in temperature, rainfall, and solar radiation as well as the presence of various man-induced contaminants and pollutants.

Ecological base-line monitoring stations should be located in the following biomes: deciduous forest, conifer forest, tropical forest, savanna, thorn scrub, grasslands, desert, tundra, estuaries, and various ocean areas. About twenty terrestrial stations should be located over the various continents with at least two stations for each biome type. There should probably be about ten estuary and ten open ocean stations.

The terrestrial base-line stations should be permanent, protected natural areas with a long record of ecological and meteorological study. The monitoring stations should be composed of an intensively studied site with a large surrounding area relatively free from human intervention and direct contamination.

A meteorological station and a contaminant sampling station should be located near the base-line station to permit correlation of biological results with physical and chemical measurements.

Global monitoring and surveying require a number of global base-line stations that measure the backgrounds and variations of environmental parameters and biota. It is essential to measure those factors that characterize the heat and water balance of the atmosphere and earth's surface. It is necessary to measure changes in physical and chemical composition of the atmosphere, water, and soil, particularly those changes with marked biological effects.

Detailed ecological studies should be continued in the base-line stations, and regular intensive surveys and censuses should be taken of the population, reproductive successes, and health or condition of the biota. No major man-caused changes should be permitted in the surrounding area or of the watershed in which the station occurs. The results of this type of station can be compared with impact stations designed to study direct biological effects of man including urban- and industrial-caused changes.

Study of multiple species or communities or organisms is of great value in biological monitoring of pollution, especially the species composition, diversity of species, and change in species. A special case of importance is the predation of accumulators by predators, often at the end of a long food chain. The death of predators may be of great importance in calling attention to the presence of pollutants not otherwise discovered.

Monitoring of ecological base-line stations involves detailed studies of species composition, species diversity, abundance, density, genetic changes, pathology, mortality, and reproductive success. These should be correlated with measurements of the physical and chemical environment. Any major changes should be studied carefully such as the following: drastic increase in the population of some unexpected species component; reduction in species diversity; or decrease in numbers or disappearance of typical species or groups of species for the particular ecosystem. In short, any marked change in ecosystem structure should be taken as a warning signal that some new factor (for example, pollution) may be influencing the system.

Base-line studies and monitoring of the biome stations are made difficult by the requirement for minimum disturbance. In the base-line or bench-mark control area few plants or lower animals should be collected and no vertebrate animals should be killed. This makes quantitation of many species, particularly insects, very difficult. Light-trap collectors are known to decimate local populations of insects in an area. The control area should be studied by maximum field observation and minimum collection.

In the surrounding area at a suitable distance, some collections can be made of particular plants, invertebrates, and other animals. No larger vertebrates should be killed which have wide territories or ranges such as ungulates, hawks and owls, and most carnivores. All dead animals should be collected, especially those in higher trophic levels, and assayed for presence of biocides and cause of death.

Various indigenous organisms are highly sensitive detectors or sentinels while others are bioaccumulators. These should be observed more frequently and studied in greater detail for determining low concentrations and effects of pollutants.

Various plants are known to be highly sensitive to low levels of different air pollutants. Over 200 species of wild and cultivated plants have been shown to be sensitive and damaged by various pollutants. Specific plants such as lichens, orchids, and certain grasses and legumes are known to be highly sensitive and cosmopolitan in distribution. These and other selected plants should receive special study in the base-line control area.

Biocides, particularly insecticides, are most effective against certain sensitive insect species and predatory insects which are exposed to higher levels through biological magnification. Selected groups of insects should be studied in detail in the area surrounding the control area. Predatory fish, birds, and mammals are particularly subject to being killed by pesticides since they are high in the trophic levels of food chains and webs.

The use of species diversity as a measure of the perturbation of an ecosystem shows the following: (1) There is a drastic increase in the population of some unexpected species component of the system. (2) There is a reduction in species diversity. (3) Typical species or groups of species for the particular ecosystem decrease in numbers or disappear. Any marked change in ecosystem structure should be taken as a warning signal that some new factor (for example, pollution) is influencing the system.

In aquatic situations, the aquatic population that occurs in a given area is a representation or indicator of environmental conditions which have prevailed during the life history of the organisms comprising the population. It is this property indicating past environmental conditions, especially the extreme conditions of brief duration, that make aquatic populations valuable indicators of pollution. The best indicators are diatoms, various aquatic insects, mollusks, and fish. Diatoms occur throughout the world and are sensitive to change and have been used effectively for monitoring pollutants and environmental changes.

Bottom organisms have proved a sensitive indicator of pollution in the Great Lakes. In Lake Erie, in the period of 1930 to 1961, mayfly larvae decreased to 1 percent of their original numbers. Other indicator organisms showed comparable decreases.

Because each species has its own requirements for oxygen, nitrogen, and other organic and inorganic compounds in the water, certain algae, water insects, and some microscopic organisms are reliable indicators of the degree and nature of pollution. The identification of species present in a particular location provides a rapid and inexpensive monitoring system for detecting pollution.

Diatoms, which are unicellular plants, are very useful in monitoring pollution in lakes, rivers, and estuaries. They would also be useful in monitoring pollution in the open sea. Diatoms belong to the large group of plants known as algae. They are a very important food source for most forms of aquatic life that feed upon plants. They also carry out the process of photosynthesis which is so important in the generation of oxygen needed by all organisms in order to carry out metabolic processes.

Monitoring of the marine environment is an enormous and difficult problem. It is necessary to select the critical environmental and biotic parameters, establish priorities, and develop baseline information on present contamination of the marine environment and effects on the biota, and establish monitoring stations.

The major environmental parameters of global importance which should be monitored are:

1. Hydrocarbons—Chlorinated hydrocarbon insecticides including DDT; aldrin group—dieldrin, toxaphene, chlordane, endrin, and heptachlor; Polychlorinated biphenyl plasticizers; crude oil and petroleum products.
2. Toxic metals—Hg, Pb, As, Cd.

The most important of these environmental factors requiring monitoring are DDT and aldrin-group insecticides. Determination of the presence and amount of these parameters in seawater is very difficult. Biological monitoring is considered to be the most effective and efficient monitoring method due to bioaccumulation of chlorinated hydrocarbons and toxic metals.

Biological monitoring would include study of species that are known to be sensitive or killed at low levels of pesticides such as sea trout, shrimp, and jack mackerel. Other monitoring studies would include the bioaccumulators such as mollusks, oysters, clams, and predatory fish and birds that show high levels of the insecticide or toxic metal.

Monitoring stations should be located in strategic places characteristic of the oceans and estuaries that will permit extrapolation to the world marine environment. Oceanic base-line stations should be established in critical areas in open ocean such as upwelling areas, major fishing regions, and representative of the different oceans and seas. The stations should be located at sites that have been studied extensively for a period of time such as ocean stations, ocean buoys, and sites of extensive oceanographic cruise studies. The Ocean Station sites PAPA off Alaska, the CALCOFI, International Indian Oceanographic Program, IGOSS, GARP, and other sites and study areas offer suitable initial sites.

Some stations should be located in coastal shore areas, seas, and larger estuaries if these are not too polluted to consider as protected natural areas. Much more study is necessary before final selection of oceanic and coastal stations.

Acknowledgments

It is a pleasure to acknowledge the help of many scientists who are sincerely concerned with determining the present and future effects of the presence and activities of man on the life and environment of the world. This report is an attempt to organize the ideas on biological monitoring generated by several of my colleagues at the Summer Study of Critical Environmental Problems sponsored by the Massachusetts Institute of Technology during July 1970. It also summarizes the biological discussions of various committees and the task force on the Global Network for Environmental Problems of the International Biological Program, National Academy of Science, National Research Council during 1969 and 1970. I would like to especially thank Dr. Bengt Lundholm of Sweden and Dr. Frank Blair, Mr. Robert Citron, Mr. William Gusey, Dr. Glenn Hilst, Dr. Ruth Patrick, and Dr. Robert Rowland.

23
Identification of Globally Distributed Wastes in the Marine Environment

Edward D. Goldberg and M. Grant Gross

A base-line set of measurements is proposed to identify the distributions and concentrations of man-introduced materials in the atmosphere and ocean. Such a program can define present or potential threats to the continued survival and well-being of marine ecosystems.

Attention is focused on atmospherically transported materials for several reasons:
1. River-carried solids are extremely diverse in type and are deposited primarily near the river mouth.
2. Background concentrations of particles introduced by rivers are high and can mask the contributions of man.
3. Airborne materials are widely dispersed over oceanic areas within days or weeks after entry into the atmosphere.

To simplify the difficult problems of designating those materials that may pose problems to our environment, we have grouped them into three categories:
1. Minerals and solid phases from the combustion of fossil fuels and from industrial activities.
2. Elements introduced to the atmosphere in amounts that are of the same order of magnitude as those brought to the world's rivers by natural weathering processes.
3. Synthetic organic chemicals.

Table 23.1 makes it clear that a modern industrial society now competes with natural processes as sources of materials to the environment. Although the data are poor in quality, they suggest that the burning of the fossil fuel will have a significant impact on the environment. Since energy production is accompanied by the emission of many other materials that may be detrimental to the environment, some measures of fossil-fuel combustion seems essential. Further, by using a tracer of the paths of materials introduced to the atmosphere by energy generation, we can more readily ascertain the fates of other materials simultaneously introduced.

Elemental carbon (soot, cokey balls) formed by fossil-fuel

Prepared during SCEP.

Table 23.1 Particles Mobilized into the Atmosphere (Grams/year)

A. Natural processes	
Continental rock debris	$10^{14} - 10^{15}$
Sea salt	10^{15}
Volcanic emanations	10^{14}
B. Industrial, agricultural, and social releases (U.S.)	
Total	10^{14}
Incineration	1×10^{12}
Transportation	1×10^{12}
Fuel combustion	9×10^{12}
Industrial	8×10^{12}
Agricultural	10×10^{12}
Miscellaneous	10×10^{12}

Source: Study of Critical Environmental Problems, 1970.

combustion are disseminated over the earth's surface by winds. Rains over England contain appreciable amounts of elemental carbon which covaries with their sulfate contents; both materials were presumed to have an origin in the burning of coal (Gorham, 1955). Atmospheric dusts collected over the Atlantic Ocean contain elemental carbon, attributed to fossil-fuel burning in the eastern United States (Parkin, Phillips, and Sullivan, 1970).

Elemental carbon can be determined in atmospheric dusts and rains as part of the base-line program. Further, the historical record of fossil-fuel burning most probably is held by the annual ice strata preserved in glaciers.

The simultaneous measurement of sulfate and carbon would give a broad picture of the fates of fossil-fuel emissions in the atmosphere. Sulfur initially introduced as sulfur dioxide and later oxidized to sulfate may have a somewhat different path in the atmosphere than that of carbon. To confirm the fossil-fuel origins of the elemental carbon and sulfate isotopic analyses of the sulfur and carbon could also be used.

The relative contribution of gas and oil to that of coal may be found in assays for the aluminosilicate mineral mullite made during the burning of coals. Mullite is abundant in fly ash and is not normally found in rocks. Being rare in the environment, it may well be useful as a tracer of coal combustion.

Most of the particles generated by the processes included in Table 23.1 are large and therefore fall out close to their sources.

Smaller particles with dimensions of microns or less may have long residence times in the atmosphere and may travel long distances before entering the ocean, where they can further affect plant and animal populations.

Another group of chemicals encompasses the elements that are injected into the atmosphere at annual rates comparable to those in the major sedimentary cycle. In the latter case, compounds of the element are brought to the oceans by rivers. The chemical forms introduced by man may be different from those naturally present, and they may have an unusual size distribution or shape if they come in as solids. In any case, they can create new chemical surroundings for the marine organisms. Five metals are given in Table 23.2 that fall within this classification.

Lead emitted to the atmosphere from the combustion of lead tetraethyl and lead tetramethyl in gasoline constituent accounts for about 20 percent of the total world lead consumption. The vanadium and nickel come from burning of fuel oil, in which both elements are naturally bound in porphyrin complexes. Mercury is introduced to the atmosphere primarily through stack gases from caustic/chlorine plants where it is used as an electrode in the dissociation of saline solutions. Some mercury escapes to our immediate surroundings through its used as a biocide, in agricultural and industrial applications.

It is estimated that about half of man-induced emissions of cadmium to the atmosphere result from processing cadmium ores;

Table 23.2 Mobilization of Elements at the Earth's Surface (Grams/year) Man-made emissions for mercury and lead are on a worldwide basis. For the other metals, U.S. values are given.

Element	Natural emission (a)	Industrial emission	Approximate value (f)
Cd	3.1×10^8	2.0×10^9 (b)	$10,000,000
Hg	5.0×10^9	4.0×10^9 (c)	50,000,000
Ni	2.0×10^{11}	7.0×10^9 (b)	15,000,000
Pb	2.1×10^{11}	4.0×10^{11} (d)	20,000,000
V	2.8×10^{11}	5.0×10^9 (e)	2,000,000

Sources:
(a) Goldberg, forthcoming
(b) NAPCA, 1970
(c) Klein and Goldberg, 1970
(d) D. D. Patterson, personal communication
(e) SCEP, 1970
(f) U.S. Bureau of Mines, 1967.

cadmium is also lost in the mining and processing of zinc, lead, and copper ores. Another source of cadmium to the U.S. atmosphere is incineration of waste products containing cadmium. This is thought to be equal to the loss resulting from the original processing.

The amounts of sulfur may be sensed from the following argument, 95 percent of the synthetic organic chemicals are produced from petroleum. This requires that 5 percent of the annual world production of petroleum, 50 million tons per year in the United States, goes into such an activity. In addition there is a substantial loss of hydrocarbons through fuel combustion, transportation, industrial processes, solid waste disposal, and so on. Such dispersions are given in Table 23.3.

Several criteria are suggested for use in selecting synthetic organic substances for inclusion in a base-line program:
1. The material is produced in large quantities.
2. The material interferes with metabolic activities of one or more forms of life.

Table 23.3 Organic Chemicals—Some Production and Potential Emission Data for United States

Substance	Present Production Rate (grams/year)	Emission Rate (grams/year)
Total amounts of synthetic organics	5×10^{13}	
DDT	2×10^{11}	
PCBs	5×10^{9}	
Dry-cleaning solvents, perchloroethylene, and trichloroethylene	2×10^{11}	2×10^{11}
Freon-12 (CCl_2F_2)	1×10^{11}	1×10^{11}
Gasoline into atmosphere		2×10^{12}
Hydrocarbons lost in transportation, fuel combustion, industrial processes, solid waste disposal, and so forth		3×10^{13}
Petroleum introduced directly into marine environment by man's activities		3×10^{12}

Source: Estimated by E. D. Goldberg from unpublished data.

3. The material is concentrated by one form of life upon which it may have no discernible effect, but which is food for another form of life upon which it can have a most deleterious effect. The passage of DDT and its residues through the food chain and their subsequent effect upon bird and fish populations illustrates such a case.

4. Its physical form interferes with an organism's growth activity or wellbeing. For example, asbestos fibers and siliceous particles cause damage to man's lungs.

5. The material, or its decomposition products, is alien to the marine environment but similar to substances that have known teratogenic, mutagenic, toxic, or carcinogenic effects. For example, the dry-cleaning solvent, trichloroethylene, may affect enzyme activities as do the heavier chlorinated hydrocarbons.

A sense of priority for the inclusion of synthetic organic chemicals in a base-line program may be established with such criteria combined with the chemical and biochemical characteristics of a given substance. This suggests special attention be given towards halogenated organics.

The 3 million tons per year of petroleum introduced directly into the marine environment provide a special case of a substance to be measured. Much of this material, introduced in coastal waters, estuaries, harbors, and shipping lanes, may accumulate in the air-water interface to form slicks. Such a veneer over the ocean may be a site of uptake of nonpolar organic materials transported about the atmosphere or the surface ocean.

Hydrocarbons appear to be a major organic emission to the environment and can be globally dispersed by wind systems. Since Northern Hemisphere emissions vastly exceed those from the Southern Hemisphere and since most industrial activity is at midlatitudes, it is necessary to focus attention upon mid-latitudinal oceanic areas in the Northern Hemisphere. With halogenated hydrocarbons, entry through the atmosphere to the ocean likely involves a reservoir at the air-water interface. The natural substances forming such a surface film include fatty acids and their alcohols, while the man-introduced materials consists of oil from leakages and spills. Such a surface film may pick up many of the other organic compounds translocated from the continents to the oceans via the atmosphere. Thus, we should direct attention to the analysis of a variety of organic compounds in surface films, such as DDT and its residues, freon, and gasoline hydrocarbons.

References

Goldberg, E. D., forthcoming. *Chemical Oceanography*, vol. 1, edited by J. P. Riley and G. Skirrow (New York: Academic Press).

Gorham, E., 1955. On the acidity and salinity of rain, *Geochimica et Cosmochimica Acta, 1:* 231–239.

Klein, D., and Goldberg, E. D., 1970. Mercury in the marine environment, *Environmental Science and Technology, 4:* 765–768.

National Air Pollution and Control Administration (NAPCA), 1970. Unpublished NAPCA document (Durham, North Carolina).

Parkin, D. W., Phillips, D. R., and Sullivan, R. A. L., 1970. Airborne dust collections over the North Atlantic, *Journal of Geophysical Research, 75:* 1782–1792.

Study of Critical Environmental Problems (SCEP), 1970. *Man's Impact on the Global Environment* (Cambridge, Massachusetts: The M.I.T. Press).

U.S. Bureau of Mines, 1967. *Minerals Yearbook* (Washington, D.C.: U.S. Government Printing Office).

24
Proposal for a Base-Line Sampling Program

Edward D. Goldberg,
Geirmundur Arnason,
M. Grant Gross,
Frank G. Lowman, and
Joseph L. Reid

Introduction

There exists a paucity of data on the present concentrations of potentially toxic pollutants in the marine environment. Such base-line data are essential in identifying needed research. While perhaps desirable in the future, a global monitoring program requiring great expenditures and technological advances is an inexpeditious approach to lacunae requiring immediate attention. A simplified base-line sampling program, however, meets that need within reasonable financial and technological bounds. Moreover, such a program can serve as a foundation for designing a more ambitious monitoring system.

Guided by such concerns, we propose a one-year program to gather approximately one thousand samples from the following components of the environment: wind systems; current systems; organisms; rivers; glaciers, rain, and deep-ocean sediments.

Wind Systems

Introduction

To design a network of sampling stations at which to measure concentrations of gaseous and particulate pollutants in the atmosphere, one must consider the prevailing global wind systems in relation to the locations of major sources of airborne pollutants. One must also identify these mechanisms by which pollutants are transported laterally and vertically and estimate the thickness of the layer within which appreciable vertical mixing occurs.

It is to be noted that most sources are located on the North American, European, and Asian (including Japan) continents and are situated in a predominantly westerly flow. Furthermore, waste-bearing effluents are usually released near the ground at

Prepared during SCEP.

low speeds and are therefore unable to penetrate far up into the atmosphere, unlike volcanic eruptions that inject matter into the stratosphere. Finally, it should be borne in mind that we are primarily concerned here with the atmospheric part of the biosphere, a very thin layer bounded by the earth's surface.

Prevailing Wind Systems

1. The tradewinds (30° N to 30° S): These are easterly winds near the earth's surface but *decrease* in strength with increasing height and become westerly except near the equator (where they retain the same direction at all heights).

2. The westerlies (30° N to 70° N, 30° S to 65° S): The westerly winds increase in strength up to a height of 10 km and more.

3. The polar easterlies (70° N to 90° N, 65° S to 90° S): These are shallow winds which decrease with height and reverse their direction at about 3 km.

4. The major wind systems are modified by seasons. Warming and cooling of the continents cause the monsoon winds. North-south oriented mountain ranges cause large-scale meandering of the otherwise zonal flow.

Lateral and Vertical Transport Mechanisms

Among those processes that mix airborne wastes in the atmosphere are the following:

1. Small-scale turbulence;
2. Local circulations such as ordinary convection and severe storms;
3. Extratropical cyclones within the westerlies; and easterly waves, tropical storms, and hurricanes within the tropical easterlies.

Thickness of the Mixing Layers

Toba (1965) reports that the number of sea salt particles in the size range 1- to 10-μ radius decreases quite rapidly with height; for the smallest of these particles (1-μ radius), the decrease is approximately tenfold from the sea surface up to a height of 6,000 m. For submicron-size particles one may expect a somewhat smaller decrease with height.

For this base-line study it is suggested that one may not generally need to make measurements above 5 km in the westerlies. The corresponding height is somewhat lower in the transitional zone and the polar easterlies.

Proposal for a Base-Line Sampling Program

Table 24.1 Number of Sampling Stations

Location	Surface	500 m	1,500 m	5,000 m	Frequency	Number
North America	15	7	4	2	2	56
Europe and Asia	20	10	5	3	2	76
Africa	3		3	3	1	9
South America	3		3	3	1	9
North Atlantic	5		3	3	1	11
South Atlantic	3		3	3	1	9
North Pacific	7		4	4	1	15
South Pacific	3			3	1	6
Australia	1	1	1		1	3
Arctic	1			1	1	2
Antarctic	1			1	1	2
Subtropical high-pressure cells	2			2	1	4
Total						202

Geographic Location of Airborne Pollutant Sources
1. United States—a major source, located within the westerlies;
2. Western Europe, USSR, China, Japan—major sources located within the westerlies;
3. Africa—a minor source, located within the trade winds and to a lesser extent within the westerlies;
4. South America—a minor source, located within the trade winds and the westerlies;
5. Australia—a minor source, within the trade winds.

Where and When to Sample?
Based on the foregoing, the following sample network is suggested:
1. Frequency: 1 to 2 samples per year
2. Geographic distribution of the sampling stations (see Table 24.1).

Irregular Observations
It is suggested that additional samples be obtained near the sea surface by means of research and commercial ships, and at greater heights by means of special-purpose or commercial aircraft. Aircraft observations may provide the only observations needed above 5,000 m.

Ocean Current Systems

Pollutants of potential importance in the ocean occur in a variety of forms (gases, solutes, particles, oils) and may enter through various paths (atmosphere, rivers, ships, and so on). It is suggested therefore that a base-line set of water samples be collected. These must be spaced both horizontally and vertically in such a way as to detect any of these wastes. Both the physical circulation of the ocean and the distribution effected by biological transport must be taken into account, so that a proper framework for assessment of the distributions in the oceans, by whatever paths and transformations, can be established.

Sampling methods, means of treatment for preservation, and transport for analysis are not discussed here. When such problems are confronted, and the suggestions for sampling of other parts of the environment compared, some modifications can be expected.

A tentative framework for base-line sampling of ocean water is outlined here. This framework involves three kinds of vertical spacing: deep array, shallow array, and surface film. These require seven, three, and one samples, respectively. In reverse order, the surface film involves one sample at the surface; the shallow array involves samples from the surface film, the mixed layer, and the pycnocline; the deep array involves samples as for the shallow array plus samples from about 1,000 m, 2,000 m, 3,000 m, and near the bottom.

For the deep array, samples should be taken in the centers of the major gyres and in the major mediterranean seas (Arctic, Gulf of Mexico, Mediterranean, Okhotsk, Bering, and Sea of Japan). With a central equatorial array for each ocean, the total is 20 deep arrays of 7 samples each, or 140 samples.

For the shallow array, samples should be taken in the major eastern and western boundary currents (about 16 positions) and at the eastern and western edges of the three oceans at the equator. This totals 22 arrays with 3 samples each, yielding 66 samples.

In addition to the 42 surface film samples indicated for the deep and shallow arrays combined, a set of perhaps 50 samples should be collected, spaced generally over the ocean but with emphasis on the major zones of precipitation and evaporation, and along the major shipping routes (especially for oil).

The program outlined here is of necessity intuitive. Before attempting to collect and analyze the 256 samples, it might be worthwhile to look at a much smaller number as soon as possible. With the data from such a survey, a more effective and less expensive base-line grid might be developed.

Organisms

The expense of marine sampling limits the number of organisms that can be collected for a worldwide program designed to provide a base-line measurement of the amounts of pollutants in the marine biosphere. The following considerations are significant for planning a sampling program for marine organisms:
1. Geographical distribution of primary biological productivity
2. Geographical distribution of world fisheries
3. Sites of major river discharges
4. General air circulation patterns
5. Sites of large-scale vertical water movements, especially upwelling
6. Sites of centripetal centers of major water gyres
7. Desired coverage of marine populations of organisms
8. Ease of sampling
9. Cost

Collection of adequate samples of marine organisms is important in areas of high primary productivity and major fisheries since several of these areas (1) are near cities and sites of pollution production, (2) support relatively large marine biomasses that may function as reservoirs with slow turnover rates for various wastes, and (3) provide pathways for wastes leading to man through his utilization of marine fishery products as food.

The world catch of marine food exceeds 50 million metric tons per year, is unevenly distributed throughout the ocean, and includes a great variety of marine species. Many pollutants toxic to man are concentrated by many marine organisms. Furthermore, pollutants present at extremely low concentrations often exhibit log-normal frequency distributions in apparently homogeneous populations of marine organisms. To minimize the effect of such variable concentrations, edible parts of ten to twenty individuals of a selected species collected at a given site should be combined to form a homogenized, composite sample. Composite samples will be collected for demersal and pelagic fish, mollusks, and crustacea. Also, net plankton will be collected at selected sites.

Major areas of high productivity (Figure 24.1) include the antarctic oceans, the North Atlantic banks, other shallow areas and of less importance, the outflows of rivers. The landing of fish in millions of metric tons (1967) were as indicated in Table 24.2.

The fisheries of the N. E. Atlantic, W. Central Pacific and S. E. Pacific accounted for 61 percent of the 1967 marine fishery and, with the N. Pacific and N. W. Atlantic, for more than 80 percent of the total. More than 65 percent of the Atlantic catch was taken north of 30° N. latitude, while a significant part of the total sources of pollution for this ocean was also located north of the same parallel.

Most of the major fisheries are on or near continental shelves, close to the outflows of the world's major rivers. The stations of the sampling pattern shown in Figure 24.2 are located mainly on the continental shelves at sites of high productivity, major fisheries, the outflows of major rivers and at the sites of down and upwelling of water. In addition, sampling stations are placed in the Mediterranean Sea and off the southern shores of Africa,

Table 24.2 Landing of Fish, 1967 (millions of metric tons)

	10^6 tons	Percent
N. W. Atlantic	4.0	7.6
N. E. Atlantic	10.2	19.5
W. Central Atlantic	1.3	2.5
E. Central Atlantic	1.6	3.1
S. W. Atlantic	1.3	2.5
S. E. Atlantic	2.5	4.8
N. Pacific	6.4	12.2
W. Central Pacific	10.5	20.0
E. Central Pacific	0.7	1.3
S. E. Pacific	11.2	21.5
S. W. Pacific	0.4	0.8
W. Indian Ocean	1.3	2.5
E. Indian Ocean	0.8	1.5

Source: Holt, 1969.

383 Proposal for a Base-Line Sampling Program

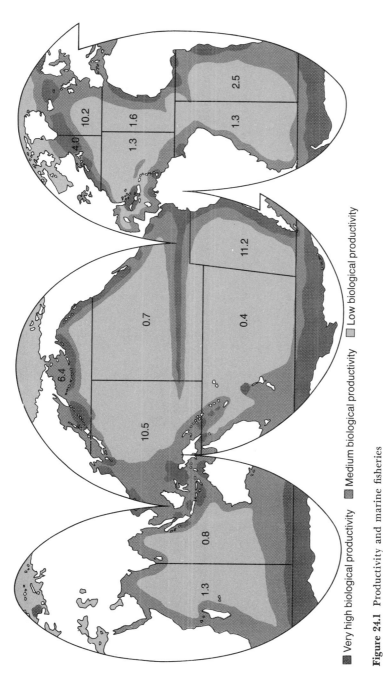

Figure 24.1 Productivity and marine fisheries
Source: Derived from Holt, 1969; Isaacs, 1969. The number that appears within each oceanic subdivision indicates the quantity (in metric tons) of fishes landed in 1967.

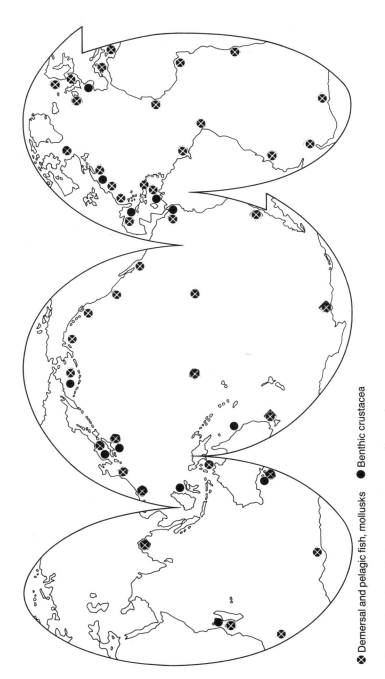

Figure 24.2 Fish, mollusk, and crustacea sampling stations

⊗ Demersal and pelagic fish, mollusks ● Benthic crustacea

Australia, and South America. This plan calls for the collection of 42 composite samples each of demersal fish, mollusks, and pelagic fish and 12 composite samples of benthic crustacea—a total of 138 samples.

Except for two sampling stations in the Central Pacific, downcurrent from the upwelling area off the Pacific Coast of South America, central oceanic areas are not sampled. The great average depth of these open ocean sites precludes sampling of demersal organisms, and the populations of pelagic fishes are greatly reduced. Although tunas and other fishes are sometimes taken in these areas, the lack of appropriate catch data from fishing boats operating in the central oceans, and the limited knowledge of the migration habits of these fishes complicate the interpretation of data derived from these samples.

Two types of samples may be collected from the central oceanic regions, to include net plankton and flying fish. Plankton may be taken on a predetermined collection grid, and sufficient flying fish normally come aboard research vessels operating in these regions at night to supply adequate samples for the central ocean areas. A total of 142 plankton samples and 40 flying fish (or more, if available) should be collected from the Atlantic, Pacific, and Indian oceans and the Mediterranean Sea (see Figure 24.3).

In summary, the suggested program for base-line studies of marine organisms consists of the following:

	No. of samples
Fish	124
Mollusks	42
Crustacea	12
Plankton	142
Total	320

Rivers, Estuaries, and Continental Shelves

Rivers are important routes by which waterborne wastes reach coastal oceans. Identification of major routes and reservoirs of such wastes requires that river water, riverborne sediment, and related sediment deposits be included in base-line studies. For the

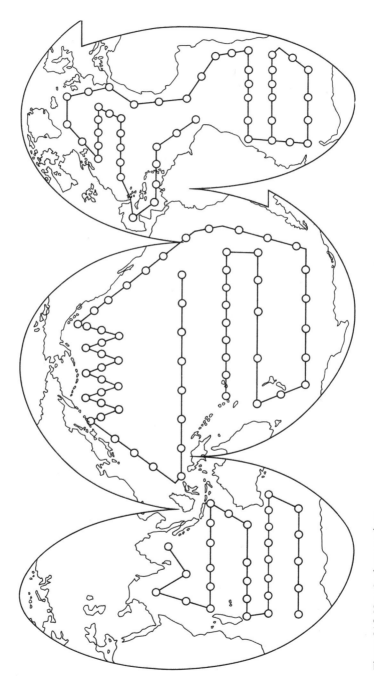

Figure 24.3 Net plankton stations

initial study, a limited number of areas are suggested. Criteria used for their selection are as follows:
1. Rivers
    a. large river with large drainage basin
    b. extensive human activity
    (1) industrial
    (2) agricultural
    c. different climatic zones
    d. different stages of industrial development
2. Continental Shelf Areas
    a. accumulation areas for modern sediments
    b. areas used for other activities, such as fisheries
    c. near rivers affected by man's activities, industrial or agricultural

Water and sediment samples should be composited in space and time. This procedure permits an estimate of concentrations for quasi-steady-state releases and should work well for dissolved (or dispersed) constituents. For rivers, two sampling periods are suggested, each of approximately six months' duration. One period would sample water and sediment moving during low-flow conditions, the other during high-flow conditions.

Sediment transport by rivers is extremely variable. In most rivers, sediment moves primarily during the few weeks of maximum river flow. Hence composite samples provide minimal data on annual sediment-transport phenomena.

River water and sediment samples should be collected in volumes proportional to river flow and composited over a six-month period. One sampler located near the river surface (perhaps on a float) would collect river water; another near the bottom would collect water and sediment.

Four samples a year from 19 river systems (Table 24.3) total 76 samples. The number of river samples could profitably be doubled to provide more information about time variability in such systems.

Special samples should be collected from recently deposited sediments at the head of the estuary for each of the major rivers. For example, a sample might be taken from the navigation channel farthest from the ocean. The sample should be typical of

Table 24.3 Rivers Included in Base-Line Study

North America	Europe
Hudson	Rhine
Mississippi	Danube
St. Lawrence	Po
Columbia	Thames
South America	**Asia**
Orinoco	Ganges-Brahmaputra
Amazon	Yellow (Hwang Ho)
Plata	Amur
	Ob
Africa	Indus
Nile	
Congo	
Niger	

riverborne sediment and wastes accumulating in the estuary (the materials most likely to be removed by dredging or washed out by floods). This provides another 19 samples.

To understand long-term movements of particulate matter out of a river, it is necessary to study deposits on the adjacent continental shelf or associated delta. Some areas are suggested for study because they are associated with major rivers and others (such as the Japan Sea or two Japanese bays) because of industrial activities in their regions. The limited number of samples (approximately 50) should be composited to permit estimates of concentrations in recently deposited sediment. Eleven continental shelf areas, deltas, and harbors would therefore be included in the study:

Osaka Bay
Tokyo Bay
Ganges-Brahmaputra Delta
Mississippi Delta
Nile Delta
Rhone Delta
Baltic Sea
Black Sea
Japan Sea
North Sea
Yellow Sea

Glaciers, Rains, and Deep-Ocean Sediment Samples

The most appropriate set of samples with which to measure the rates of atmospheric entry of pollutants to the environment over the past several hundred years is contained in the permanent snowfields (glaciers). At present, there are no well-documented techniques for making such historical studies on river influxes of materials (except possibly for varved sediment in isolated fjords). Glaciers exist over a wide range of latitudes, permitting an examination of fallout from all of the major wind systems. Time of layer formation can be determined by radiometric dating (Pb-210), firn stratigraphy, or oxygen isotopic stratigraphy. The variations in oxygen isotopic compositions of glacial waters allow yearly ice accumulation to be subdivided into summer and winter layers. Such a subdivision may be useful in isolating sources of introduced materials, as wind directions vary seasonally. The examples shown in Table 24.4 are glaciers that might be assayed (Windom, 1969; reprinted by permission of the Geological Society of America).

A sampling program of these six zones is proposed using samples of glacial ice that accumulated at known times. For ex-

Table 24.4 Glacial Compositions for Sampling Program

Glacier	Wind System Involved	Rates of Ice Accumulation (cm/year)
Greenland	Northern Hemisphere polar easterlies	63
Yukon Territory or Mount Olympus, Washington	Northern Hemisphere westerlies	40–65
Mount Orizaba or Mount Popocatepetl, Mexico	Northern Hemisphere trades	6–11
Andean Glaciers	Southern Hemisphere trades	
Tasman Glaciers, New Zealand	Southern Hemisphere westerlies	
Antarctica	Southern Hemisphere polar easterlies	10

ample, one such set might include ice from 1970 (summer and winter), 1969 (summer and winter), 1968 (summer and winter), 1965, 1960, 1953, 1943, 1930, 1900, and 1800. This permits study of recent variations and long-term trends in airborne materials.

For each of the trade and westerly systems in the Northern and Southern hemispheres, composite rain samples (perhaps a year's collection of rainwater) collected at single locations in both the Pacific and Atlantic oceans would be most useful to compare with the glacial results as well as to give a measure of precipitation washout.

The total number of samples proposed:

78 permanent snowfields
48 rain samples
———
126

Deep sea sediment accumulates at extremely slow rates, ranging from fractions of a millimeter up to centimeters per thousand years. These deposits are composed of rock debris from continents, volcanic materials and their degradation products, minerals precipitated from seawater (such as the ferromanganese minerals and barite) as well as animal and plant remains.

Sediment deposits are often disturbed by burrowing organisms and near-bottom currents. Such phenomena disturb (smear) the record, diminishing its resolution. Thus, the utility of an extensive sampling program for deep-ocean sediment is questionable.

On the other hand, the biological remains, calcareous and siliceous tests, do provide a measure of the removal of materials from surface waters to the sediments by biological agencies. A suite of 4 samples each of siliceous and calcareous deep-ocean sediment (ooze) from the Pacific, Indian, and Atlantic oceans can provide a most reasonable set of base-line materials. Total sediment samples: 12.

References

Holt, S. J., 1969. The food resources of the world, *Scientific American, 221*, September, pp. 180–181.

Isaacs, J. D., 1969. The nature of oceanic life, *Scientific American, 221*, September, pp. 152–153.

Toba, Y., 1965. On the giant sea-salt particles in the atmosphere. *Tellus, 17:* 131–145.

Windom, H. L., 1969. Atmospheric dust records in permanent snowfields: implications to marine sedimentation. *Geological Society of America Bulletin, 80.*

25
International Environmental Monitoring Programs
Robert Citron

Introduction

This report includes outlines of the major international environmental monitoring programs that utilize intergovernmental or international nongovernmental organizational machinery to coordinate their activities. There are a number of national and multinational environmental monitoring programs that are regional or global in scope but do not operate through international organizational structures. These programs are not included in this report but will be described in another document (Smithsonian Institution, 1970).

For the past several years a number of national and international groups and organizations have discussed the need for a global network for environmental monitoring.* It has become clear that if man is to improve the quality of life on this planet, he must have a much better understanding of his impact on his environment. A global environmental monitoring network would provide base-line information and a continuous flow of data from which man-modified ecosystem changes could be detected, measured, and assessed.

The purpose of this report is to outline the characteristics and status of the major international environmental monitoring programs that are now operational or are in their planning stages. While these preliminary outlines are not comprehensive, they do give some idea of the magnitude and scope of current and planned international monitoring programs. These activities range from specific environmental monitoring and research programs that cover relatively small regional areas to very comprehensive biosphere monitoring and research programs that are global in scope and multidisciplinary in character.

Environmental Monitoring Program sheets contained in this report include information on program purpose, status, coverage, duration, participating countries, coordinating and cooperating organizations, principal disciplines involved, principal phenomena monitored, number of monitoring stations, physical

Prepared for SCEP.
* Listed in References.

sphere, time response, sampling frequency, data acquisition techniques, data storage and retrieval facilities, communications facilities, and research center facilities. This report also includes outlines of developing proposals for the establishment of a Global Network for Environmental Monitoring (GNEM), a World Environmental Institute (WEI), an International Center for the Environment (ICE), and a Global Network for Monitoring the Biosphere (MABNET). Monitoring programs in operation are found in Table 25.1; monitoring programs planned, in Table 25.2; and proposed monitoring programs, in Table 25.3. Finally, there is a list of abbreviations for all these organizations.

Operational Monitoring Programs
1. World Weather Watch (WWW)
2. Monitoring Sun-Earth Environment (MONSEE)
3. International Biological Programme (IBP)
4. International Hydrological Decade (IHD)
5. Environmental Health Surveillance and Monitoring Programmes
6. Endangered Ecosystem Monitoring Programs
7. Cooperative Investigations of the Caribbean and Adjacent Regions (CIGAR)
8. Antarctic Research Programs (ARP)
9. Tsunami Warning System (TWS)
10. Cooperative Study of Kuroshio and Adjacent Regions (CSK)
11. Monitoring of Basic Environmental Climate (MBEC)
12. Monitoring Background Air Pollution (MBAP)
13. WMO Research Programme, Monitoring Solar Radiation (MSR)
14. WMO Research Programme, Monitoring Isotopes in Precipitation (MIP)
15. WMO Research Programme, Monitoring Atmospheric Electricity (MAE)
16. WMO Research Programme, Monitoring Atmospheric Ozone (MAO)
17. Hydrometeorological Monitoring for Hydrological Purposes (HMH)
18. Meteorological Services to Air Transport (MSAT)

Table 25.1 Operational Monitoring Programs

INTERNATIONAL MONITORING ACTIVITIES PROGRAM INFORMATION SHEET			
NAME OF PROGRAM		WORLD WEATHER WATCH (WWW)	
PURPOSE OF PROGRAM		A global observing, telecommunication and processing system to make available to each member country of WMO the basic meteorological and related geophysical environmental information they require. It is the basic system providing support for the WMO Programme on the Interaction of Man and his Environment and the WMO Research Programme.	
STATUS	FULLY OPERATIONAL		
COVERAGE	GLOBAL, REGIONAL AND NATIONAL		
DURATION	LONG-TERM, INDEFINITE		
NO. OF COUNTRIES PARTICIPATING	133		
COORDINATING ORGANIZATION		WORLD METEOROLOGICAL ORGANIZATION	
COOPERATING ORGANIZATION(S)		FAO (See also IGOSS)	
PRINCIPAL DISCIPLINES		METEOROLOGY AND HYDROMETEOROLOGY	
PRINCIPAL PHENOMENA MONITORED		At various levels in atmosphere, over land and sea; temperature, pressure, geopotential, moisture, wind, precipitation, hydrometeors, sea-surface temperature, sea sub-surface temperature as relevant to atmospheric forecasting models.	
NO. OF MONITORING STATIONS	THOUSANDS (GOS)		
PHYSICAL SPHERE	ATMOSPHERE: AIR-LAND, AIR-SEA INTERFACE		
TIME RESPONSE	REAL AND DELAYED		
SAMPLING FREQUENCY	INTERVALS OF 3,6 OR 12 HOURS		
DATA ACQUISITION TECHNIQUES		Land stations (direct and remote reading), balloons, rockets, ships, buoys, aircrafts, satellites.	
DATA STORAGE AND RETRIEVAL FACILITIES	EXTENSIVE (NATIONAL REGIONAL AND WORLD WEATHER CENTERS (GDPS)		
COMMUNICATIONS FACILITIES		EXTENSIVE NATIONAL, REGIONAL, AND GLOBAL (GTS)	
RESEARCH CENTER(S)		NONE: NATIONAL METEOROLOGICAL SERVICES AND RESEARCH CENTRES.	
REMARKS		Existing scheme is being continuously developed in the light of new technology and scientific developments - Major attention is being given to assisting developing centres.	

REFERENCES
1. *The Essential Elements of the World Weather Watch*, World Meteorological Organization, October, 1966.
2. *Activities and Plans of the World Meteorological Centres*, World Meteorological Organization, March 1967.
3. *World Weather Watch: The Plan and Implementation Programme*, World Meteorological Organization, May 1967.
4. *World Weather Watch: First Status Report on Implementation*, World Meteorological Organization, July 1968.
5. *Catalogue of Meteorological Data for Research*, World Meteorological Organization, 1965.

Additional information may be obtained from: World Meteorological Organization, 1211 Geneva 20, Case Postale No. 1, Switzerland.

Table 25.1 (continued)

Abbreviation or symbol	Meaning	Number of stations
AGRIMET	Agrometeorological station.	55
ATMEL	Atmospheric electricity measurements	15
ATMOS	Atmospherics location by narrow-sector direction-finder	7
AUT	Automatic station or observation made by automatic equipment	13
AUR	Visual aurora.	43
CLIMAT (C)	Station for which monthly climatological means of surface elements are transmitted.	984
CLIMAT (T)	Station for which monthly climatological means of upper-air elements are transmitted.	93
CLIMAT (CT)	Station for which monthly climatological means of both surface and upper-air elements are transmitted.	306
EVAP	Evaporation measurements.	1,117
H	Hourly observations — The letters are followed by figures showing the hours during which the observations are made (e.g. H 00-24 or S 0630-1830)	
S	Half-hourly observations	3,275
HU/FC	Hurricane, tropical cyclone or typhoon forecast centre.	21
ICE	Ice observations.	79
IONOS	Ionospheric observations.	7
LIT	Lightning counter	18
MAGNET	Magnetic observations.	17
METAR	Aviation routine weather report.	105
M/B	Station making reports of sudden changes	1,252
MONT	Observations of cloud below the level of the station	92
NEPH	Nephoscope observations	414
NLC	Noctilucent cloud.	58
NOCTRA	Nocturnal radiation measurements	10
OZONE	Ozone observations.	32
PH	Phenological observations	252
RAD	Radiation measurements.	52
RAREP	Weather radar report.	9
RECCO	Aircraft reconnaissance flights	6
ROCOB	Rocket-sonde observations.	9
RSD	Radar storm and meteorological phenomena detection.	234
SEA	State-of-sea observations.	275
SEA/SWELL	Sea and swell observations.	75
SEATEMP	Sea temperature measurements.	106
SEISMO	Seismological observations.	175
SFERIC	Atmospherics detection by cathode-ray direction-finder.	29
SKYRA	Sky radiation measurements	48
SNOW	Snow survey.	105
SOILTEMP	Soil temperature measurements.	792
SOLRA	Solar radiation measurements.	231
SPECI	Aviation selected special weather reports	193
SUNDUR	Sunshine duration measurements	1,610
SWELL	Swell observations.	8
TIDE	Tide observations.	92
TI/WA/FC	Tidal wave forecast centre	7
TOTRA	Total radiation measurements.	211

Table 25.1 (continued)

INTERNATIONAL MONITORING ACTIVITIES PROGRAM INFORMATION SHEET		
NAME OF PROGRAM		MONITORING SUN-EARTH ENVIRONMENT (MONSEE)
PURPOSE OF PROGRAM		Planning and Coordination of observations and data exchange on monitoring the solar-terrestrial environment.
STATUS	OPERATIONAL	
COVERAGE	GLOBAL	
DURATION	INDEFINITE	
NO. OF COUNTRIES PARTICIPATING	71	
COORDINATING ORGANIZATION		Inter-Union Commission on Solar-Terrestrial Physics (IUCSTP)
COOPERATING ORGANIZATION(S)		CIG, IUGG, IUPAP, URSI, IAU, COSPAR, WMO*
PRINCIPAL DISCIPLINES		SOLAR-TERRESTRIAL PHYSICS
PRINCIPAL PHENOMENA MONITORED		Solar and interplanetary phenomena, ionospheric phenomena, flare associated phenomena, geomagnetic phenomena, aurora, cosmic radiation, air glow.
NO. OF MONITORING STATIONS	HUNDREDS	
PHYSICAL SPHERE	SUN, SPACE, ATMOSPHERE	
TIME RESPONSE	REAL & DELAYED	
SAMPLING FREQUENCY	CONTINUOUS & PERIODIC	
DATA ACQUISITION TECHNIQUES		Satellites, rockets, ground stations.
DATA STORAGE AND RETRIEVAL FACILITIES	MODERATE (WORLD DATA CENTERS)	
COMMUNICATIONS FACILITIES		NONE: USE MILITARY, COMMERCIAL, NASA & GTS NETWORKS
RESEARCH CENTER(S)		NONE: NATIONAL AND REGIONAL RESEARCH CENTERS
REMARKS	The World Data Centers also collect data on glaciology, meteorology, oceanography, seismology, paleomagnetism, volcanology, geochemistry and geothermics.	
REFERENCES 1. STP Notes No. 6: Guide for International Exchange of Data in Solar-Terrestrial Physics, IUCSTP Secretariat, October 1969. 2. Annals of the IQSY: Sources and Availability of IQSY Data, International Council of Scientific Unions. 3. World Data Center A: Catalogue of Data, March 1970. 4. World Data Center A: Catalogue of Data, Feb. 1968. 5. World Data Center A: Catalogue of Data on Solar-Terrestrial Physics, June 1969. 6. Guide to International Data Exchange (IQSY Disciplines), IQSY Committee, International Council of Scientific Unions, 1963. 7. Shapley, A. H. Chairman, Working Group on "Monitoring of the Solar-Terrestrial Environment" of the Inter-Union Commission on Solar-Terrestrial Physics (IUCSTP), Letter of 23 June 1970. Additional information may be obtained from: Dr. A.H.Shapley, Environmental Science Services Administration, Research Laboratories, Boulder, Colorado, 80302. *Studies are going on about how WMO may support this monitoring system.		

Table 25.1 (continued)

INTERNATIONAL MONITORING ACTIVITIES PROGRAM INFORMATION SHEET			
NAME OF PROGRAM			INTERNATIONAL BIOLOGICAL PROGRAMME (IBP)
PURPOSE OF PROGRAM			World-wide study of the biological basis of productivity and human welfare; organic production on land, sea and fresh water, potential uses of natural resources, and human adaptability to change.
STATUS		OPERATIONAL	^
COVERAGE		GLOBAL	^
DURATION		1964-1972	
NO. OF COUNTRIES PARTICIPATING			58
COORDINATING ORGANIZATION			SPECIAL COMMITTEE FOR THE IBP (SCIBP)
COOPERATING ORGANIZATION(S)			IUCN, UNESCO, FAO, IGU, IUBS, IUGG, WHO, IUPS, IUB, IUNS
PRINCIPAL DISCIPLINES			ECOLOGY, BIOLOGY, GENETICS, ANTHROPOLOGY, DEMOGRAPHY, SYSTEMS ANALYSIS.
PRINCIPAL PHENOMENA MONITORED			Terrestrial productivity, production processes, terrestrial conservation, freshwater productivity, marine productivity, and human adaptability.
NO. OF MONITORING STATIONS		HUNDREDS	^
PHYSICAL SPHERE		BIOSPHERE	^
TIME RESPONSE		DELAYED	
SAMPLING FREQUENCY		CONTINUOUS, PERIODIC	
DATA ACQUISITION TECHNIQUES			Survey and Sampling and experimentation.
DATA STORAGE AND RETRIEVAL FACILITIES			MODERATE: PRIVATE & NATIONAL CENTERS. FRESHWATER AND MARINE DATA HELD BY FAO FISHERY DATA CENTRE, ROME.
COMMUNICATIONS FACILITIES			NONE
RESEARCH CENTER(S)			NONE: NATIONAL FACILITIES
REMARKS			Plans call for the establishment of data centres, development of plans for a Global Network for Environmental Monitoring, and the continuation of important themes through the UNESCO Man and the Biosphere Programme, SCOPE, IUCN, and other agencies.
REFERENCES			
1. *IBP News No. 2: The Scientific Plan of IBP*, International Council of Scientific Unions, Special Committee for the International Biological Programme, February, 1965. 2. *IBP News No. 9:* International Council of Scientific Unions, Special Committee for the International Biological Programme, June 1967. 3. *International Biological Programme 1970 Review*, International Council of Scientific Unions, Special Committee for the International Biological Programme, September 1970. 4. Series of 15 handbooks on methodology published - others in press. Additional information may be obtained from: International Biological Programme, Central Office, 7 Marylebone Road, London N.W. 1, England.			

Table 25.1 (continued)

INTERNATIONAL MONITORING ACTIVITIES PROGRAM INFORMATION SHEET			
NAME OF PROGRAM		INTERNATIONAL HYDROLOGICAL DECADE (IHD)	
PURPOSE OF PROGRAM		Improve understanding of global hydrologic cycle to strengthen scientific base for water use, management, and conservation.	
STATUS	OPERATIONAL		
COVERAGE	GLOBAL		
DURATION	1965-1974		
NO. OF COUNTRIES PARTICIPATING		105	
COORDINATING ORGANIZATION		IHD Co-ordinating Council (UNESCO)	
COOPERATING ORGANIZATION(S)		ICSU (IASH, IAH, IBP) WMO, FAO, IAEA, WHO	
PRINCIPAL DISCIPLINES		Hydrology	
PRINCIPAL PHENOMENA MONITORED		River discharge, lake level changes, evaporation, ground water fluctuations.	
NO. OF MONITORING STATIONS	HUNDREDS		
PHYSICAL SPHERE	WATER SYSTEMS		
TIME RESPONSE	DELAYED		
SAMPLING FREQUENCY	CONTINUOUS, PERIODIC, SPORADIC		
DATA ACQUISITION TECHNIQUES		Monitoring water flow and water level changes.	
DATA STORAGE AND RETRIEVAL FACILITIES	NONE		
COMMUNICATIONS FACILITIES	NONE		
RESEARCH CENTER(S)	NONE (NATIONAL AND REGIONAL CENTERS)		
REMARKS	Standardizing methodology		

REFERENCES: 1. IHD Secretariat Report-Contribution of UNESCO to the Decade Programme (1965-1969) Doc. SC/HYMIDEC/25 2. Final Report of the International Conference on the Practical and Scientific Results of the International Hydrological Decade and on International Cooperation in Hydrology, Paris 8-16 December 1969, Doc. SC/MD/18. 3. Information on the International Hydrological Decade, U. S. National Committee for the International Hydrological Decade, Division of Earth Sciences, U. S. National Academy of Sciences, National Science Council, January 1968. 4. Your Contribution to the IHD, U. S. National Committee for the IHD (as above). 5. Water, Man, and the IHD, U. S. National Committee for the IHD (as above). 6. International Field Year for the Great Lakes, U. S. National Committee for the IHD and Canadian National Committee for the IHD, National Research Council of Canada. 6. Heindl, L.A. Executive Secretary, U.S. National Committee for the IHD, The U.S/IHD Program: The Half-Way Mark, IHD Bulletin, No. 11, in EOS, Vol. 50, No. 10, pp 566-570, October 1969. Additional information may be obtained from: U.S. National Committee for the IHD, National Research Council, National Academy of Sciences, 2101 Constitution Ave., Washington, D. C. 20418.

Table 25.1 (continued)

INTERNATIONAL MONITORING ACTIVITIES PROGRAM INFORMATION SHEET			
NAME OF PROGRAM			ENVIRONMENTAL HEALTH SURVEILLANCE AND MONITORING PROGRAMMES
PURPOSE OF PROGRAM			1. Communicable Diseases 2. Adverse Effects of Drugs 3. River Basin Pollution 4. Air Pollution 5. Radiation
STATUS	OPERATIONAL		
COVERAGE	REGIONAL		
DURATION	INDEFINITE		
NO. OF COUNTRIES PARTICIPATING		30	
COORDINATING ORGANIZATION			WORLD HEALTH ORGANIZATION (WHO)
COOPERATING ORGANIZATION(S)			IAEA, VARIOUS NATIONAL AUTHORITIES
PRINCIPAL DISCIPLINES			ENVIRONMENTAL HEALTH, TOXICOLOGY, COMMUNICABLE DISEASES, COMMUNICATIONS SCIENCE
PRINCIPAL PHENOMENA MONITORED			1. Communicable Diseases 2. Adverse Drug Effects 3. River Basin Monitoring (water) 4. Human Bone Strontium-90 Content 5. Suspended Dust, SO_2
NO. OF MONITORING STATIONS		DOZENS	
PHYSICAL SPHERE		MAN, ATMOSPHERE RIVER BASINS	
TIME RESPONSE		DELAYED	
SAMPLING FREQUENCY		PERIODIC SPORADIC	
DATA ACQUISITION TECHNIQUES		Sampling	
DATA STORAGE AND RETRIEVAL FACILITIES		MODEST; (INTERNATIONAL REFERENCE CENTRE, HQ COMPUTER)	
COMMUNICATIONS FACILITIES		RADIO, CABLES	
RESEARCH CENTER(S)		DOZENS	
REMARKS			Plans call for the establishment of a global health monitoring and information network.

REFERENCES
1. *The Problems of the Human Environment*, memos on Global Environmental Health Monitoring, Office of Science and Technology, WHO, March and June 1970 (restricted).
2. *Water Pollution Control in Developing Countries*, Report of a WHO Expert Committee, World Health Organization, Technical Report Series No. 404, 1968.
3. *Water Pollution Control*, Report of a WHO Expert Committee, Technical Report Series No. 318, World Health Organization, 1966.
4. *Measurement of Air Pollutants, Guide to the Selection of Methods*, M. Katz, World Health Organization, 1969.
5. *WHO Annual Reports*

Additional information may be obtained from: World Health Organization, 1211 Geneva 27, Switzerland.

400   Robert Citron

Table 25.1 (continued)

INTERNATIONAL MONITORING ACTIVITIES PROGRAM INFORMATION SHEET Suggested alternate name: Conservation of Nature and Natural Resources			
NAME OF PROGRAM			ENDANGERED ECOSYSTEM MONITORING PROGRAMS
PURPOSE OF PROGRAM			Population monitoring of threatened species (flora & fauna), ecosystem status of National parks and reserves; record environmental information on unique ecosystems.
STATUS	OPERATIONAL		^
COVERAGE	GLOBAL		^
DURATION	INDEFINITE		^
NO. OF COUNTRIES PARTICIPATING		DOZENS	^
COORDINATING ORGANIZATION			INTERNATIONAL UNION FOR THE CONSERVATION OF NATURE (IUCN)
COOPERATING ORGANIZATION(S)			UNESCO, FAO, ECOSOC, ICBP, IWRW
PRINCIPAL DISCIPLINES			ECOLOGY, CONSERVATION
PRINCIPAL PHENOMENA MONITORED			Endangered flora, fauna, and ecosystems.
NO. OF MONITORING STATIONS		DOZENS	^
PHYSICAL SPHERE		FLORA, FAUNA	^
TIME RESPONSE		DELAYED	^
SAMPLING FREQUENCY		PERIODIC, SPORADIC	^
DATA ACQUISITION TECHNIQUES			Survey and sampling.
DATA STORAGE AND RETRIEVAL FACILITIES		SMALL	
COMMUNICATIONS FACILITIES		NONE	
RESEARCH CENTER(S)		NONE: PRIVATE OR NATIONAL	
REMARKS				Red data book programs, National Park and Reserve programs. Programs developing for measurements of specific pollutants of biological significance, sociological parameters of environmental quality, and endangered landscapes.
REFERENCES			
1. This is IUCN, International Union for Conservation of Nature and Natural Resources, August 1968. Additional information may be obtained from: International Union for Conservation of Nature and Natural Resources, 1110 Morges, Switzerland.			

Table 25.1 (continued)

INTERNATIONAL MONITORING ACTIVITIES PROGRAM INFORMATION SHEET		
NAME OF PROGRAM		COOPERATIVE INVESTIGATIONS OF THE CARIBBEAN AND ADJACENT REGIONS (CICAR)
PURPOSE OF PROGRAM		Improvement of knowledge of Caribbean waters, basic structure, marine life, and harvestable food resources.
STATUS	OPERATIONAL	
COVERAGE	REGIONAL	
DURATION	1970-1972	Colombia, Cuba, Fed. Rep. of Germany, France, U.S.S.R., and Venezuela, Jamaica, Netherlands, Trinidad, Tobago, and U.S.A.
NO. OF COUNTRIES PARTICIPATING	11	
COORDINATING ORGANIZATION		INTERGOVERNMENTAL OCEANOGRAPHIC COMMISSION (IOC)
COOPERATING ORGANIZATION(S)		ICSPRO, ESPECIALLY UNESCO, FAO, WMO, and NATIONAL ORGANIZATIONS
PRINCIPAL DISCIPLINES		Physical and Biological oceanography
PRINCIPAL PHENOMENA MONITORED		Ocean processes, physical structure, and marine life.
NO. OF MONITORING STATIONS	DOZENS	
PHYSICAL SPHERE	OCEAN	
TIME RESPONSE	DELAYED	
SAMPLING FREQUENCY	CONTINUOUS AND PERIODIC	
DATA ACQUISITION TECHNIQUES		Sea-borne survey, measurements and monitoring.
DATA STORAGE AND RETRIEVAL FACILITIES		NATIONAL CENTERS NODC (U.S.A.) DESIGNATED AS REGIONAL DATA CENTRE.
COMMUNICATIONS FACILITIES		MODEST
RESEARCH CENTER(S)		NATIONAL RESEARCH CENTERS

REFERENCES
1. Intergovernmental Oceanographic Commission, Summary Report of its Sixth Session, UNESCO, June 1970.
2. Global Ocean Research, Report of a Joint Working Party of the Advisory Committee on Marine Resources Research, the Scientific Committee on Oceanic Research, and the World Meteorological Organization, June 1969.
3. Intergovernmental Oceanographic Commission, Resolutions Adopted at the VIth Session, UNESCO, September 1969.
4. International Co-ordination Group for CICAR, Summary report of the first meeting.
5. International Co-ordination Group for CICAR, Summary report of the second meeting.

The 6th Session of IOC requested FAO and WMO to cooperate with IOC in the planning of these investigations. Each of these organizations has nominated an Assistant International Coordinator on the CICAR International Coordinating Group.

Additional information may be obtained from: Intergovernmental Oceanographic Commission, Place de Fontenoy, 75 Paris 7e, France.

Table 25.1 (continued)

INTERNATIONAL MONITORING ACTIVITIES PROGRAM INFORMATION SHEET			
NAME OF PROGRAM			ANTARCTIC RESEARCH PROGRAMS (ARP)
PURPOSE OF PROGRAM			Conduct scientific and engineering studies of Antarctica and surrounding oceans and correlate these investigations with global programs. Ultimate purpose is to better understand the Antarctic region.
STATUS	OPERATIONAL		
COVERAGE	REGIONAL		
DURATION	INDEFINITE		Argentina, Australia, Belgium, Chile, France, Japan, New Zealand, Norway, So. Africa, U.K., U.S.A. and U.S.S.R.
NO. OF COUNTRIES PARTICIPATING		12	
COORDINATING ORGANIZATION			SCIENTIFIC COMMITTEE ON ANTARCTIC RESEARCH (SCAR)
COOPERATING ORGANIZATION(S)			IGU, IUBS, IUGG, IUGS, IUPAC, IUPS, URSI, WMO
PRINCIPAL DISCIPLINES			Most disciplines (see below)
PRINCIPAL PHENOMENA MONITORED			Human, terrestrial, and marine biology; geology; solid-earth geophysics; physical oceanography; glaciology; geodesy and cartography; meteorology and climatology; upper atmosphere physics; astronomy; conservation; environmental protection; polar engineering; logistics.
NO. OF MONITORING STATIONS		DOZENS	
PHYSICAL SPHERE		EARTH, ATMOSPHERE FLORA, FAUNA, SPACE	
TIME RESPONSE		REAL AND DELAYED	
SAMPLING FREQUENCY		CONTINUOUS AND PERIODIC	
DATA ACQUISITION TECHNIQUES			Ground stations, balloons, rockets, satellites, sampling, survey.
DATA STORAGE AND RETRIEVAL FACILITIES		Extensive: WORLD DATA CENTERS & NATIONAL	
COMMUNICATIONS FACILITIES			MODEST
RESEARCH CENTER(S)			NONE: NATIONAL AND REGIONAL
REMARKS			
REFERENCES 1. "SCAR Manual" published by SCAR, 1966. 2. Annual Report to SCAR, Committee on Polar Research, U.S. National Research Council, 1969. 3. Annual Information Kit, Office of Polar Programs, National Science Foundation, October 1969. Additional information may be obtained from the SCAR Secretariat, c/o Scott Polar Research Institute, Cambridge, U.K.			

Table 25.1 (continued)

INTERNATIONAL MONITORING ACTIVITIES PROGRAM INFORMATION SHEET			
NAME OF PROGRAM		TSUNAMI WARNING SYSTEM (TWS)	
PURPOSE OF PROGRAM			Operate a regional (Pacific) seismograph and tide gauge network to monitor earthquakes and ocean levels and an extensive rapid communications network to warn of impending tsunamis.
STATUS	OPERATIONAL		
COVERAGE	REGIONAL		
DURATION	LONG-TERM (INDEFINITE)		
NO. OF COUNTRIES PARTICIPATING		6: Canada,	France, Japan, Philippines, U.S.A., U.S.S.R.
COORDINATING ORGANIZATION			INTERGOVERNMENTAL OCEANOGRAPHIC COMMISSION (IOC)
COOPERATING ORGANIZATION(S)			WMO, IUGG
PRINCIPAL DISCIPLINES			SEISMOLOGY, PHYSICAL OCEANOGRAPHY
PRINCIPAL PHENOMENA MONITORED			Seismicity, ocean level, wind-wave motions.
NO. OF MONITORING STATIONS		Seismic - 21 Tidal - 44	
PHYSICAL SPHERE		OCEAN	
TIME RESPONSE		REAL	
SAMPLING FREQUENCY		CONTINUOUS	
DATA ACQUISITION TECHNIQUES			Tide stations, seismograph stations
DATA STORAGE AND RETRIEVAL FACILITIES		MODEST; NATIONAL CENTERS	
COMMUNICATIONS FACILITIES			NEAR REAL TIME PACIFIC TELECOMMUNICATIONS NETWORK
RESEARCH CENTER(S)			USE OF NATIONAL CENTERS
REMARKS			

REFERENCES

1. Tsunami, The Seismic Sea-Wave Warning System, U. S. Department of Commerce Coast & Geodetic Survey.

2. International Tsunami Information Center, Newsletter, Volume III, No. 2, 25 June 1970.

   For additional information contact: International Tsunami Information Center, P. O. Box 3887, Honolulu, Hawaii 96812.

Table 25.1 (continued)

INTERNATIONAL MONITORING ACTIVITIES PROGRAM INFORMATION SHEET		
NAME OF PROGRAM		COOPERATIVE STUDY OF KUROSHIO AND ADJACENT REGIONS (CSK)
PURPOSE OF PROGRAM		Synoptic surveys of Kuroshio system; study of fishery resources of the Kuroshio.
STATUS	OPERATIONAL	
COVERAGE	REGIONAL	
DURATION	1965-1970	
NO. OF COUNTRIES PARTICIPATING		China, Indonesia, Japan, Korea, Philippines, Singapore, Thailand, U.K., USA, USSR, Vietnam.
COORDINATING ORGANIZATION		INTERGOVERNMENTAL OCEANOGRAPHIC COMMISSION (IOC)
COOPERATING ORGANIZATION(S)		NATIONAL ORGANIZATIONS
PRINCIPAL DISCIPLINES		PHYSICAL AND BIOLOGICAL OCEANOGRAPHY
PRINCIPAL PHENOMENA MONITORED		Short-term fluctuation of ocean processes; physical structure and marine life.
NO. OF MONITORING STATIONS	DOZENS	
PHYSICAL SPHERE	OCEAN	
TIME RESPONSE	DELAYED	
SAMPLING FREQUENCY	CONTINUOUS & PERIODIC	
DATA ACQUISITION TECHNIQUES		Sea-borne survey, measurement and monitoring.
DATA STORAGE AND RETRIEVAL FACILITIES		NATIONAL CENTERS. NODC (JAPAN) DESIGNATED AS REGIONAL DATA CENTRE.
COMMUNICATIONS FACILITIES	MODEST	
RESEARCH CENTER(S)		NATIONAL RESEARCH CENTERS
REMARKS		

REFERENCES
1. Intergovernmental Oceanographic Commission, Summary Report of its Sixth Session, UNESCO, June 1970.
2. Global Ocean Research, Report of a Joint Working Party of the Advisory Committee on Marine Resources Research, the Scientific Committee on Oceanic Research, and the World Meteorological Organization, June 1969.
3. Intergovernmental Oceanographic Commission, Resolutions Adopted at the VIth Session, UNESCO, September 1969.
4. Summary Report of the Vth and VIth Meetings of the International Co-ordination Group for the CSK.
5. A Proposal Plan for a Synoptic Survey of the South China Sea under the CSK Program, National Coordinator of Thailand.
   Additional information may be obtained from: Intergovernmental Oceanographic Commission, Place de Fontenoy, 75 Paris 7e, France.

Table 25.1 (continued)

INTERNATIONAL MONITORING ACTIVITIES PROGRAM INFORMATION SHEET			
NAME OF PROGRAM:	WMO PROGRAMME ON THE INTERACTION OF MAN AND HIS ENVIRONMENT	MONITORING OF BASIC ENVIRONMENTAL CLIMATE (MBEC)	
PURPOSE OF PROGRAM		To monitor parameters for studies of long-term changes in the climate of the world and for application to economic activities and human environment problems	
STATUS	OPERATIONAL		
COVERAGE	GLOBAL		
DURATION	PERMANENT		
NO. OF COUNTRIES PARTICIPATING	132		
COORDINATING ORGANIZATION		WMO	
COOPERATING ORGANIZATION(S)			
PRINCIPAL DISCIPLINES		METEOROLOGY	
PRINCIPAL PHENOMENA MONITORED		Temperature, humidity of air wind precipitation and snow cover evaporation soil temperature	
NO. OF MONITORING STATIONS	THOUSANDS		
PHYSICAL SPHERE	ATMOSPHERE AND SOIL		
TIME RESPONSE	DELAYED		
SAMPLING FREQUENCY	AT LEAST DAILY		
DATA ACQUISITION TECHNIQUES		Various meteorological instruments.	
DATA STORAGE AND RETRIEVAL FACILITIES	NATIONAL: COMPUTERS		
COMMUNICATIONS FACILITIES	NONE; PARTLY BY GTS AND PARTLY BY MAIL		
RESEARCH CENTER(S)	NONE: NATIONAL METEOROLOGICAL SERVICES AND RESEARCH CENTRES		
REMARKS Data published nationally or internationally and exchanged internationally (monthly or annually) between WMO Members			

REFERENCES
    1. Guide to Climatological Practices - WMO No. 100 TP 44, Geneva, 1960
 	
    Additional information may be obtained from World Meteorological Organization, 1211 Geneva 20, Case Postale No. 1, Switzerland.			

Table 25.1 (continued)

INTERNATIONAL MONITORING ACTIVITIES PROGRAM INFORMATION SHEET		
NAME OF PROGRAM	WMO PROGRAMME ON THE INTERACTION OF MAN AND HIS ENVIRONMENT	MONITORING BACKGROUND AIR POLLUTION (MBAP)
PURPOSE OF PROGRAM		To document long-term changes in atmospheric environment parameters of particular significance to weather and climate
STATUS	OPERATIONAL	
COVERAGE	GLOBAL	
DURATION	PERMANENT	
NO. OF COUNTRIES PARTICIPATING	12	
COORDINATING ORGANIZATION		WMO
COOPERATING ORGANIZATION(S)		
PRINCIPAL DISCIPLINES		METEOROLOGY, CHEMISTRY
PRINCIPAL PHENOMENA MONITORED		- Turbidity - Constituents of precipitation and dry fallout - CO_2) - CO and methane) at selected stations - SO_2 and H_2S) - oxides of nitrogen) - total precipitable water
NO. OF MONITORING STATIONS	20	
PHYSICAL SPHERE	ATMOSPHERE	
TIME RESPONSE	DELAYED	
SAMPLING FREQUENCY	DAILY, AT LEAST	
DATA ACQUISITION TECHNIQUES		- air samplers, laboratory analyses - turbidity metre at land stations
DATA STORAGE AND RETRIEVAL FACILITIES	UNDER CONSIDERATION	
COMMUNICATIONS FACILITIES	NONE: BY MAIL	
RESEARCH CENTER(S)	NONE: NATIONAL METEOROLOGICAL SERVICES AND RESEARCH CENTRES	
REMARKS Number of Stations will increase gradually in order of magnitude of 100-200 Collection centers and publication arrangements to be established. Expansion over oceanic areas under investigation.		

REFERENCES
  1. WMO Circular letter MC-1558 - Establishment of a WMO network on background pollution stations - Geneva July 1970
  2. Resolution 11 (EC-XXI) - Establishment of a Network of stations to measure background pollution
  3. Resolution 14(EC-XXI) - Marine Pollution
  4. A brief survey of the activities of the World Meteorological Organization relating to human environment.
    Additional information may be obtained from World Meteorological Organization, 1211 Geneva 20, Case Postale No. 1, Switzerland.		

Table 25.1 (continued)

INTERNATIONAL MONITORING ACTIVITIES PROGRAM INFORMATION SHEET		
NAME OF PROGRAM		WMO RESEARCH PROGRAMME MONITORING SOLAR RADIATION (MSR)
PURPOSE OF PROGRAM		To monitor, at the surface of the earth, incoming and outgoing radiative energy
STATUS	OPERATIONAL	
COVERAGE	GLOBAL	
DURATION	INDEFINITE	
NO. OF COUNTRIES PARTICIPATING	74	
COORDINATING ORGANIZATION		WMO
COOPERATING ORGANIZATION(S)		
PRINCIPAL DISCIPLINES		METEOROLOGY, SOLAR PHYSICS
PRINCIPAL PHENOMENA MONITORED		Solar and Terrestrial Radiation
NO. OF MONITORING STATIONS	HUNDREDS	
PHYSICAL SPHERE	ATMOSPHERE	
TIME RESPONSE	DELAYED	
SAMPLING FREQUENCY	DAILY AND CONTINUOUS	
DATA ACQUISITION TECHNIQUES		Pyrheliometer, Pyranometer
DATA STORAGE AND RETRIEVAL FACILITIES	NATIONAL-COMPUTER	
COMMUNICATIONS FACILITIES	NONE: BY MAIL	
RESEARCH CENTER(S)		NONE: NATIONAL METEOROLOGICAL SERVICES AND RESEARCH CENTRES
REMARKS	published monthly by Main Geophysical Observatory, Leningrad, for international exchange	
REFERENCES	1. Solar Radiation and Radiation Balance Data A. I. Voejkov- Main Geophysical Observatory Leningrad - USSR - published monthly Additional information may be obtained from World Meteorological Organization, 1211 Geneva 20, Case Postale No. 1, Switzerland.	

Table 25.1 (continued)

\multicolumn{4}{	c	}{INTERNATIONAL MONITORING ACTIVITIES PROGRAM INFORMATION SHEET}

NAME OF PROGRAM		\multicolumn{2}{l	}{WMO RESEARCH PROGRAMME MONITORING ISOTOPES IN PRECIPITATION (MIP)}
PURPOSE OF PROGRAM			To collect and analyse samples of precipitation on a global basis in order to determine the circulation of the following isotopes:- - tritium - deuterium - oxygen - 18
STATUS	OPERATIONAL		
COVERAGE	GLOBAL		
DURATION	INDEFINITE		
NO. OF COUNTRIES PARTICIPATING		65	
COORDINATING ORGANIZATION			IAEA/WMO
COOPERATING ORGANIZATION(S)			
PRINCIPAL DISCIPLINES			METEOROLOGY, ISOTOPE CHEMISTRY
PRINCIPAL PHENOMENA MONITORED			Concentration in precipitation of isotopes of: -tritium -deuterium -oxygen -18
NO. OF MONITORING STATIONS		155	
PHYSICAL SPHERE		ATMOSPHERE	
TIME RESPONSE		DELAYED	
SAMPLING FREQUENCY		CONTINUOUS OVER 1 MONTH PERIODS	
DATA ACQUISITION TECHNIQUES			Precipitation collection, laboratory analysis
DATA STORAGE AND RETRIEVAL FACILITIES		INTERNATIONAL COMPUTER (IAEA)	
COMMUNICATIONS FACILITIES		NONE: BY MAIL	
RESEARCH CENTER(S)		\multicolumn{2}{l	}{IAEA LABORATORY VIENNA, AND CO-OPERATING NATIONAL LABORATORIES}
REMARKS	\multicolumn{3}{l	}{Collected and made available by IAEA Laboratory, Vienna}
REFERENCES	\multicolumn{3}{l	}{Technical Report Series No. 96 Environmental Isotope Data No. 1 World Survey of Isotope concentration in Precipitation (1953-1963) International Atomic Energy Agency - Vienna: 1969 Additional information may be obtained from World Meteorological Organization, 1211 Geneva 20, Case Postale No. 1, Switzerland.}

Table 25.1 (continued)

	INTERNATIONAL MONITORING ACTIVITIES PROGRAM INFORMATION SHEET	
NAME OF PROGRAM		WMO RESEARCH PROGRAMME MONITORING ATMOSPHERIC ELECTRICITY (MAE)
PURPOSE OF PROGRAM		To monitor changes in the earth's electrical field and to relate these to atmospheric processes.
STATUS	OPERATIONAL	
COVERAGE	GLOBAL	
DURATION	INDEFINITE	
NO. OF COUNTRIES PARTICIPATING	8	
COORDINATING ORGANIZATION		WMO
COOPERATING ORGANIZATION(S)		IAMAP/IAGA
PRINCIPAL DISCIPLINES		METEOROLOGY
PRINCIPAL PHENOMENA MONITORED		- hourly means of potential gradients of atmospheric electrical fields - atmospheric electrical conductivity - density of air-earth current
NO. OF MONITORING STATIONS	19	
PHYSICAL SPHERE	ATMOSPHERE	
TIME RESPONSE	REAL-TIME	
SAMPLING FREQUENCY	CONTINUOUS	
DATA ACQUISITION TECHNIQUES		Electronic sensors
DATA STORAGE AND RETRIEVAL FACILITIES	NATIONAL	
COMMUNICATIONS FACILITIES	NONE: BY MAIL	
RESEARCH CENTER(S)	NONE: NATIONAL METEOROLOGICAL SERVICES AND RESEARCH CENTRES	
REMARKS Atmospheric electricity may be related to: atmospheric pollutants, orographic wind systems, lightning, precipitable water, sferics, fog		

REFERENCES
    1. Results of Ground Observations of Atmospheric Electricity
 	
       A.I. Voejkov - Main Geophysical Observatory, Leningrad, USSR - Published monthly
       Additional Information may be obtained from World Meteorological Organization, 1211 Geneva 20, Case Postale No. 1, Switzerland.		

Table 25.1 (continued)

INTERNATIONAL MONITORING ACTIVITIES PROGRAM INFORMATION SHEET			
NAME OF PROGRAM			WMO RESEARCH PROGRAMME MONITORING ATMOSPHERIC OZONE (MAO)
PURPOSE OF PROGRAM			To monitor variations in concentration of ozone in the atmosphere for studies of troposphere-stratoshpere energy exchange processes and other energy balance research
STATUS	OPERATIONAL		
COVERAGE	GLOBAL		
DURATION	INDEFINITE		
NO. OF COUNTRIES PARTICIPATING		22	
COORDINATING ORGANIZATION			WMO
COOPERATING ORGANIZATION(S)			IAMAP
PRINCIPAL DISCIPLINES			METEOROLOGY
PRINCIPAL PHENOMENA MONITORED			Concentration of ozone in the atmosphere
NO. OF MONITORING STATIONS		32	
PHYSICAL SPHERE		ATMOSPHERE	
TIME RESPONSE		DELAYED	
SAMPLING FREQUENCY		DAILY	
DATA ACQUISITION TECHNIQUES			- balloon-borne chemical sensors - surface optical instruments
DATA STORAGE AND RETRIEVAL FACILITIES		NATIONAL: COMPUTER	
COMMUNICATIONS FACILITIES			NONE: BY MAIL
RESEARCH CENTER(S)			NONE: NATIONAL METEOROLOGICAL SERVICES AND RESEARCH CENTRES
REMARKS			

REFERENCES
1. WMO Technical Note 36 - Ozone Observations and their Meteorological Applications by Dr. H. Taba, 1961

2. Ozone Data for the World - Meteorological Service of Canada - published bi-monthly for international exchange.
  Additional information may be obtained from World Meteorological Organization, 1211 Geneva 20, Case Postale No. 1, Switzerland.

International Environmental Monitoring Programs

Table 25.1 (continued)

	INTERNATIONAL MONITORING ACTIVITIES PROGRAM INFORMATION SHEET	
NAME OF PROGRAM	WMO PROGRAMME ON THE INTERACTION OF MAN AND HIS ENVIRONMENT	HYDROMETEOROLOGICAL MONITORING FOR HYDRO-LOGICAL PURPOSES (HMH)
PURPOSE OF PROGRAM		Promoting international co-operation in the operational aspects of hydrology; developing and improving methods procedures and techniques for: hydrometeorological networks; data processing; standardization of instruments; methods and terminology; hydrological forecasting; hydrometeorological design data for water resources projects.
STATUS	OPERATIONAL	
COVERAGE	GLOBAL AND REGIONAL	
DURATION	INDEFINITE	
NO. OF COUNTRIES PARTICIPATING	122	
COORDINATING ORGANIZATION		WMO
COOPERATING ORGANIZATION(S)		NONE
PRINCIPAL DISCIPLINES		HYDROMETEOROLOGY, HYDROLOGY
PRINCIPAL PHENOMENA MONITORED		Precipitation, snow cover, evaporation and evaportranspiration, water level of lakes and streams, streamflow and storage, sediment discharge, river and lake ice, water temperature, soil moisture
NO. OF MONITORING STATIONS	THOUSANDS	
PHYSICAL SPHERE	WATER	
TIME RESPONSE	REAL & DELAYED	
SAMPLING FREQUENCY	CONTINUOUS, PERIODIC, SPORADIC	
DATA ACQUISITION TECHNIQUES		Hydrometeorological stations, aircraft, satellites.
DATA STORAGE AND RETRIEVAL FACILITIES	REGIONAL AND NATIONAL CENTERS	
COMMUNICATIONS FACILITIES	NONE: USE OF GTS	
RESEARCH CENTER(S)	NONE: NATIONAL CENTERS	
REMARKS	Interdependence of meteorology and hydrology covered by close co-ordination with other relevant WMO activities.	
REFERENCES	1. Weather and Water. WMO 1966 2. Report of the Third Session of the Commission for Hydrometeorology, WMO 1968 3. Guide to Hydrometeorological Practices, 2nd edition, WMO 1970 Additional information may be obtained from World Meteorological Organization, 1211 Geneva 20, Case Postale No. 1, Switzerland.	

Table 25.1 (continued)

	INTERNATIONAL MONITORING ACTIVITIES PROGRAM INFORMATION SHEET	
NAME OF PROGRAM:	WMO PROGRAMME ON THE INTERACTION OF MAN AND HIS ENVIRONMENT	METEOROLOGICAL SERVICES TO AIR TRANSPORT (MSAT)
PURPOSE OF PROGRAM		To monitor the atmosphere with a view to providing forecasts and statistical information in support of air transport planning (aerodrome design included) and operations; sub-programs include aeronautical climatological summaries and memoranda.
STATUS	OPERATIONAL	
COVERAGE	GLOBAL AND REGIONAL	
DURATION	INDEFINITE	
NO. OF COUNTRIES PARTICIPATING	132	
COORDINATING ORGANIZATION		WMO
COOPERATING ORGANIZATION(S)		ICAO
PRINCIPAL DISCIPLINES		METEOROLOGY
PRINCIPAL PHENOMENA MONITORED		Observations and forecasts: atmospheric pressure patterns, wind and temperature fields, water vapour, visibility, clouds, turbulence, significant weather. Statistics: aeronautical climatological summaries and aeronautical descriptive climatological memoranda for particular areas and air routes
NO. OF MONITORING STATIONS	THOUSANDS	
PHYSICAL SPHERE	ATMOSPHERE	
TIME RESPONSE	REAL TIME AND DELAYED	
SAMPLING FREQUENCY	CONTINUOUS AND PERIODIC	
DATA ACQUISITION TECHNIQUES		Aeronautical meteorological stations (surface and the upper-air) - North Atlantic Ocean ships (NAOS) - satellites, aircrafts
DATA STORAGE AND RETRIEVAL FACILITIES	NATIONAL AND REGIONAL	
COMMUNICATIONS FACILITIES		AFTN - MOTNE - VOLMET (EXTENSIVELY ASSISTED BY WWW, GTS)
RESEARCH CENTER(S)		NONE
REMARKS	An Area Forecast System encompassing 18 area forecast centres relying heavily on the WWW for basic meteorological and processed data and their exchange	
REFERENCES	1. WMO Technical Regulations, Vo. II - WMO Geneva, 1970 edition 2. Aviation aspects of mountain waves - WMO Geneva, T.N. No. 18, 1958 edition (Reprinted in 1967) 3. Aviation hail problem - Turbulence in clear air and in cloud - Ice formation on aircraft -occurrence and forecasting of cirrostratus clouds WMO Geneva TN Nos. 37,38,39, and 40, 1961 edition (reprinted 1968) 4. Meteorological problems in the design and operation of supersonic aircraft, WMO Geneva, TN No. 89, 1967 edition. 5. Manual on meteorological observing in transport aircraft, WMO Geneva, Publication No. 197 TP 102, 1966 edition. Additional information may be obtained from World Meteorological Organization, 1211 Geneva 20, Case Postale No. 1, Switzerland.	

Planned Monitoring Programs
1. Global Atmospheric Research Programme (GARP)
2. Integrated Global Ocean Station System (IGOSS)
3. Man and the Biosphere (MAB)
4. Long-Term and Expanded Program of Oceanic Exploration and Research (LEPOR)
5. International Decade of Ocean Exploration (IDOE)
6. Cooperative Investigations of the Northern Part of the Eastern-Central Atlantic (CINECA)
7. World Health Monitoring and Information Network
8. International Cooperative Studies of the Mediterranean (ICSM)

Table 25.2 Monitoring Programs Planned

INTERNATIONAL MONITORING ACTIVITIES PROGRAM INFORMATION SHEET			
NAME OF PROGRAM		GLOBAL ATMOSPHERIC RESEARCH PROGRAMME (GARP)	
PURPOSE OF PROGRAM		Study the physical processes in the stratosphere and at the ocean-atmosphere and land-atmosphere interfaces as well as the statistical properties of the general circulation of the atmosphere in order to ascertain the inherent predictability of the large scale feature of the general circulation over a two to three weeks period and to improve the understanding of the physical basis of climate.	
STATUS	PLANNING		
COVERAGE	GLOBAL AND REGIONAL		
DURATION	1973-1976		
NO. OF COUNTRIES PARTICIPATING		132	
COORDINATING ORGANIZATION			INTERNATIONAL COUNCIL OF SCIENTIFIC UNIONS (ICSU) WORLD METEOROLOGICAL ORGANIZATION (WMO)
COOPERATING ORGANIZATION(S)			
PRINCIPAL DISCIPLINES			METEOROLOGY
PRINCIPAL PHENOMENA MONITORED			Wind, temperature, pressure, geopotential, soil moisture, cloud, snow or ice cover, precipitation, radiation fluxes, sea temperatures and salinity profiles, currents in surface layers.
NO. OF MONITORING STATIONS		THOUSANDS	
PHYSICAL SPHERE		ATMOSPHERE, OCEAN	
TIME RESPONSE		REAL	
SAMPLING FREQUENCY		CONTINOUS PERIODIC	
DATA ACQUISITION TECHNIQUES			GARP Global Observing System (GGOS): Satellites, aircrafts, balloons, ships, buoys, land stations.
DATA STORAGE AND RETRIEVAL FACILITIES		ELABORATE, NATIONAL, REGIONAL WORLD CENTERS	
COMMUNICATIONS FACILITIES			NONE - GLOBAL TELECOMMUNICATIONS SYSTEM (GTS)
RESEARCH CENTER(S)			NATIONAL
REMARKS: Plan first tropical experiment 1973-1974; plan first global experiment 1975-1976			

REFERENCES
1. An Introduction to GARP, GARP Publications Series No. 1, International Council of Scientific Unions, World Meteorological Organization.
2. Systems Possibilities for an Early GARP Experiment COSPAR Working Group VI report to JOC, GARP Publications Series No. 2. (same as above)
3. The Planning of the First GARP Global Experiment GARP Publications Series No. 3 (Same as above)
4. The Planning of GARP Tropical Experiments, GARP Publications Series No. 4, (same as above)
5. Comments on the Observing and Data Processing Systems for FGGE, as given in GARP Publication No 3, COSPAR Working Group 6, February 1970.
6. Report of Planning Conference on GARP, GARP Special Report No. 1, Brussels, March 1970.
7. Report of Interim Planning Group on GARP Tropical Experiment in the Atlantic, GARP Special Report No. 2, London, July 1970. Additional information may be obtained from World Meteorological Organization, 1211 Geneva 20, Case Postale No. 1, Switzerland.

Table 25.2 (continued)

INTERNATIONAL MONITORING ACTIVITIES PROGRAM INFORMATION SHEET		
NAME OF PROGRAM		INTEGRATED GLOBAL OCEAN STATION SYSTEM (IGOSS)
PURPOSE OF PROGRAM		Develop a global observation, radio communication, and data processing system, together with the WWW and other meteorological systems, to provide real-time oceanographic and meteorological information which will improve the understanding of oceanic phenomena and permit improved and more comprehensive forecast of ocean conditions.
STATUS	INTERMEDIATE PLANNING	
COVERAGE	GLOBAL	
DURATION	INDEFINITE	
NO. OF COUNTRIES PARTICIPATING	DOZENS	
COORDINATING ORGANIZATION		INTERGOVERNMENTAL OCEANOGRAPHIC COMMISSION
COOPERATING ORGANIZATION(S)		WORLD METEOROLOGICAL ORGANIZATION (WMO)
PRINCIPAL DISCIPLINES		PHYSICAL OCEANOGRAPHY, METEOROLOGY
PRINCIPAL PHENOMENA MONITORED		In conjunction with WWW and other meteorological systems, temperature and salinity at surface and as a function of depth, wind waves and swell, current, wind speed and direction, atmospheric pressure, dew point and net solar radiation.
NO. OF MONITORING STATIONS	HUNDREDS	
PHYSICAL SPHERE	OCEAN	
TIME RESPONSE	REAL	
SAMPLING FREQUENCY	CONTINUOUS, PERIODIC	
DATA ACQUISITION TECHNIQUES		IGOSS plus WWW facilities, namely: manned and unmanned fixed platforms, weather ships and research ships, ships of opportunity, satellites and aircraft, coastal and island stations, buoys
DATA STORAGE AND RETRIEVAL FACILITIES		NATIONAL AND REGIONAL FACILITIES
COMMUNICATIONS FACILITIES		EXTENSIVE: IGOSS SYSTEM PLUS WWW, GTS
RESEARCH CENTER(S)		NATIONAL AND REGIONAL RESEARCH CENTERS
REMARKS		Heavy participation by World Weather Watch, global observing, data processing and telecommunications systems.

REFERENCES
1. 2nd Joint Meeting IOC W/C for IGOSS and WMO EC Panel on Meteorological Aspects of Ocean Affairs, September, 1969. 2. Intergovernmental Oceanographic Commission Summary Report of its Sixth Session, UNESCO, 1 June 1970. 3. Global Ocean Research, Report of Joint Working Party of the Advisory Committee on Marine Resources Research, The Scientific Committee on Oceanic Research, and the World Meteorological Organization, June 1969. 4. Resolution 17-(EC-XXI) - Integrated Planning of IGOSS with the World Weather Watch. 5. General Plan and Implementation Programme of IGOSS for Phase I, Intergovernmental Oceanographic Commission, UNESCO, Paris, 27 October 1969. Further information may be obtained from: Intergovernmental Oceanographic Commission, UNESCO, Place de Fontenoy, 75 Paris 7e, France.

At the request of the Governments meeting under the aegis of IOC, and with the agreement of the governments meeting under the aegis of WMO, IGOSS is being developed through joint action between IOC and WMO. By definition there should be no duplication between IGOSS and WWW; the two systems "must be looked at in conjunction with a view to the ultimate development of complete data gathering, data processing and prediction services which avoid duplication of efforts and means" (extract from the Plan of IGOSS for PHASE I as approved by IOC and WMO). It is through this conjunction of IGOSS and WWW that "during Phase I of Igoss" certain oceanographic and meteorological parameters will be monitored.

Table 25.2 (continued)

INTERNATIONAL MONITORING ACTIVITIES PROGRAM INFORMATION SHEET		
NAME OF PROGRAM		MAN AND THE BIOSPHERE (MAB)
PURPOSE OF PROGRAM		Develop a global program to inventory and assess the resources of the biosphere, including systematic observations and monitoring and research into the structure and functioning of terrestrial and aquatic (non-marine) ecosystems and research into changes in the biosphere brought about by man and the effects of these changes on man. Focus on non-oceanic areas.
STATUS	ADVANCED PLANNING	
COVERAGE	GLOBAL	
DURATION	LONG TERM - INDEFINITE	
NO. OF COUNTRIES PARTICIPATING	HUNDRED OR MORE	
COORDINATING ORGANIZATION		United Nations Educational, Scientific, and Cultural Organization (UNESCO)
COOPERATING ORGANIZATION(S)		FAO, WHO, IUCN, IBP, WMO, ICSU
PRINCIPAL DISCIPLINES		ECOLOGY, BIOLOGY, BIOGEOCHEMISTRY
PRINCIPAL PHENOMENA MONITORED		Natural, modified, and managed ecosystems; quantitative and qualitative environmental changes in the biosphere.
NO. OF MONITORING STATIONS	HUNDREDS	
PHYSICAL SPHERE	BIOSPHERE	
TIME RESPONSE	REAL AND DELAYED	
SAMPLING FREQUENCY	CONTINUOUS AND PERIODIC	
DATA ACQUISITION TECHNIQUES		Sampling, survey, ground station instrumentation, aircraft and satellite remote sensing.
DATA STORAGE AND RETRIEVAL FACILITIES	MODERATE: USE OF NATIONAL & REGIONAL FACILITIES.	
COMMUNICATIONS FACILITIES		MODERATE: INITIALLY USE OF EXISTING FACILITIES.
RESEARCH CENTER(S)		SPECIAL NATIONAL AND REGIONAL BIOSPHERE RESEARCH CENTERS, DESIGNATED BY INDIVIDUAL COUNTRIES.
REMARKS		The "Man and the Biosphere" program proposal will be submitted to the UNESCO General Conference in October 1970.
REFERENCES		1. Intergovernmental Conference of Experts on the Scientific Basis for Rational Use and Conservation of the Resources of the Biosphere, UNESCO, 6 January 1969. 2. Plan for a Long-Term Intergovernmental and Interdisciplinary Program on Man and the Biosphere, Doc. 16/C/78-UNESCO, Paris, Sept. 1970. 3. Dasmann, Raymond F., Man and the Biosphere--A Challenge for UNESCO in a New International Program, Conservation Foundation, 16 September 1969. 4. "Man and the Biosphere", Working Group Reports I-V, Natural Resources Research Division, UNESCO, November 1969. 5. Citron, Robert. An Approach to the Organization of the UNESCO "Man and the Biosphere" Program, Smithsonian Institution, 5 January 1970. 6. Citron, Robert, MABNET, The Establishment of a Global Network of Ecostations for the "Man and the Biosphere" Program, Smithsonian Institution, 2 February 1970. Additional information may be obtained from: UNESCO, Division of Natural Resources, Place de Fontenoy, 75 Paris 7e, France.

International Environmental Monitoring Programs

Table 25.2 (continued)

	INTERNATIONAL MONITORING ACTIVITIES PROGRAM INFORMATION SHEET	
NAME OF PROGRAM		LONG-TERM AND EXPANDED PROGRAM OF OCEANIC EXPLORATION AND RESEARCH (LEPOR)
PURPOSE OF PROGRAM		To increase knowledge of the ocean, its contents, interfaces with land, the atmosphere, and the ocean floor; to improve the understanding of marine environmental processes to enhance utilization of the ocean and its resources. Develop a world-wide system of pollution monitoring at vital points.
STATUS	PRELIMINARY PLANNING	
COVERAGE	GLOBAL	
DURATION	LONG TERM - INDEFINITE	
NO. OF COUNTRIES PARTICIPATING	67	
COORDINATING ORGANIZATION		INTERGOVERNMENTAL OCEANOGRAPHIC COMMISSION
COOPERATING ORGANIZATION(S)		ICSPRO (U.N.) IMCO, UNESCO/IOC, WMO, FAO, ICSU (SCOR)
PRINCIPAL DISCIPLINES		MARINE GEOLOGY, PHYSICAL, CHEMICAL AND BIOLOGICAL OCEANOGRAPHY
PRINCIPAL PHENOMENA MONITORED		1. Ocean circulation and ocean-atmosphere interactions, variability, and tsunamis. 2. Marine biological production. 3. Marine pollution. 4. Dynamics of the ocean floor.
NO. OF MONITORING STATIONS	HUNDREDS	
PHYSICAL SPHERE	WORLD OCEAN	
TIME RESPONSE	REAL & DELAYED	
SAMPLING FREQUENCY	CONTINUOUS AND PERIODIC	
DATA ACQUISITION TECHNIQUES		Initially contributions from World Weather Watch, Global Atmospheric Research Program, Integrated Global Ocean Station System, and on-going IOC sponsored cooperative investigations of ocean.
DATA STORAGE AND RETRIEVAL FACILITIES		
COMMUNICATIONS FACILITIES	TO BE DETERMINED	
RESEARCH CENTER(S)	TO BE DETERMINED	

REFERENCES 1. Intergovernmental Oceanographic Commission Summary Report of its Sixth Session, UNESCO, 1 June 1970. 2. Global Ocean Research, Report of Joint Working Party of the Scientific Committee on Oceanic Research, and the World Meteorological Organization, June, 1969. 3. U. N. Resolution 2414, 2467 4. Resolution 13 (EC-XXI)Participation of WMO in the Long-term Coordinated Programme of Scientific Research and exploration relating to the Ocean. Additional information may be obtained from: Intergovernmental Oceanographic Commission, Place de Fontenoy, 75 Paris 7e, France. LEPOR is a multidisciplinary programme deriving from resolutions of the UN. The General Assembly invited IOC to coordinate the scientific aspects of the LEPOR and to act as the focal point for the development of its comprehensive outline, acting within its terms of reference. It is to put IOC in a position to act in this coordinating job that its base has been broadened so as to enable it better to interact with the U. N. and the specialized agencies involved that it receives their support. According to an agreeemnt reached by the Inter-Secretariat Committee on Scientific Programmes relating to Oceanography (ICSPRO) as established between the Executive Heads of the U.N, UNESCO, WMO, FAO and IMCO, much of the detailed planning and implementation of specific projects of LEPOR will be the responsibility of the organizations best suited and equipped for the purpose. Much of the same remarks apply to the International Decade of Ocean Exploration which is regarded as the initial acceleration phase of LEPOR.

Table 25.2 (continued)

\multicolumn{3}{	l	}{INTERNATIONAL MONITORING ACTIVITIES PROGRAM INFORMATION SHEET}
NAME OF PROGRAM		INTERNATIONAL DECADE OF OCEAN EXPLORATION (IDOE)
PURPOSE OF PROGRAM		To obtain a comprehensive knowledge of ocean processes which will lead to more efficient management of the ocean and its resources.
STATUS	PLANNING	
COVERAGE	GLOBAL	
DURATION	1970-1980	
NO. OF COUNTRIES PARTICIPATING	DOZENS	
COORDINATING ORGANIZATION		INTERGOVERNMENTAL OCEANOGRAPHIC COMMISSION (IOC) UNDER LEPOR
COOPERATING ORGANIZATION(S)		WMO, FAO, ICSU (SCOR)
PRINCIPAL DISCIPLINES		PHYSICAL, BIOLOGICAL, CHEMICAL, AND GEOLOGICAL OCEANOGRAPHY, AND METEOROLOGY
PRINCIPAL PHENOMENA MONITORED		Phenomena associated with seabed assessment, environmental quality, and environmental forecasting.
NO. OF MONITORING STATIONS	HUNDREDS	
PHYSICAL SPHERE	WORLD OCEANS	
TIME RESPONSE	DELAYED	
SAMPLING FREQUENCY	PERIODIC	
DATA ACQUISITION TECHNIQUES		Research ships, ocean stations, and others.
DATA STORAGE AND RETRIEVAL FACILITIES	USE OF NATIONAL FACILITIES	
COMMUNICATIONS FACILITIES	MODEST	
RESEARCH CENTER(S)	MAJOR OCEANOGRAPHIC LABORATORIES	
REMARKS		

REFERENCES
1. United Nations General Assembly Resolution 2467.
2. An Oceanic Quest, National Academy of Sciences, 1969.
3. Intergovernmental Oceanographic Commission, Summary Report of its Sixth Session, UNESCO, June 1970.
4. Global Ocean Research, Report of a Joint Working Party of the Advisory Committee on Marine Resources Research, the Scientific Committee on Oceanic Research, and the World Meteorological Organization, June 1969.

Additional information may be obtained from: Head, Office of International Decade of Ocean Exploration, National Science Foundation, 1800 G. Street, N.W., Washington, D. C. 20550.

International Environmental Monitoring Programs

Table 25.2 (continued)

INTERNATIONAL MONITORING ACTIVITIES PROGRAM INFORMATION SHEET			
NAME OF PROGRAM			COOPERATIVE INVESTIGATIONS OF THE NORTHERN PART OF THE EASTERN-CENTRAL ATLANTIC (CINECA)
PURPOSE OF PROGRAM			Synoptic survey of the northern part of East Central Atlantic; study fishery resources of the East-Central Atlantic.
STATUS	PRELIMINARY PLANNING		
COVERAGE	REGIONAL		
DURATION	YEARS		
NO. OF COUNTRIES PARTICIPATING		SEVERAL	
COORDINATING ORGANIZATION			INTERGOVERNMENTAL OCEANOGRAPHIC COMMISSION AND ICES
COOPERATING ORGANIZATION(S)			FOOD AND AGRICULTURE ORGANIZATION FCECA AND WMO.
PRINCIPAL DISCIPLINES			OCEANOGRAPHY - MARINE LIVING RESOURCES
PRINCIPAL PHENOMENA MONITORED			Fluctuation of ocean processes, and air-sea interaction, physical structure; and marine life. Possible support from and coordination with GARP
NO. OF MONITORING STATIONS		DOZENS	
PHYSICAL SPHERE		OCEAN AND ATMOSPHERE	
TIME RESPONSE		DELAYED	
SAMPLING FREQUENCY		CONTINUOUS, PERIODIC	
DATA ACQUISITION TECHNIQUES			Shipboard instrumentation, sampling and survey
DATA STORAGE AND RETRIEVAL FACILITIES		NATIONAL: FAO FISHERY DATA	CENTRE HOLDS FISHERY AND BIOLOGICAL DATA.
COMMUNICATIONS FACILITIES			MODEST: USE EXISTING FACILITIES
RESEARCH CENTER(S)			NATIONAL, REGIONAL AND PLANNED
REMARKS			

REFERENCES 1. Intergovernmental Oceanographic Commission, Summary Report, UNESCO, Paris, 2-13 September 1969, published 1 June 1970. 2. Global Ocean Research, Report of a Joint Working Party of the Advisory Committee on Marine Resources Research, the Scientific Committee on Oceanic Research, and the World Meteorological Organization, Ponza and Rome, MCMLXIX 27 April to 7 May 1969, published 1 June 1969. 3. Intergovernmental Oceanographic Commission, Resolutions Adopted at the VIth Session, UNESCO, Paris, 2-13 September 1969, published 23 September 1969. 4. CINECA Newsletter, Report of the First Session of the Coordinating Group for the Planning and Execution of CINECA, Paris, UNESCO House, 27-30 April 1970. Additional information may be obtained from: Intergovernmental Oceanographic Commission, Place de Fontenoy, 75 Paris 7e, France. Many of the countries intending to participate in CINECA are the same as those intending to participate in the Tropical Experiment of the GARP. The area to be covered by the Tropical Experiment while larger, overlaps the area to be investigated by the CINECA and there is little difference in the proposed timing of the experiments. Consideration is being given to coordination of the two activities.

Table 25.2 (continued)

INTERNATIONAL MONITORING ACTIVITIES PROGRAM INFORMATION SHEET			
NAME OF PROGRAM			WORLD HEALTH MONITORING AND INFORMATION NETWORK
PURPOSE OF PROGRAM			To establish a global health information network including a world-wide health monitoring programme, early warning system, a global health intelligence center, a computerized network of national and regional centres, and additional health research centres.
STATUS		PRELIMINARY PLANNING	^
COVERAGE		GLOBAL	^
DURATION		LONG-TERM - INDEFINITE	^
NO. OF COUNTRIES PARTICIPATING		DOZENS	
COORDINATING ORGANIZATION			WORLD HEALTH ORGANIZATION
COOPERATING ORGANIZATION(S)			UNITED NATIONS AND SPECIFIED AGENCIES, NATIONAL AND LOCAL AUTHORITIES
PRINCIPAL DISCIPLINES			PUBLIC HEALTH, MEDICAL AND COMMUNICATIONS
PRINCIPAL PHENOMENA MONITORED			Environmental pollution, epidemiology applied to ecology, waste disposal, communicable diseases, mental health, nutrition, population fluctuations, sanitation, occupational health, cancer and other degenerative diseases, adverse drug reactions, maternal and child health
NO. OF MONITORING STATIONS		DOZENS	^
PHYSICAL SPHERE		MAN AND HIS ACTIVITIES	^
TIME RESPONSE		DELAYED	^
SAMPLING FREQUENCY		PERIODIC	
DATA ACQUISITION TECHNIQUES			Sampling, survey, ground-based instrumentation.
DATA STORAGE AND RETRIEVAL FACILITIES			LARGE COMPUTER
COMMUNICATIONS FACILITIES			MODEST: INITIALLY USE OF EXISTING FACILITIES
RESEARCH CENTER(S)			MODERATE: ESTABLISHMENT OF ADDITIONAL INTERNATIONAL HEALTH RESEARCH CENTERS.
REMARKS Development of proposal for presentation to the World Health Assembly, May 1971, and the UN Human Environment Conference in Stockholm, June 1972.			

REFERENCES
1. The Problems of the Human Environment, memos on Global Environmental Health Monitoring, Office of Science and Technology, WHO, March and June 1970 (restricted).
2. Water Pollution Control in Developing Countries, Report of a WHO Expert Committee, World Health Organization, Technical Report Series No. 404, 1968.
3. Water Pollution Control, Report of a WHO Expert Committee, Technical Report Series No. 318, World Health Organization, 1966.
4. Measurement of Air Pollutants, Guide to the Selection of Methods, M. Katz, World Health Organization, 1969.
5. WHO Annual Reports

For further information contact: World Health Organization, 1211 Geneva 27, Switzerland.

Table 25.2 (continued)

INTERNATIONAL MONITORING ACTIVITIES PROGRAM INFORMATION SHEET		
NAME OF PROGRAM		INTERNATIONAL COOPERATIVE STUDIES OF THE MEDITERRANEAN (ICSM)
PURPOSE OF PROGRAM		Synoptic Surveys of the Mediterranean; study of fishery resources of Mediterranean
STATUS	PLANNING AND IMPLEMENTATION	
COVERAGE	REGIONAL	
DURATION	1970-1975	
NO. OF COUNTRIES PARTICIPATING	MEDITERRANEAN	COUNTRIES AND USA, USSR, AND FED. REP. OF GERMANY
COORDINATING ORGANIZATION		INTERGOVERNMENTAL OCEANOGRAPHIC COMMISSION (IOC)
COOPERATING ORGANIZATION(S)		WITH THE GFCM OF FAO AND CIESMM NATIONAL ORGANIZATIONS
PRINCIPAL DISCIPLINES		PHYSICAL AND BIOLOGICAL OCEANOGRAPHY
PRINCIPAL PHENOMENA MONITORED		Fluctuations of ocean processes; physical structure; and marine life.
NO. OF MONITORING STATIONS	DOZENS	
PHYSICAL SPHERE	OCEAN	
TIME RESPONSE	DELAYED	
SAMPLING FREQUENCY	CONTINUOUS & PERIODIC	
DATA ACQUISITION TECHNIQUES		Sea-borne survey, sampling, measurement, and monitoring
DATA STORAGE AND RETRIEVAL FACILITIES	NATIONAL CENTERS -WDC(B)	MOSCOW DESIGNATED AS REGIONAL DATA CENTRE FOR TIME BEING
COMMUNICATIONS FACILITIES		MINIMAL-OPERATIONAL UNIT LOCATED AT MONACO.
RESEARCH CENTER(S)	NATIONAL RESEARCH CENTERS	
REMARKS		
REFERENCES		

1. Intergovernmental Oceanographic Commission, Summary Report, UNESCO, Paris 2-13 September, 1969, published 1 June 1970. (Sixth Session)
2. Global Ocean Research, Report of a Joint Working Party of the Advisory Committee on Marine Resources Research, the Scientific Committee on Oceanic Research, and the World Meteorological Organization, Ponza and Rome, MCMLXIX 27 April to 7 May 1969, published 1 June 1969.
3. Intergovernmental Oceanographic Commission, Resolutions adopted at the VIth Session, UNESCO, Paris, 2-13 September, 1969, published 23 September 1969.

Additional information may be obtained from Intergovernmental Oceanographic Commission, Place de Fontenoy, 75 Paris 7e, France

Proposed Monitoring Programs
1. International Center for the Environment (ICE)
2. Global Network for Environmental Monitoring (GNEM)
3. World Environmental Institute (WEI)
4. Global Network for Monitoring the Biosphere (MABNET)

Table 25.3 Monitoring Programs Proposed

INTERNATIONAL MONITORING ACTIVITIES PROGRAM INFORMATION SHEET			
NAME OF PROGRAM		INTERNATIONAL CENTER FOR THE ENVIRONMENT (ICE)	
PURPOSE OF PROGRAM			
STATUS	PROPOSED	Establish a global environmental monitoring program, an international research center on global environmental problems, and a central "intelligence service" or clearinghouse on environmental knowledge.	
COVERAGE	GLOBAL		
DURATION	LONG-TERM (INDEFINITE)		
NO. OF COUNTRIES PARTICIPATING	DOZENS		
COORDINATING ORGANIZATION		SCIENTIFIC COMMITTEE ON PROBLEMS OF THE ENVIRONMENT (SCOPE)	
COOPERATING ORGANIZATION(S)		UNITED NATIONS	
PRINCIPAL DISCIPLINES		ECOLOGY, SOCIOLOGY, TECHNOLOGY	
PRINCIPAL PHENOMENA MONITORED		Human population; atmospheric carbon dioxide and turbidity, pollution of the oceans and coastal waters, radioactivity in the biosphere pressure on water resources, soil erosion, noise, pollution in air, soils, and water; thermal pollution, loss of gene pools, urban environment.	
NO. OF MONITORING STATIONS	HUNDREDS		
PHYSICAL SPHERE	BIOSPHERE		
TIME RESPONSE	REAL AND DELAYED		
SAMPLING FREQUENCY	CONTINUOUS, PERIODIC		
DATA ACQUISITION TECHNIQUES		Sampling, survey, ground stations, remote sensing systems.	
DATA STORAGE AND RETRIEVAL FACILITIES	TO BE DETERMINED		
COMMUNICATIONS FACILITIES	TO BE DETERMINED		
RESEARCH CENTER(S)		COORDINATE ENVIRONMENTAL RESEARCH ACTIVITIES	
REMARKS	Proposal to be submitted to the U. S. General assembly following the U. N. Conference on the Human Environment in 1972.		
REFERENCES	1. Report of the Ad Hoc Committee of ICSU on "Problems of the Human Environment", May 1970 Additional information may be obtained from: International Council of Scientific Unions, 7 via Cornelio Celso, Rome 00161, Italy.		

Table 25.3 (continued)

INTERNATIONAL MONITORING ACTIVITIES / PROGRAM INFORMATION SHEET			

NAME OF PROGRAM		GLOBAL NETWORK FOR ENVIRONMENTAL MONITORING (GNEM)	
PURPOSE OF PROGRAM		Develop a global capability for the measurement of biological and phhsical parameters to establish ecological baselines which will lead to a better understanding of man's impact on the biosphere.	
STATUS	REPORT NEAR COMPLETION	^	
COVERAGE	GLOBAL	^	
DURATION	INDEFINITE	^	
NO. OF COUNTRIES PARTICIPATING	DOZENS		
COORDINATING ORGANIZATION		ICSU/SCIBP/POSSIBLY U.N.	
COOPERATING ORGANIZATION(S)		TO BE DETERMINED UPON COMPLETION OF FEASIBILITY STUDY	
PRINCIPAL DISCIPLINES		ECOLOGY, BIOLOGY	
PRINCIPAL PHENOMENA MONITORED		Water; air; flora; fauna; ecosystems: CO, O, SO, CH, NO, NO; Aerosols and particulates; Radioactivy Species diversity Biological stimulants Mutations Biological toxins Radioisotopes Biological indicators Population numbers.	
NO. OF MONITORING STATIONS	25 - 30	^	
PHYSICAL SPHERE	BIOSPHERE	^	
TIME RESPONSE	REAL & DELAYED	^	
SAMPLING FREQUENCY	CONTINUOUS, PERIODIC, SPORADIC	^	
DATA ACQUISITION TECHNIQUES		Sampling, ground instrumentation, remote sensing	
DATA STORAGE AND RETRIEVAL FACILITIES	TO BE DEVELOPED		
COMMUNICATIONS FACILITIES	TO BE DEVELOPED		
RESEARCH CENTER(S)		NONE: USE OF NATIONAL AND REGIONAL CENTERS	
REMARKS	Ad hoc committee of SCIBP near final report on feasibility of a Global Network for Environmental Monitoring; U. S. Task Force of this ad hoc committee near final report on design of a baseline station for environmental monitoring.		

REFERENCES
1. Blair, Frank, U. S. National Committee for the International Biological Program, National Academy of Sciences, Washington, D. C. 20418
2. Oliver, Richard, Division of Biology and Agriculture, U. S. National Research Council, National Academy of Sciences.
3. Hilst, Glenn, Chairman, GNEM Task Force, U. S. National Committee for the IBP.
4. Jenkins, Dale, Vice Chairman, GNEM Task Force, U. S. National Committee for the IBP.

For further information contact: U. S. National Academy of Sciences, 2101 Constitution Avenue, Washington, D. C. 20418.

Table 25.3 (continued)

INTERNATIONAL MONITORING ACTIVITIES PROGRAM INFORMATION SHEET			
NAME OF PROGRAM		WORLD ENVIRONMENTAL INSTITUTE (WEI)	
PURPOSE OF PROGRAM		Establish a non-political World Environmental Institute (WEI) to act as a global environmental research center and clearinghouse for environmental information, independent of existing international organizations.	
STATUS	CONCEPTUAL PROPOSAL		
COVERAGE	GLOBAL		
DURATION	LONG-TERM- INDEFINITE		
NO. OF COUNTRIES PARTICIPATING		All countries that wish to	of the world participate
COORDINATING ORGANIZATION		A NEW INDEPENDENT ORGANIZATION OUTSIDE U. N. FAMILY	
COOPERATING ORGANIZATION(S)		TO BE DECIDED	
PRINCIPAL DISCIPLINES		HUMAN ECOLOGY	
PRINCIPAL PHENOMENA MONITORED		World-wide pollution, environmental phenomena resulting from technology and population growth.	
NO. OF MONITORING STATIONS	Possibly HUNDREDS (?)		
PHYSICAL SPHERE		BIOSPHERE	
TIME RESPONSE		REAL AND DELAYED?	
SAMPLING FREQUENCY		CONTINUOUS & PERIODIC (?)	
DATA ACQUISITION TECHNIQUES		To be decided	
DATA STORAGE AND RETRIEVAL FACILITIES		TO BE DECIDED	
COMMUNICATIONS FACILITIES		TO BE DECIDED	
RESEARCH CENTER(S)		TO BE DECIDED	
REMARKS	U.S.Senate Resolution 399 urges U.S.representatives to the First International Conference on the Human Environment propose the creation of a World Environmental Institute to the Conference.		
REFERENCES	1. Magnuson, W.G., "A world View of the Environment", Remarks before the Second Annual International Geoscience Electronics Symposium, Washington, D. C., April 16, 1970. 2. Senator Magnuson's Plan for a World Environmental Institute, The Congressional Record, Vol. 116, No. 64, 23 April 1970 3. U.S. Senate Resolution 399-Resolution to Create A World Environmental Institute, Vol.116, No. 65, 27 April 1970. Further information may be obtained from The Honorable Warren G. Magnuson, The United States Senate, Washington, D. C.		

Table 25.3 (continued)

INTERNATIONAL MONITORING ACTIVITIES PROGRAM INFORMATION SHEET			
NAME OF PROGRAM		GLOBAL NETWORK FOR MONITORING THE BIOSPHERE (MABNET)	
PURPOSE OF PROGRAM		To establish a global network of environmental monitoring ecostations, national and global aircraft and satellite remote sensing systems, a global biosphere communications network, extensive data storage and retrieval systems, and an international Biosphere Research Center to develop ecosystem models for the purposes of evaluating man's impact on his environment and developing plans for the rational management and conservation of the natural resources of the biosphere.	
STATUS	PROPOSAL DEVELOPMENT		
COVERAGE	GLOBAL		
DURATION	LONG TERM - INDEFINITE		
NO. OF COUNTRIES PARTICIPATING		DOZENS	
COORDINATING ORGANIZATION			TO BE DECIDED
COOPERATING ORGANIZATION(S)			ALL NATIONAL AND INTERNATIONAL PROGRAMS INVOLVED IN BIOSPHERE MONITORING AND RESEARCH
PRINCIPAL DISCIPLINES			BIOSPHERE DYNAMICS
PRINCIPAL PHENOMENA MONITORED			
NO. OF MONITORING STATIONS		THOUSANDS	MABNET would establish monitoring facilities to obtain data and information where these were not provided by established programs but considered essential to the understanding of the fundamental processes of the biosphere.
PHYSICAL SPHERE		BIOSPHERE	
TIME RESPONSE		REAL & DELAYED	
SAMPLING FREQUENCY		CONTINUOUS AND PERIODIC	
DATA ACQUISITION TECHNIQUES			Sampling, survey, ground-based ecostations, aircraft and satellite remote sensing systems.
DATA STORAGE AND RETRIEVAL FACILITIES		NETWORK OF BIOSPHERE DATA CENTERS	
COMMUNICATIONS FACILITIES			GLOBAL BIOSPHERE TELECOMMUNICATIONS SYSTEM (GBTS)
RESEARCH CENTER(S)			BIOSPHERE RESEARCH CENTER (BRC)
REMARKS			

REFERENCES
1. MABNET: The Establishment of a Global Network of Ecostations for the "Man and the Biosphere" Program, Smithsonian Institution, February 1970.
2. Batisse, Michael, Director, Natural Resources Research Division, UNESCO, personal conversations, 1969, 1970.

For further information contact: UNESCO, Division of Natural Resources, Place de Fontenoy, 75 Paris 7e, France.

List of Abbreviations

ACMRR	Advisory Committee on Marine Resources Research, FAO Cooperative Investigations of the Caribbean and Adjacent Regions
CIG	Comité International de Geophysique
COSPAR	Committee on Space Research of the ICSU
CSK	Cooperative Study of the Kurochio and Adjacent Regions
ECOSOC	Economic and Social Council of the United Nations
FAGS	Federation of Astronomical and Geophysical Services
FAO	Food and Agriculture Organization of the United Nations
GARP	Global Atmospheric Research Program
GNEM	Global Network for Environmental Monitoring (IBP)
GTS (of WWW)	Global Telecommunication System
IABO	International Association of Biological Oceanography
IAEA	International Atomic Energy Agency
IAPSO	International Association for the Physical Sciences of the Ocean
IBP	International Biological Program
ICE	International Center for the Environment (ICSU)
ICES	International Council for the Exploration of the Sea
ICSM	International Cooperative Studies of the Mediterranean
ICSU	International Council of Scientific Unions
IDOE	International Decade of Ocean Exploration
IGOSS	Integrated Global Ocean Station System
IGU	International Geographical Union, ICSU
IHD	International Hydrological Decade
IMCO	Intergovernmental Maritime Consultative Organization
IOC	Intergovernmental Oceanographic Commission

IUBS	International Union of Biological Sciences, ICSU
IUCN	International Union for the Conservation of Nature
IUCSTP	Inter-Union Commission on Solar-Terrestrial Physics
IUGG	International Union of Geodesy and Geophysics, ICSU
IUGS	International Union of Geological Sciences, ICSU
MABNET	Global Network for Monitoring the Biosphere (UNESCO)
SCAR	Scientific Committee on Antarctic Research of the ICSU
SCIBP	Special Committee on the International Biological Program
SCOR	Scientific Committee on Ocean Research of the ICSU
UNESCO	United Nations Educational, Scientific and Cultural Organization
WEI	World Environment Institute
WMO	World Meteorological Organization
WWW	World Weather Watch

References

Blair, W. Frank, 1969. *Need for a Global Network for Environmental Monitoring* (Washington, D.C.: National Academy of Sciences, SCIBP Ad Hoc Committee on Global Monitoring, the International Council of Scientific Unions).

Citron, Robert, 1970a. *An Approach to the Organization of the UNESCO "Man and the Biosphere Program"* (Washington, D.C.: Smithsonian Institution).

Citron, Robert, 1970b. *MABNET, The Establishment of a Global Network of Ecostations for the "Man and the Biosphere" Program* (Washington, D.C.: Smithsonian Institution).

Dasmann, Raymond F., 1970. *Man and the Biosphere—A Challenge for UNESCO in a New International Program* (Washington, D.C.: The Conservation Foundation).

International Council of Scientific Unions, 1969. *Report of the Ad Hoc Committee of ICSU on Problems of the Human Environment.*

Joint Group of Experts on the Scientific Aspects of Marine Pollution, 1970. *Report of the Second Session,* Intergovernmental Maritime Consultative Organization, Food and Agricultural Organization, U.N. Educational, Scientific and Cultural Organization, World Meteorological Organization, World Health

Organization, International Atomic Energy Agency (New York: United Nations).

Lundholm, Bengt, 1968. *Global Baseline Stations* (Stockholm: Swedish Ecological Research Committee).

Report of the Secretary-General, 1969. *Problems of the Human Environment* (New York: United Nations).

Smithsonian Institution, 1970. *Current and Planned National, Regional, and Global Environmental Monitoring Programs* (Cambridge, Massachusetts).

U.N. Educational, Scientific and Cultural Organization (UNESCO), 1968. *Intergovernmental Conference of Experts on the Scientific Basis for Rational Use and Conservation of the Resources of the Biosphere; Final Report,* 1969 (Paris: UNESCO).

U.N. Educational, Scientific and Cultural Organization, 1970. *Proposals for a Long-term Intergovernmental and Interdisciplinary Program on Man and the Biosphere* (draft), Natural Resources Research Division (Paris: UNESCO).

World Health Organization (WHO), 1970. *Global Environmental Health Monitoring* (Geneva: WHO).

Part VI   Modeling: A Tool for
                           Understanding and
                           Management

Until recently, the extensive and fruitful use of mathematical models of environmental processes has been confined mainly to the field of meteorology where they are used to predict weather and climate conditions. In principle, these same models can be developed for ocean prediction, but lack of synoptic data from the world's oceans has hampered progress. The state-of-the-art of estuary and ocean modeling is reviewed in the following three papers prepared for SCEP.

In the first paper, Dr. Arnason explains how mathematical models can be established for estuaries where the problems of lack of data are not as severe as in the case for the open oceans. In addition to providing a conceptual and mathematical framework for these models, the paper also notes the types of management problems which these models might help resolve. Since estuaries are usually the target for much of man's discharge of wastes and are also of critical importance in the development of the world's fisheries and in coastal urban living, information for effective management is seriously needed.

The analysis and prediction of the movements of two coupled fluids, the hydrosphere and the atmosphere, have provided a focus of study for mathematically inclined earth scientists. Forecasts of wind and current patterns have been useful in the management of such enterprises as shipping, air transport, and agriculture. A new type of question is now being posed by students of air and water circulation: What will be the spatial distribution of materials introduced by man at one location after various periods of time? The paper by Dr. Reid reviews the present literature on general circulation patterns in the oceans, and the paper by Dr. Bryan is a report on the present status of qualitative and quantitative models of oceanic circulation. In the other volume in this series, *Man's Impact on the Climate*, edited by Drs. William H. Matthews, William W. Kellogg, and G. D. Robinson, there are several papers that review the present state of mathematical modeling of atmospheric processes. Modeling of the atmosphere-ocean system is also treated in the Report of the Study of Men's Impact on Climate (SMIC), *Inadvertent Climate Modification* (The M.I.T. Press, 1971).

26 Estuary Modeling

Geirmundur Arnason

Introduction

Following Pritchard (1968), one may define an estuary as "a semi-enclosed coastal body of water which has a free connection with the open sea and within which seawater is measurably diluted with fresh water derived from land drainage." In simple terms, the main characteristics of an estuary are therefore a freshwater river at one end and a free connection with the open sea at the other which allows the ocean-generated tidal wave to enter the estuary and saltwater to intrude (see Figure 26.1).

Physiographically, estuaries vary considerably. An important type is the so-called *coastal-plain estuary* that is quite common along the Atlantic and Gulf coasts of the North American continent. It is typified by the drowned-out mouths of rivers or large valleys and includes water bodies such as Chesapeake Bay and the mouth of the Mississippi River. Salinity typically drops from around thirty parts per thousand at the mouth of the estuary to about 0.1 part per thousand at the head. Above this limit the estuary is usually a stretch of freshwater river that is still subject to tidal currents and is called the tidal section of the river.

Another common type is the *bar-built estuary*. It is an area enclosed by barrier beaches and is generally elongated and parallel to the coastline. A bar-built estuary might be considered a com-

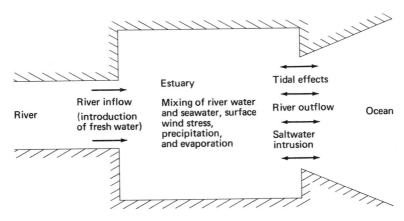

Figure 26.1 Schematic representation of an estuary and the dynamic processes

Prepared for SCEP.

posite system, part being an outer bayment partially enclosed by barrier beaches, and part being a drowned river valley. Because the inlets connecting the bar-built estuary with the ocean are usually small, tidal action is considerably reduced in such estuaries. These systems are often shallow, and the wind provides an important mixing mechanism. Albemarle Sound and Pamlico Sound in North Carolina are examples of bar-built estuaries.

A third type of estuary is the *fjord*. Its characteristics include a U-shaped cross section, considerable depth (300 to 400 meters), and a shallow sill formed by terminal glacial deposits at the mouth. Examples of this type are the fjords of Norway and British Columbia.

In this paper I shall describe devices, or models, by which processes taking place in an estuary may be suitably reproduced. The primary objective of such a model is to simulate the effects of tides, salinity intrusion, and wind stress in sufficient detail, and to thereby predict the motion, water height, and salinity within the estuary over a tidal cycle. There are three types of models presently being used for these purposes: the electric analog, the hydraulic model, and the dynamic (mathematical) model. The first one, as the name indicates, is an analog device by which the tides, the flow, and water height in a natural estuary are simulated by an alternating current and voltage in an electric circuit. Because of the very limited usefulness of this type of model, further discussion of it will be omitted. Applications of this model to one-dimensional flow are described by Einstein and Harder (1959) and by Harder and Masch (1961). The hydraulic model is a scaled-down replica where tides and river flow are generated mechanically. It has been in use in Europe since before the turn of the century and used extensively in this country since around 1930 by the U.S. Army Corps of Engineers. The third type of model—the dynamic model—is the most recent one but potentially the most promising of the three for future studies of estuarine processes. It is with this model I shall be mainly concerned in this article. Here, the physical laws governing estuarine processes are expressed mathematically, by a set of differential equations and appropriate initial and boundary conditions, and then solved numerically as an initial-value problem. A model of this type is by no means restricted to simulation of dynamic processes alone. It is equally capable, in principle, of reproducing other physical processes as well as known chemical and biological processes. Within this expanded framework, I shall refer to the model as a mathematical model. With the availability of high-speed computers and the impressive advances in the use of finite-difference techniques, mathematical models are becoming increasingly attractive and preferred as an alternative to both analog and hydraulic models. In principle, a mathematical model can simulate all processes known to be of importance and provide a computer-stored collective representation of a variety of estuaries. Through the internal logic of the computer program, provisions can be made for variability in the geometrical configuration, size, amount of freshwater inflow, and other parameters distinguishing one estuary from another.

Until recently, an extensive and fruitful use of dynamic models was mainly confined to meteorology. While such models are equally applicable to oceanic prediction, lack of synoptic data from the world oceans has so far hampered similar progress in oceanic prediction over large areas. This restriction, however, does not apply in the same degree to estuaries. Here, the knowledge of initial state is less important. Future states are essentially determined by topographic features and by atmospheric and oceanic conditions at the boundaries, all of which are either known or predictable.

In the following section, I shall briefly describe the use and limitations of hydraulic models and then devote the remaining sections to mathematical models, with emphasis on simulation of the dynamic processes.

Hydraulic Models
A hydraulic model, also called a physical model, is a suitably scaled reproduction of a natural estuary. In principle, such a model should satisfy the laws of geometric, kinematic, and dynamic similarities (Keulegan, 1966); in practice, almost all hydraulic models are geometrically distorted, and only in some cases it is possible to satisfy exactly the law of dynamic similarity (Simmons, 1966). The most common geometric distortion is brought about by the practical necessity of using different linear scales for vertical and horizontal dimensions.

Since about 1930, a great variety of hydraulic models has been built and operated by the U.S. Army Corps of Engineers at their Waterways Experiment Station in Vicksburg, Mississippi, and used extensively to simulate conditions in natural estuaries. Problems studied by means of such models include shoaling, saltwater intrusion, the hydraulics of river estuaries, tidal flooding by storm surges, and proper disposal of dredge spoils (Simmons and Lindner, 1965).

In studies of shoaling, both movable-bed and fixed-bed types of models are used to simulate natural conditions depending upon the physical mechanism affecting the particular shoaling process (Simmons, 1969). The movable bed is generally used when wave action as well as tides and littoral currents cause movement and deposition of ocean sand; fixed-bed models are often used in studies of shoaling of inner-harbor navigation channels.

Studies of saltwater intrusion necessitate the simulation of tides at the mouth of the estuary and of freshwater inflow of significant tributaries. This is achieved by means of tide generators and pumping plants attached to a water supply. A model of Narragansett Bay has been used to simulate the effects of the great hurricane of September 1938 which crossed the New England coast just west of Narragansett Bay and caused the water level at Providence, Rhode Island, to rise 16 feet above mean sea level. Tides and tidal currents were reproduced by a tide generator, and a hurricane surge was simulated by a special surge generator that consisted of a motorized movable bulkhead, located in a basin adjacent to, and connected with, the Rhode Island Sound portion of the model. One of the objects of this study was to show how various barriers would affect current velocities and water heights throughout the bay.

Hydraulic models representing many of the better-known estuaries in the United States have been found useful in a number of studies of the types mentioned. While these models will continue to be useful, it is not economically feasible to build them at the rate required by rapidly growing needs for comprehensive studies of a variety of estuaries. Other limitations to the use of hydraulic models are as follows:

1. The impossibility of satisfying the laws of dynamic similarity except for very simple models; according to Birkhoff (1960):

"in practice, theoretical considerations are seldom invoked in hydraulic model studies of rivers and harbors. Reliance is placed on reproducing various aspects of the observed behavior under actual conditions. It is hoped that variation in behavior due to altered conditions will then also be reproduced to scale—even though there is no rational argument to support this hope."

2. The problem that verification often calls for painstaking modification in the bottom roughness of the model in order to obtain results in satisfactory agreement with measurements gathered from the prototype.

3. Inherent physical limitations in reproducing the effects of wind and Coriolis force.

4. High cost of construction and operation.

Until quite recently, hydraulic models were the main tool for simulating physical conditions in natural estuaries.

A Conceptual Framework for Mathematical Modeling

Within the last decade, mathematical models have been used for simulating storm surges and tidal effects on flow and water height in rivers and estuaries. Experience with these models has been accumulating rapidly over the past few years and holds out high hopes of useful simulation of a variety of estuarine processes.

Figure 26.2 represents an attempt to identify the main building blocks of a mathematical model complex that would simulate a variety of estuarine processes. The first column lists the essential steps in the required mathematical procedures while the second distinguishes a variety of models on the basis of geometrical, physical, and other constraints. Each of these models may itself be capable of providing an answer to a practical question. More important, perhaps, the models may be considered essential building blocks in a model complex required to facilitate management decisions. The third column names the types of water bodies being considered, and the fourth column lists the main processes to be modeled.

It is not presently feasible, for a number of reasons, to make a model of the complexity outlined in Figure 26.2. Among these are incomplete quantitative knowledge of the chemical and biological processes, of turbulent diffusion, and of the laws governing the population dynamics of the various species that inhabit

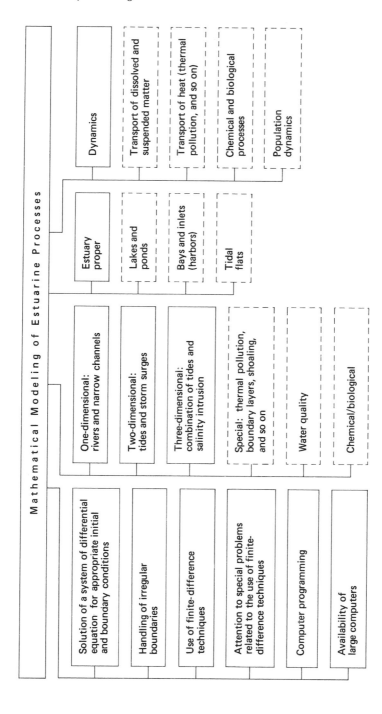

Figure 26.2 Conceptual building blocks in model simulation

the estuary. Further, the very extensive computer program required for such a model would be well beyond present-day computer capacities.

For these reasons, and because of my very limited knowledge of the chemical and biological processes, I shall mainly confine my further discussion to the modeling of the dynamic processes as indicated by the solid blocks in Figure 26.2. A dynamic model of this scope is, in my opinion, presently feasible, although I know of no three-dimensional model yet in existence for an estuary. Within the community of researchers and practitioners specifically concerned with estuaries, doubts have been expressed about the feasibility of designing and operating such a model; these are exemplified by a recent report by the Committee on Tidal Hydraulics of the Corps of Engineers (1969) and another report (relying heavily on the first one) by the Systems Analysis Group (Civil Functions) Office, Secretary of the Army and the Office of Chief Engineers (1969).

The more optimistic view expressed in this article is based on advances in modeling in other fields such as the general circulation of the atmosphere (Manabe, 1969a, 1969b, Bryan, 1969b), numerical weather prediction (Shuman and Hovermale, 1968), atmospheric convection (Arnason, Greenfield, and Newberg, 1968; Arnason, 1969), and in other branches of oceanography and fluid dynamics. As examples, I mentioned specifically the impressive general circulation numerical experiments presently conducted at the Geophysical Fluid Dynamics Laboratory of ESSA and the six-level operational weather prediction model of the National Meteorological Center of the U.S. Weather Bureau, which, in complexity, equal or exceed the three-dimensional estuary model just mentioned.

Dynamic Models

One of the first contributions on the dynamics of an estuary is an article published in 1899 by the renowned Swedish oceanographer, von Walfrid Ekman (1899). The subject of Ekman's study is the density current induced by saltwater intrusion at the mouth of the river Götaälv in southwestern Sweden. Observations show that the river water flowing into the sea near the surface is compensated by a denser (saltier) counterflow along the river bed, resulting in upwelling motion from below. This motion pattern was reflected in the salinity distribution and could be shown experimentally by drift buoys. Wind-produced surface stress was shown to modify the flow pattern importantly. Ekman succeeded in explaining the observed behavior theoretically, and this is, perhaps, the first well-conceived dynamic model.

The foremost recent contributor to the description and understanding of estuarine processes is D. W. Pritchard, who, in a series of articles (1952, 1956, 1958) has given a comprehensive treatment of estuaries, including their dynamics. The governing equations, however, are rather complex and defy general solutions by classical mathematical tools. Special solutions, however, may be obtained by analytic means applicable to some important features of a given estuary such as Ekman's solutions for the flow near the mouth of Götaälv. Other special solutions are those of Takano (1954), who studied the ocean flow pattern of light surface water off the mouth of a river, and the solutions of Rattray and Hansen (1962, 1965) for flow and salinity. Common to these solutions is that they all represent steady-state conditions, are restricted to two space dimensions, and include only some of the dynamic processes known to be of importance. Such simplifications are often justified by geometrical and physical restraints applicable to a given estuary. Narrowness, for example, may so restrict lateral variation of momentum and salinity that a lateral space coordinate is not required in a model. Another case is that of vigorous tidal flushing that may obliterate vertical gradients in the physical variable and thereby eliminate the need for a vertical coordinate in the dynamic model.

To obtain time-dependent solutions for estuaries of arbitrary geometric configuration, without suppressing physical processes that may be of importance, it is necessary to resort to numerical techniques for solving the appropriate set of equations. The generally best approach in this case is what is commonly known as the "initial value" approach. This procedure entails formulation of the physical problem at hand, which is expressed in terms of one or more governing differential equations, appropriate initial and boundary conditions, and other pertinent constraints. Provided the initial state is sufficiently known, later states are obtained by a stepwise procedure involving successive short-time extrapolations. For requirements of accuracy, and for computational stability, the time step may not exceed a certain threshold value determined by numerical analysis of the problem.

Adequate simulation of the dynamic processes may require the inclusion of the following physical factors:
1. River discharge
2. Astronomical tides
3. Meteorological tides
4. Salinity distribution
5. Turbulent transport
6. Wind stress at the surface
7. Others (such as evaporation and precipitation).

It is customary to assume the fluid to be incompressible and its pressure to be hydrostatic.

Existing dynamic models of rivers and estuaries, including storm surge models, are confined to one or two space dimensions (Hansen, 1956; Freeman, Baer, and Jung, 1957; Svansson, 1959; Fisher, 1959; Rossiter and Lennon, 1965; Jelsnianski, 1967; Leendertse, 1967; Reid and Bodine, 1968; Garrison, 1968; Banks, 1968) and may be steady state or time dependent. As indicated in the previous section models embracing three space dimensions have not yet appeared in the published literature and are considered by some (Committee on Tidal Hydraulics, 1969; Systems Analysis Group, 1969) beyond the state-of-the-art of modeling and the capacities of present computers.

One- and Two-Dimensional Models

For many rivers, including their estuarine portions, a one-dimensional model may provide an adequate simulation of flow, water height, and salinity within a tidal cycle. Such models assume that the flow is directed along the course (axis) of the river, that the slopes of the river bed are small, and that density is essentially uniform. With these assumptions, the equations of motion and mass continuity take the form

$$\frac{\partial u}{\partial t} + u \frac{\partial u}{\partial x} + g \frac{\partial h}{\partial x} + \frac{g}{C^2 R} u|u| + \frac{qu}{\sigma} = 0 \qquad (26.1)$$

$$\frac{\partial \sigma}{\partial t} + \frac{\partial}{\partial x}(Q) - q = 0 \qquad (26.2)$$

$$\frac{\partial \sigma S}{\partial t} + \frac{\partial}{\partial x} Q_1 S = \frac{\partial}{\partial x} \sigma A \frac{\partial S}{\partial x}. \qquad (26.3)$$

Here, t and x are time and space coordinates, u is velocity and $|u|$ its absolute value, g is acceleration due to gravity, h is height of the free water surface, C is a friction coefficient, R is hydraulic radius, σ is cross section, q is volume transport from tributaries, Q is volume transport through a cross section, S is salinity, Q_1 is salt transport, and A is a coefficient of turbulent diffusion. The foregoing system is readily solved by means of finite-difference techniques and by appropriately specifying the values of u, σ, S, and $\partial S/\partial x$ at the boundaries (Boicourt, 1969).

The system of equations governing simulation of a two-dimensional horizontal flow is

$$\frac{\partial u}{\partial t} + u\frac{\partial u}{\partial x} + v\frac{\partial u}{\partial y} - fv + g\frac{\partial h}{\partial x} = F_x \tag{26.4}$$

$$\frac{\partial v}{\partial t} + u\frac{\partial v}{\partial x} + v\frac{\partial v}{\partial y} + fu + g\frac{\partial h}{\partial y} = F_y \tag{26.5}$$

$$\frac{\partial h}{\partial t} + \frac{\partial}{\partial x}(Hu) + \frac{\partial}{\partial y}(Hv) = 0 \tag{26.6}$$

$$\frac{\partial S}{\partial t} + \frac{\partial}{\partial x}(Su) + \frac{\partial}{\partial y}(Sv) = \frac{\partial}{\partial x}A\frac{\partial S}{\partial x} + \frac{\partial}{\partial y}B\frac{\partial S}{\partial y}, \tag{26.7}$$

where t, x, and y denote time and the two space coordinates u and v are the two velocity components, f is the Coriolis parameter, H is the depth, F_x and F_y are the x- and y-components of turbulent friction, and A and B are coefficients of turbulent salinity diffusion; the remaining symbols were explained previously.

A two-dimensional, time-dependent model in which one of the space coordinates refers to the vertical is in certain cases more appropriate than the preceding one; but, in either case, the system of governing equations is quite similar.

Three-Dimensional Models

Three-dimensional time-dependent flow and salinity distribution is governed by the following set of equations

$$\frac{\partial u}{\partial t} + u\frac{\partial u}{\partial x} + v\frac{\partial u}{\partial y} + w\frac{\partial u}{\partial z} - fv + \frac{1}{\rho}\frac{\partial p}{\partial x} = F_x \tag{26.8}$$

$$\frac{\partial v}{\partial t} + u\frac{\partial v}{\partial x} + v\frac{\partial v}{\partial y} + w\frac{\partial v}{\partial z} + fu + \frac{1}{\rho}\frac{\partial p}{\partial y} = F_y \tag{26.9}$$

$$\frac{\partial p}{\partial z} + g\rho = 0 \tag{26.10}$$

$$\frac{\partial u}{\partial x} + \frac{\partial v}{\partial y} + \frac{\partial w}{\partial z} = 0 \tag{26.11}$$

$$\rho = \rho_0(1 + cS) \tag{26.12}$$

$$\frac{\partial S}{\partial t} + \frac{\partial}{\partial x}(Su) + \frac{\partial}{\partial y}(Sv) + \frac{\partial}{\partial z}(Sw)$$
$$= \frac{\partial}{\partial x} A \frac{\partial S}{\partial x} + \frac{\partial}{\partial y} B \frac{\partial S}{\partial y} + \frac{\partial}{\partial z} C \frac{\partial S}{\partial z}, \tag{26.13}$$

where w is the vertical velocity component, p is pressure, ρ is the density of the estuarine water, ρ_0 is the density of fresh water, c is a constant, and A, B, and C are coefficients of turbulent diffusion. It may be shown that, because of the greatly variable depth of an estuary and the variable height of its free surface, it is preferable, before solving the foregoing system, to replace the independent variable by a new independent variable σ defined as

$$\sigma = \frac{p}{p_b}, \tag{26.14}$$

where p_b is the bottom pressure, and p is assumed to be zero at the free surface. It is to be noted that in this new coordinate system, both the bottom and the free water surface are coordinate surfaces where σ has the values 1 and 0, respectively. For the sake of brevity, the transformed set of equations is not given here.

Limitations

There are at least two major limitations to a realistic simulation of the dynamic processes. The first, which is physical in nature, is incomplete knowledge of turbulent transport processes. The usual expression for turbulent fluxes of momentum, energy, and matter is the product of a transfer coefficient and a gradient of the element under consideration. By assuming this to be a proper form, the problem is reduced to that of a correct formulation of the transfer coefficient which, in general, varies both in space and time in a manner that is not well understood. The subject of turbulent diffusion in estuaries has been given some attention, in particular by staff of the Chesapeake Bay Institute (Kent and Pritchard, 1959; Pritchard, 1960; Pritchard and Carpenter, 1960; Okubo, 1962; and Pritchard, Okubo, and Carter, 1966), but in actual model experiments for a given estuary, one may have to rely upon indirect determination of the transfer coefficients by measurements of flow and other pertinent elements (Boicourt, 1969).

The other major limitation which, however, is more easily overcome, is that of cost and computer capacity. Sufficiently fine spatial resolution may require thousands of data points in any given horizontal plane of an estuary, and this figure is then to be multiplied by the number of vertical data points and by the number of elements (flow components, water height, and salinity, depth, and so forth) pertinent to the model. The finer the spatial resolution, the smaller the time step that is permissible without violating computational stability. For many estuaries, however, computers such as the CDC 6600, Sperry Rand 1108, and IBM System 360 Model 91 are capable of handling a three-dimensional dynamic model. While detailed simulation of a complex estuary such as the entire Chesapeake Bay, however, is now beyond the capacity of these computers, this is not likely to be more than a passing limitation. Rapid advances in computer design make it probable that within the span of a few years computer capacity will increase, perhaps, a thousandfold.

Other Models

One may consider the dynamic model of primary importance in estuarine simulation because its output—flow information—provides input to special-purpose models such as those simulating temperature changes due to thermal wastes and the distribution of water quality elements. Knowledge of the elements predicted by these models is not required by the dynamic model since these elements have little or no effect upon the dynamic forces. An exception, perhaps, is temperature, which may change the density distribution sufficiently to modify the flow. Such, however, is not the case with dissolved oxygen, biochemical oxygen demand, nutrients, and other water quality elements.

Both thermal pollution and other water quality models are governed by a transport equation of the general form

$$\frac{\partial s_i}{\partial t} + \frac{\partial}{\partial x}(s_i u) + \frac{\partial}{\partial y}(s_i v) + \frac{\partial}{\partial z}(s_i w)$$
$$= \frac{\partial}{\partial x} A_i \frac{\partial s_i}{\partial x} + \frac{\partial}{\partial y} B_i \frac{\partial s_i}{\partial y} + \frac{\partial}{\partial z} C_i \frac{\partial s_i}{\partial z} + \sum_j S_j, \qquad (26.15)$$

where u, v, and w are the flow components (provided by a dynamic model), and s_i is temperature, dissolved oxygen, biochemical oxygen demand, a nutrient, or any other element one may wish to predict. Terms A_i, B_i, and C_i are coefficients of turbulent diffusion, and S_j is a source term. Equation (26.15) may involve time-variable elements other than the particular s_i being predicted, in which case prediction entails solving a closed set of governing equations containing all the elements pertinent to the processes at hand. Prediction of nutrients and other water quality parameters involves source terms often reflecting complex chemical and biological processes that, in some cases, are poorly known or not readily formulated.

In the case of predicting temperature changes in response to waste heat effluent, $\sum_j S_j$ includes the magnitude of the heat source as well as the following source terms:
1. Incoming solar (short-wave) radiation
2. Reflected solar radiation
3. Incoming long-wave radiation
4. Outgoing long-wave radiation
5. Reflected long-wave radiation
6. Heat transport due to evaporation
7. Heat transport due to conduction.

Since these terms depend on atmospheric conditions, a temperature prediction model, in principle, requires prediction of these conditions. In practice, these source terms may be determined from appropriate climatological and empirical data.

Mathematical models have been developed (Edinger and Geyer, 1965, 1968; Seaders and Delay, 1966; Jaske, 1968; and Edinger, Brady, and Graves, 1968) by which temperatures are evaluated downstream of sites of heavy thermal pollution. These models are still somewhat crude and need refinement. A more precise temperature prediction may be done by air-sea interaction models, in which detailed temperature and flow predictions are made for two adjacent layers of air and water. A model of this type has been developed and tested by Pandolfo (1969) and could be adapted to thermal pollution problems.

Change in dissolved and suspended matter is governed by equation (26.15). Examples of biologically important elements are nutrients, such as nitrogen and phosphorus, coliform bacteria, biochemical oxygen demand, and dissolved oxygen. The concentration of dissolved oxygen is perhaps the best single indicator of the health of a water course, that is, its capacity to support marine life. Oxygen enters principally by aeration and by photosynthetic activity of phytoplankton. It is consumed by respiration and by bacteria in reducing organic matter, so that its rate of consumption is proportional to the rate of organic matter present.

Prediction of the concentration of dissolved oxygen and of biochemical oxygen demand requires, in order to determine the appropriate sources in equation (26.15), the knowledge of

1. Aeration
2. Waste discharged along the watercourse
3. Bacterial oxidation of organic matter
4. Photosynthesis
5. Plant respiration
6. Oxygen demand of the benthal layer.

It is beyond the scope of this article to touch more than briefly upon the important area of water quality modeling. Processes that it would be desirable to simulate include extraction of nutrients by plants, the carbon-oxygen photosynthetic cycle, and various influencing factors such as illumination and turbidity. A number of relatively simple models have been designed to deal with various measures of water quality (Stommel, 1953; Preddy and Webber, 1968; Holly and Harleman, 1965; and Grill, 1970) including nutrients (Riley, 1965) and plankton (Riley, 1967).

The important area of population dynamics is not treated here.

Potential Applications

Mathematical modeling is of great interest in its own right, in that it furthers the understanding of the estuarine processes and displays in considerable detail their interactions. More important, perhaps, models are potentially important tools for a variety of practical decisions and for marine resources management in general. Dynamic models alone have a number of applications, such as

1. Determining the effects of upstream changes of river flow— for example, by dam and reservoir construction and operation

2. Determining effects of other engineering projects on a river
3. Experimenting with alternate navigation routes and various channel depths
4. Obtaining views of the effects of navigation projects on the physical structure and function of the lower estuary
5. Estimating the effects of storm surges
6. Aiding in the selection of best location and design of terminals, piers, and seawalls
7. Simulating the flow characteristics in a canal connecting two bodies of water
8. Simulating the effects of proposed engineering structures on flow, water height, and salinity
9. Predicting areas of maximum shoaling
10. Providing information of biological importance such as data on salinity and areas of upwelling
11. Conducting water outfall and intake studies to prevent recirculation of wastes in plant processing systems.

Thermal pollution models will be helpful in:

1. Determining ability of an estuary to provide cooling water for industry without serious ecological effects
2. Selecting proper sites for nuclear power plants
3. Determining maximum waste heat output that will not lead to violation of given standards of temperature increase
4. Deciding whether to concentrate waste heat at the surface or to mix it completely with the receiving water.

Applications of water quality models would be in the general area of pollution abatement. More specifically, such models could provide input to management charged with deciding whether a contemplated scheme for waste disposal will meet required standards for water quality.

References

Arnason, G., Greenfield, R. S., and Newburg, E. A., 1968. A numerical experiment in dry and moist convection including the rain stage, *Journal of Atmospheric Sciences*, 25(3): 404–415.

Arnason, G., et al., 1969. Numerical simulation of the macrophysical and microphysical processes of moist convection, *Proceedings of the World Meteorological Organization/International Union of Geodesy and Geophysics Symposium on Numerical Weather Prediction in Tokyo, November 26–December 4, 1968* (Tokyo: Japan Meteorological Agency).

Banks, J. E., 1968. *A Numerical Model to Study Tides and Surges in a River-Sea*

Combination, internal paper of the Tidal Institute and Observatory, The University of Liverpool, England.

Birkhoff, G., 1960. *Hydrodynamics: A Study in Logic, Fact and Similitude,* rev. ed. (Princeton, New Jersey: Princeton University Press).

Boicourt, W. C., 1969. *A Numerical Model of the Salinity Distribution in Upper Chesapeake Bay,* CBI Reference 69–7(May), Report No. 54.

Bryan, K., 1969. Climate and the ocean circulation: III. The ocean model, *Monthly Weather Review, 97*(11): 806–827.

Committee on Tidal Hydraulics, 1969. *Special Analytic Study of Methods for Estuarine Water Resources Planning,* Corps of Engineers, U.S. Army, Technical Bulletin No. 15 (March).

Edinger, J. E., Brady, D. K., and Graves, W. L., 1968. The variation of water temperatures due to steam electric cooling operations, *Journal of Water Pollution Control Federation, 40L9:* 1632–1639.

Edinger, J. E., and Geyer, J. C., 1965. *Heat Exchange in the Environment,* Cooling water studies of Edison Electric Institute Publ. No. 49, 253 pp.

Edinger, J. E., and Geyer, J. C., 1968. Analyzing steam electric power plant discharges, *Journal of Sanitary Engineering Division, Proceedings of the American Society of Civil Engineers,* SA4: 611–623.

Einstein, H. A., and Harder, J. A., 1959. An electric analog model of a tidal estuary, *Journal of Waterways and Harbors Division, American Society of Civil Engineers, 85:* 153–165.

Ekman, V. W., 1899. Ein Beitrag zur Erklärung und Berechnung des Stromverlaufes in Flussmündungen, *Ofers. Kgl. Vet. Akad. Handl. Stockh.,* No. 5, p. 469.

Fisher, G., 1959. Ein numerisher Verfahren zur Errechnung von Windstau und Gezeiten in Randmeeren, *Tellus, 11:* 60–76.

Freeman, J. C., Jr., Baer, L., and Jung, G. H., 1957. The bathystrophic storm tide, *Journal of Marine Research, 16*(1): 12–22.

Garrison, J. M., et al., 1968. Unsteady flow simulation in rivers and reservoirs, *American Society of Civil Engineers, Hydraulic Division, Specialty Conference,* Cambridge, Massachusetts.

Grill, E. V., 1970. A mathematical model for the marine dissolved silicate cycle, *Deep-Sea Research, 17*(2): 245–266.

Hansen, W., 1956. Theorie zur Berechnung des Wasserstandes und der Strömungen in Randmeeren nebst Anwedungen, *Tellus, 8*(3): 287–300.

Hansen, D. V., and Rattray, M., Jr., 1965. Gravitational circulation in straits and estuaries, *Journal of Marine Research, 23*(2): 104–122.

Harder, J. A., and Masch, F. D., 1961. Non-linear tidal flows and electric analogs, *Journal of Waterways and Harbors Division, American Society of Civil Engineers, 87:* 27–39.

Holley, E. R., Jr., and Harleman, D. R. F., 1965. *Dispersion of Pollutants in Estuary Type Flows.* Hydrodynamics Laboratory Report No. 74 (Cambridge, Massachusetts: Massachusetts Institute of Technology).

Jaske, R. T., 1968. The use of a digital simulation system for the modeling and prediction of water quality, *Water Research, 2* (New York: Pergamon Press).

Jelsnianski, C. P., 1967. Numerical computations of storm surges with bottom stress, *Monthly Weather Review, 95*(11): 740–756.

Kent, R. E., and Pritchard, D. W., 1959. A test of mixing length theories in a coastal plain estuary, *Journal of Marine Research, 18*(1): 62–72.

Keulegan, G. H., 1966. Model laws for coastal estuarine models, *Estuary and Coastline Hydrodynamics, Eng. Soc. Monog.*, pp. 691–710.

Lauff, G. H., 1967. *Estuaries*, No. 83 (Washington, D.C.: American Association for the Advancement of Science).

Leendertse, J. J., 1967. *Aspects of a Computational Model for Long-Period Water-Wave Propagation*, Memo RM-5294-PR (Santa Monica: The RAND Corporation).

Manabe, S., 1969a. Climate and the ocean circulation: I. The atmospheric circulation and the hydrology of the earth's surface, *Monthly Weather Review, 97*(11): 739–774.

Manabe, S., 1969b. Climate and the ocean circulation: II. The atmospheric circulation and the effect of heat transfer by ocean currents, *Monthly Weather Review, 97*(11): 775–805.

Okubo, A., 1962. A review of theoretical models for turbulent diffusion in the sea, *Journal of Oceanographic Society of Japan*, 29th Anniversary Volume, pp. 286–320.

Pandolfo, J. P., 1969. A numerical model of the atmosphere-ocean planetary boundary layer, *Proceedings of the World Meteorological Organization/International Union of Geodesy and Geophysics Symposium on Numerical Weather Prediction in Tokyo, November 26–December 4, 1968* (Tokyo: Japanese Meteorological Agency).

Preddy, W. S., and Webber, B., 1968. The calculation of pollution of the Thames estuary by a theory of quantized mixing, *International Journal of Air and Water Pollution, 7:* 829–843.

Pritchard, D. W., 1952. *Estuarine Hydrography. Advances in Geophysics* (New York: Academic Press).

Pritchard, D. W., 1956. The dynamic structure of a coastal plain estuary, *Journal of Marine Research, 15*(1): 33–42.

Pritchard, D. W., 1958. The equations of mass continuity and salt continuity in estuaries, *Journal of Marine Research, 17:* 412–423.

Pritchard, D. W., 1960. *The Movement and Mixing of Contaminants in Tidal Estuaries: Waste Disposal in the Marine Environment* (Proceedings of the First International Conference, July 22–25, 1959) (New York: Pergamon Press), pp. 512–525.

Pritchard, D. W., 1968. Dispersion and flushing of pollutants in estuaries, *Journal of Hydraulics Division of the American Society of Civil Engineers, 95:*NY1, Proc. Paper 6344 (January 1969), pp. 115–124.

Pritchard, D. W., and Carpenter, J. H., 1960. Measurements of turbulent diffusion in estuarine and inshore waters, *Bulletin of the International Association of Sci. Hydrol.*, No. 20: 37–50.

Pritchard, D. W., Okubo, A., and Carter, H. H., 1966. *Observations and Theory of Eddy Movement and Diffusion of an Introduced Tracer Material in the Surface Layers of the Sea*, Symposium on the Disposal of Radioactive Wastes into Seas, Oceans, and Surface Waters, Vienna, Austria (May 16–20), pp. 397–424.

Rattray, M., Jr., and Hansen, D. V., 1962. A similarity solution for circulation in an estuary, *Journal of Marine Research, 20:* 121–132.

Reid, R. O., and Bodine, B. R., 1968. Numerical model for storm surges in Galveston Bay, *Journal of Waterways and Harbors Division, 94:* 33–57.

Riley, G. A., 1965. A mathematical model of regional variations in plankton, *Journal of Limnology and Oceanography, 10* (Supplement), 202–215.

Riley, G. A., 1967. IV. Mathematical model of nutrient conditions in coastal waters, *Bulletin of Bingham Oceanography Collection: Aspects of Oceanography of Long Island Sound,* vol. 19, art. 2, pp. 72–80.

Rossiter, J. R., and Lennon, G. W., 1965. Computation of tidal conditions in the Thames estuary by the initial value method, *Proceedings of the Institute of Civil Engineering, 31:* 25–56.

Seaders, J., and Delay, W. H., 1966. Predicting temperatures in rivers and reservoirs, *Journal of Sanitary Engineering Division, American Society of Civil Engineers, 92:* 115–134.

Shuman, F. G., and Hovermale, J. B., 1968. An operational six-layer primitive equation model, *Journal of Applied Meteorology, 7*(4): 525–547.

Simmons, H. B., 1966. Tidal and salinity model practice, *Estuary and Coastline Hydrodynamics, Engineering Society Monograph,* pp. 711–731.

Simmons, H. B., 1969. Use of models in resolving tidal problems, *Proceedings of the American Society of Civil Engineers, Journal of Hydraulics Division, 95:* 125–146.

Simmons, H. B., and Lindner, C. P., 1965. Hydraulic model studies of tidal waterway problems, *Evaluation of Present State of Knowledge of Factors Affecting Tidal Hydraulics and Related Phenomena,* edited by C. F. Wicker, Report No. 3, LX-1–LX-21.

Stommel, H., 1953. Computation of pollution in a vertically mixed estuary, *Sewage and Industrial Wastes, 25:* 1065–1071.

Svansson, A., 1959. Some computations of water heights and currents in the Baltic, *Tellus, 11*(2): 231–238.

Systems Analysis Group (Civil Functions), 1969. *Guidelines for Evaluating Estuary Studies, Models and Comprehensive Planning Alternatives,* U.S. Department of the Army (August).

Takano, K., 1954. On the velocity distribution off the mouth of a river, *Journal of the Oceanographic Society of Japan, 10:* 60.

27
General Circulation Patterns in the World Ocean
Joseph L. Reid

Mathematical investigations of the ocean circulation begin with treatments of geostrophic circulation (Bjerknes and Sandstrom, 1910; Bjerknes, 1911), of Ekman transport (Ekman, 1905), the consequences of evaporation and precipitation (Goldsbrough, 1933), Sverdrup transport (Sverdrup, 1947), and westward intensification (Stommel, 1948); a more general treatment of the wind-driven circulation was given by Munk (1950). There have been discussions by Stommel (1957). Since these basic contributions, a large number of investigators have attempted various solutions for particular processes, areas, or features: examples are the Stommel and Arons series on abyssal circulation (Wyrtki, 1961; Veronis, 1969; Welander, 1969). The most general approach is that being attempted now by Kirk Bryan and his associates at Princeton.

Setting aside the earlier studies that, in the absence of real data, attempted mathematical solutions on more or less idealized oceans, a great deal of qualitative results had been obtained through the measurements of the currents (nearly always done only at or near the surface) and measurements of the content of heat, salt, dissolved oxygen, plant nutrients, and any other measurable and possibly useful chemical concentration. Recently the concept of age dating by radioactive isotopes has been used and various measurements made, with some disagreement as to how the results should be interpreted (Bien, Rakestraw, and Suess, 1965; Munk, 1966; Craig, 1969; Kuo and Veronis, 1970).

The immediately relevant background materials for a discussion of the marine environment include those qualitative investigations based upon what is known of the distributions of heat, salt, dissolved oxygen, nutrients, and other materials, and what circulation has been directly measured. These two sets of data have led to at least a first-order scheme of general ocean circulation, including exchange with the atmosphere. It is of course quite incomplete and cannot really be tested in a quantitative way over any large areas of the ocean.

Aspects of the exchange between the ocean and atmosphere have been discussed by Jacobs (1951) and various other investiga-

Prepared for SCEP.

tors; lately Stewart (1969) has reviewed much of this work in a popular article.

Circulation
Pertinent qualitative studies of the circulation and some of its effect on the concentrations are given by Stommel (1955), Reid (1962, 1965), Wyrtki (1962), Worthington (1965), and Lynn and Reid (1968).

Surface Circulation
The major parts of the surface circulation of the world ocean are predominantly wind-driven. In low latitudes near the equator most of the surface waters move westward under the influence of the trades. In high latitudes under the influence of the westerlies most water moves eastward across the ocean. Variations in wind strength as well as the configuration of the continents cause some irregularities in this flow: we find some eastward flow near the equator and some westward flow in high latitudes. Since the continents break the oceans up into three major parts, the westerly and easterly transports feed into each other along the coasts. The West Wind Drift of the North Pacific, for example, divides when it reaches the coast of North America, part turning southward into the California Current system and then westward with the North Equatorial Current; this in turn divides as it nears the Asian continent, a part turning north to contribute to the Kuroshio Current which in turn feeds into the West Wind Drift. This sort of anticyclone is also found in the corresponding latitudes of the Atlantic Ocean and of the South Pacific, South Atlantic, and South Indian oceans. Poleward of these anticyclones cyclonic circulations are similarly imposed, the largest being the Antarctic Circumpolar Current entirely around the continent of Antarctica. This wind-driven surface circulation apparently extends at least some hundreds of meters over most of the world ocean and in the case of the Antarctic Circumpolar flow to depths as great as 4,000 meters.

Mixed Layer
The water near the surface of the ocean is stirred by the wind and by evaporative and cooling processes. The thickness of this mixed layer and its degree of homogeneity vary both with season and

area. Really deep convective overturn is limited to only two or three places in the world ocean. The thickness of this upper layer may be as great as 300 meters in the western-intensified circulation within the Sargasso Sea and the corresponding parts of the Kuroshio system, or as shallow as 15 or 20 meters along the coasts of the eastern boundary current systems. All of the exchange between the ocean and the atmosphere takes place through this mixed layer. Both the warmest and the coldest waters of the ocean are found at the surface, as well as the most saline and the least saline parts. Since convective stirring is more or less effective throughout this layer, the concentration of dissolved oxygen, for example, is very nearly at equilibrium in the layer. Significant variations from equilibrium occur only in areas of intense upwelling which make up only a very small part of the ocean's surface and in some highly productive areas where intense photosynthesis may produce oversaturation, perhaps as much as 15 or 20 percent above the equilibrium value during the summer months; these areas also appear to be fairly small in extent compared to the entire area of the world ocean.

Sinking

Downward penetration of the characteristics of the mixed layer such as heat and salt are limited by the stability of the ocean. Figure 27.1 shows three long sections drawn north-south through the Atlantic, Indian, and Pacific Oceans. The quantity plotted in Figure 27.2 is the hydrostatic stability. On the scale of these sections it is not possible to show the details at very shallow depths. Actually the values fall well below the 100 value, which is the lowest depicted near the surface. In the mixed layer, of course, which it is impossible to represent properly on this scale, the values approach zero. Since the ocean is stratified in density as well as in other characteristics, any vertical motion of the water is opposed by the vertical density gradient. These figures show the degree of stratification in various areas. The highest values of stability are found just beneath the mixed layer and especially in low and middle latitudes. The Atlantic Ocean, for example, shows three areas of conspicuously low stability where the density stratification offers the least opposition to vertical motion and thus exchange of characteristics. The first of these is the Norwegian Sea (70° to 80° N). In this basin, overturn from the top

451 General Circulation Patterns in the World Ocean

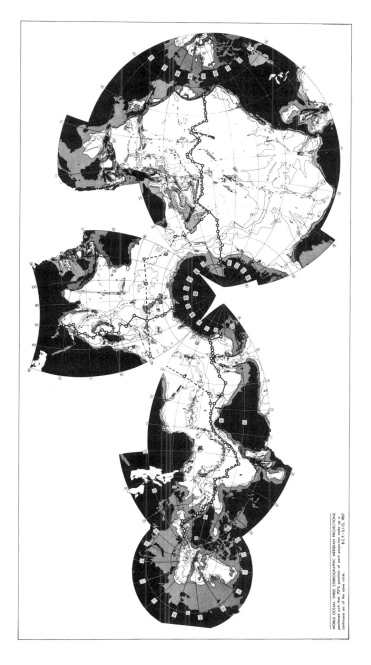

Figure 27.1 Locations of the three sections through the Atlantic, Indian, and Pacific oceans are given by the solid lines Dark shading indicates depths less than 2,000 meters; light shading less than 3,000 meters; the black line is the 4,000-meter contour.

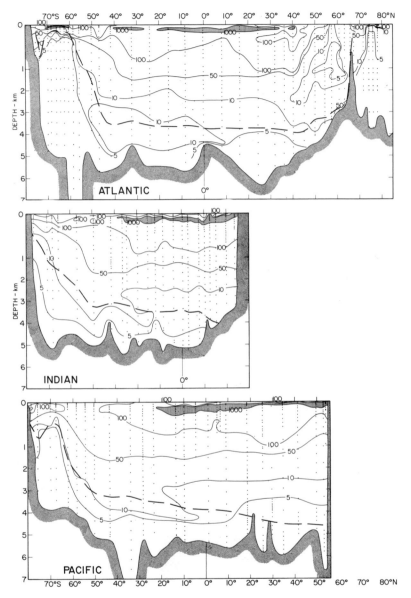

Figure 27.2 Hydrostatic stability (resistance to overturn) along north-south sections in the Atlantic, Indian, and Pacific oceans

The quantity is $E = \dfrac{\delta \rho}{\partial z} \, 10^{-8} \, \dfrac{\text{g/cm}^3}{\text{m}}$

to the bottom may actually occur in midwinter. Though this has not really been documented as well as we should like, it is generally accepted that overturn to the bottom does occur there, at least in some winters. This means that in the waters at the bottom of the Norwegian Sea, as well as throughout, the dissolved oxygen concentration should be at the saturation value, since the entire body is in effective contact with the atmosphere. The shallow sill separating the Norwegian Sea from the principal part of the Atlantic Ocean prevents these waters from moving freely into the Atlantic Ocean. They are in fact denser than any of the waters in the Atlantic Ocean, and were it not for this sill they would fill the deepest parts of the Atlantic Ocean. The flow through the narrow, shallow passage, however, imposes a high degree of turbulent mixing upon the outflowing waters. They mix with the overlying waters, which are much warmer, to such an extent that their density is decreased and that they do not reach the deepest parts of the Atlantic. Instead they penetrate to a depth of approximately 3,500 meters only; the deeper waters are denser and appear to be of antarctic origin.

A secondary region of low stability is seen at about 60° N in the Atlantic, extending down to about 2,000 meters. In this area, which is just south of Greenland, overturn to depths of 2,000 or 2,500 meters may occur in the wintertime. Downward penetration of heat, salt, dissolved oxygen, nutrients, and all other mixed-layer characteristics may take place in this area. This is the deepest downward penetration that occurs in the open ocean. (This excludes, of course, the Norwegian and Mediterranean seas, which are cut off by shallow sills from the major parts of the ocean.)

The third area of minimum stability is seen in the antarctic region. Although the stability everywhere below 500 meters appears to be very low south of 60° S, this does not imply that strong overturn does take place there. Though the stability is low, the characteristics indicate that the water is stratified (Figures 27.3 through 27.5). There are still strong gradients of temperature and salinity from 500 meters to the bottom, and of dissolved oxygen and nutrients as well. Above this 500-meter layer there is a relatively high stability beneath the mixed layer. Though the surface water is cold in this area, the high precipitation causes it to be very low in salinity, and it is very much less dense than the deeper

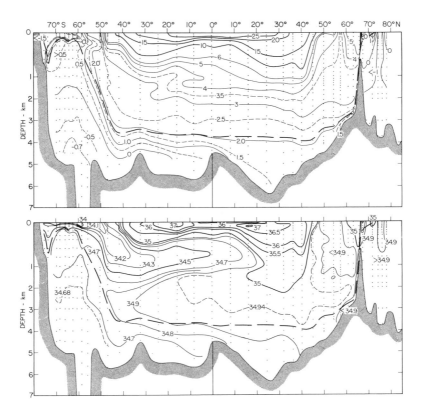

Figure 27.3 Atlantic Ocean
Potential temperature and salinity along a north-south section.

water. The extremely dense waters that are found at the bottom in the southern part of the Atlantic are formed not by convective overturn between 60° and 70° S as this figure might be taken to suggest but occur instead as a consequence of processes that take place along the continental shelf of Antarctica, particularly within the Weddell Sea. Atlantic water that is fairly saline reaches the coast of Antarctica and on these shelves is made cooler, and by contact with the ice shelves some freezing occurs which causes the salinity to increase; the water is thus made both colder and denser and flows down the slope into the bottom of the Weddell Sea area. From there it extends northward into the Atlantic and eastward around Antarctica into the deeper parts of the South Indian and South Pacific Oceans. The Weddell Sea is thus not

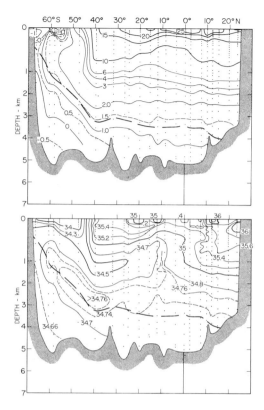

Figure 27.4 Indian Ocean
Potential temperature and salinity along a north-south section.

so effective in bringing mixed-layer characteristics to the bottom of the ocean as it would be if convective overturn did occur from the surface to the bottom in that area.

The Mediterranean Sea behaves much like the Norwegian Sea but its outflow does not reach so deep into the Atlantic Ocean. It is thus not so important to the immediate discussion and will be left out of this brief presentation.

Deep Circulation

The effect of this variation in stability can be seen on the vertical sections of temperature and salinity in the various oceans (Figures 27.3 through 27.5). The Norwegian Sea, which appears to go through severe mixing, has a small range of temperature and of salinity and in its central area appears to be almost homo-

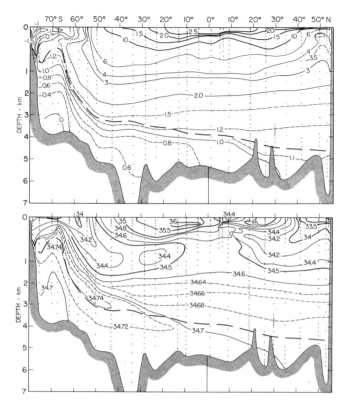

Figure 27.5 Pacific Ocean
Potential temperature and salinity along a north-south section.

geneous. Near 60° N in the Atlantic there is an immense vertical extent of water between 3° and 3.5°C and of salinity about 34.9 percent. The antarctic part of the Atlantic between 60° S and Antarctica is not very strongly stratified in density but shows a clear variation in temperature and salinity. Both temperature and salinity decrease toward the bottom. The distribution of temperature and of salinity in particular on the three sections gives some immediate notion as to the nature of the deep and bottom circulation. The Antarctic Bottom Water appears to extend northward in all oceans, becoming warmer and more saline as it moves. The immense body of highly saline water in the Norwegian Sea and North Atlantic is seen to extend southward in the Atlantic Ocean at depths from 2,000 to 4,000 meters and into the

Indian Ocean and Pacific Ocean where it has been carried by the eastward-flowing Antarctic Circumpolar Current. Above these, in all southern oceans and in the North Pacific Ocean, another stratum of low-salinity water extends from about 500 to 1,000 meters. Thus, the principal sources of deep and abyssal water throughout the world ocean appear to be the Norwegian Sea and the Weddell Sea. No other parts of the ocean produce water of such high density. The Norwegian Sea is the only one of these water bodies which appears to be filled with waters whose concentrations are characteristic of equilibrium with the atmosphere. The outflow from the Norwegian Sea into the North Atlantic is immediately diluted with other waters that have been away from contact with the atmosphere for some uncertain period.

Exchange with the atmosphere of course takes place everywhere over the surface of the ocean. Gases such as dissolved oxygen which equilibrate rapidly are very near to their equilibrium values throughout the mixed layer. The highest temperatures and the lowest temperatures are found in the mixed layer. The downward penetration of the mixed-layer characteristics by sinking is thus limited to particular areas of the ocean. The nature of the downward penetrations and the return upward of modified water has been studied mostly through distributions of temperature, salinity, dissolved oxygen, and nutrients, though recently this system has been examined through the distribution of various radioactive parameters (Bien, Rakestraw, and Suess, 1965; Bolin and Stommel, 1961; Munk, 1966; Craig, 1969). To date not many observations have been made, and the treatment of the samples has varied among investigators. From 1973 to 1975 a more comprehensive selection of water samples will be made along these three sections, with the hope of using all of the techniques of geochemistry in studying the deep flow, and in particular the time scale, of the deep ocean.

Ocean-Atmosphere Interaction
Recent studies on the interaction of the ocean and atmosphere have dealt with large-scale as well as small-scale aspects. One of the large-scale features now receiving considerable attention is the deviations of sea-surface temperature from its normal seasonal value. Areas of abnormally high temperature have been observed

to extend over areas more than 6,000 km by 4,000 km and to endure for periods of more than a year. Indeed, such warm areas (and cold areas) seem to be typical of the ocean; coherent patches of anomalous water with large areas and several months' duration are common features of the North Pacific Ocean.

References

Arons, A. B., and Stommel, Henry, 1967. On the abyssal circulation of the World Ocean—III. An advection-lateral mixing model of the distribution of a tracer property in an ocean basin, *Deep-Sea Research, 14*(4): 441–457.

Bien, G. S., Rakestraw, N. W., and Suess, H. E., 1965. Radiocarbon in the Pacific and Indian oceans and its relation to deep water movements. *Limnology and Oceanography, 10* (Redfield Anniversary Volume): R25-R37.

Bjerknes, V., 1911. Dynamic meteorology and hydrography, *Part II, Kinematics.* Carnegie Institute Washington, Pub. No. 88, 175 pp.

Bjerknes, V., and Sandstrom, J. W., 1910. Dynamic meteorology and hydrography. *Part I, Statics.* Carnegie Institute Washington, Pub. No. 88, 146 pp.

Bolin, Bert, and Stommel, Henry, 1961. On the abyssal circulation of the world ocean—IV. Origin and rate of circulation of deep ocean water as determined with the aid of tracers, *Deep-Sea Research, 8*(2): 95–110.

Craig, H., 1969. Abyssal carbon and radiocarbon in the Pacific, *Journal of Geophysical Research, 74*(23): 5491–5506.

Ekman, V. W., 1905. On the influence of the earth's rotation on ocean currents, Arkiv for matematik, astronomi, ach fysik (Stodkholm), *2*(11): 53 pp.

Goldsbrough, G. R., F.R.S., 1933. Ocean currents produced by evaporation and precipitation, *Proceedings of the Royal Society, A141:* 512–517.

Jacobs, W. C., 1951. The energy exchange between sea and atmosphere and some of its consequences, *Bulletin of the Scripps Institution of Oceanography, 6*(2): 27–122.

Kuo, Han-Hsiung, and Veronis, George, 1970. Distribution of tracers in the deep oceans of the world, *Deep-Sea Research, 17*(1): 29–46.

Lynn, Ronald J., and Reid, Joseph L., 1968. Characteristics and circulation of deep and abyssal waters, *Deep-Sea Research, 15*(5): 577–598.

Munk, W. H., 1950. On the wind-driven ocean circulation, *Journal of Meteorology, 7*(2): 79–93.

Munk, W. H., 1966. Abyssal recipes, *Deep-Sea Research, 13*(4): 707–730.

Namias, Jerome, 1970. Macroscale variations in sea-surface temperatures in the North Pacific, *Journal of Geophysical Research, 75*(3): 565–582.

Reid, Joseph L., Jr., 1962. On circulation, phosphate-phosphorus content, and zooplankton volumes in the upper part of the Pacific Ocean, *Limnology and Oceanography, 7*(3): 287–306.

Reid, Joseph L., Jr., 1965. Intermediate waters of the Pacific Ocean, *Johns Hopkins Oceanographic Studies, 2.*

Stewart, R. W., 1969. The atmosphere and the ocean, *Scientific American* (September): 76–86.

Stommel, H., 1948. The westward intensification of wind-driven ocean currents, *Transactions of the American Geophysical Union*, 29(2): 202–206.

Stommel, Henry, 1955. The anatomy of the Atlantic, *Scientific American* (January) (Reprint No. 810).

Stommel, Henry, 1957. A survey of ocean current theory, *Deep-Sea Research*, 4(3): 149–184.

Stommel, Henry, 1958. The abyssal circulation, *Deep-Sea Research*, 5(1): 80–82.

Stommel, Henry, and Arons, A. B., 1960a. On the abyssal circulation of the world ocean—I. Stationary planetary flow patterns on a sphere. *Deep-Sea Research*, 6(2): 140–154.

Stommel, Henry, and Arons, A. B., 1960b. On the abyssal circulation of the world ocean—II. An idealized model of the circulation pattern and amplitude in oceanic basins, *Deep-Sea Research*, 6(3): 217–233.

Sverdrup, H. U., 1947. Wind-driven currents in a baroclinic ocean; with application to the equatorial currents of the Eastern Pacific. *Proceedings of the National Academy of Sciences* (Washington), 33(11): 318–326.

Veronis, George, 1969. On theoretical models of the thermocline circulation, *Deep-Sea Research*, 16 (Supplement): 301–323.

Welander, Pierre, 1969. Effects of planetary topography on the deep-sea circulation, *Deep-Sea Research*, 16 (Supplement): 369–391.

Worthington, L. V., and Volkmann, G. H., 1965. The volume transport of the Norwegian Sea overflow water in the North Atlantic, *Deep-Sea Research*, 12(5): 667–676.

Wyrtki, Klaus, 1961. The thermohaline circulation in relation to the general circulation in the oceans, *Deep-Sea Research*, 8(1): 39–64.

Wyrtki, Klaus, 1962. The oxygen minima in relation to oceanic circulation, *Deep-Sea Research*, 9(1): 11–23.

28
Hydrodynamic Modeling of Ocean Systems

Kirk Bryan

Introduction

Waves and currents in the ocean can be organized into many different categories depending on horizontal dimension and the time scale of variability. Some of these categories are strongly interconnected and others almost independent. An attempt is made at classification in Table 28.1. In each case, the principal way in which the phenomena have an impact on human activities is also indicated. The emphasis in this entire outline is on ocean circulation phenomena, since this is the area with which the author is most familiar. The modeling of surface waves, tides, and storm tides is treated only briefly, although it is recognized that these are very important subjects from the standpoint of practical disaster-warning systems.

Wind Waves and Tidal Waves

The numerical models developed to predict surface waves are essentially refinements of earlier operational models developed by the U.S. Navy. These forecasting models have proved to be very useful to shipping. The new computer models allow a much

Table 28.1 Phenomena Associated with Waves and Currents that May Be Treated by Hydrodynamic Models
The arrangement is by time and space scales. The relation to applied problems is indicated in parentheses.

Time Scale	Local Interactions	Intermediate Scale	Global
Short minutes	surface waves (shipping, shore erosion, offshore drilling)		tidal waves (tsunamis) (safety of shore areas)
Intermediate hours-days	ocean turbulence and mixing (pollution, air-sea interaction)	storm surges (safety of shore areas hurricane damage)	tides (navigation)
Long months-years	near-shore circulation (pollution)	circulation of inland seas (Great Lakes pollution, polar pack ice models)	circulation in ocean basins (long-range weather forecasting, fisheries, climatic change)

Prepared for SCEP.

more detailed incorporation of the latest experimental and theoretical advances in the study of wave generation. In a short time orbiting satellites may be able to provide a good synoptic picture of the surface sea state all over the globe. Computer models would then be able to predict future sea states given an accurate weather forecast. It may turn out that the ultimate limitation to wave forecasting will involve the accuracy of the weather forecast rather than the wave prediction model itself.

Operational models for the prediction of tidal waves (tsunamis) have been developed for the Pacific where the danger of earthquakes is greatest. The model predicts the time of arrival of a tidal wave as soon as the epicenter of the earthquake is located by seismographs. Such warning systems are being developed by the National Oceanic and Atmospheric Administration (NOAA) and the Japanese Meteorological Agency. (NOAA was formerly the Environmental Science Services Administration, ESSA.)

Storm Surges and Tides
Most of the research in developing numerical models to predict storm tides has been carried out in Europe in connection with flooding in the North Sea area. In the United States the most interest has been in connection with storm surges caused by hurricanes approaching the Gulf Coast. The results of these model studies appear to be very promising. The model calculations may be used to construct graphs and charts used by Weather Bureau forecasters in making flood warnings. The models will also be very useful in the engineering design of harbor flood walls and levees. In time, computer models will probably replace the expensive and cumbersome laboratory models of harbors now in use by coastal engineers.

An Evaluation of Current Scientific Knowledge Related to Numerical Modeling of Ocean Circulation

The Data Base
Standard oceanographic data and geochemical data provide a fairly adequate data base for modeling the time-averaged mean state of the ocean. The data base for modeling the time variability

of the ocean is very limited, however. Information on the large-scale changes in ocean circulation, as well as the small-scale variability associated with mixing in the ocean, have not been gathered in any comprehensive way.

Development of Numerical Models

Over the past decade three-dimensional, numerical models have been developed by the Soviet Hydrometeorological Service and NOAA for the calculation of ocean circulation. The methods used are similar to those of numerical weather forecasting. Given the flux of heat, water, and momentum at the upper surface, the model predicts the response of the currents at deeper levels. The currents at deeper levels in turn change the configuration of temperature and salinity in the model ocean. Although active work in developing these models is being conducted at several universities, the only published U.S. calculations are based on the NOAA "box" model developed at the Geophysical Fluid Dynamics Laboratory. This model allows the inclusion of up to twenty levels in the vertical direction and a detailed treatment of the bottom and shore configuration of actual ocean basins.

A calculation of the circulation of the Indian Ocean by M. D. Cox is perhaps the most detailed application attempted with the NOAA "box" model. Using climatic data it was possible to specify the observed distribution of wind, temperature, and salinity at the surface as a function of season. In response to the changing monsoons the model was able to make an accurate prediction of the spectacular changes in currents and upwelling along the African Coast measured during the Indian Ocean Expedition of the early 1960s.

Application of the Model to Practical Problems

The same numerical models designed for studying large-scale ocean circulation problems can also be modified to study more local circulation in near-shore areas or inland seas, such as the Great Lakes. Thus, numerical models may be very useful for a very large number of problems in oceanography in which steady currents play a role. A partial list is now given:

1. Long-range weather forecasting
2. Fisheries forecasting
3. Pollution on a global or local scale
4. Transportation in the Polar Ice Pack

Requirements for Scientific Activity

Needed Scientific and Technical Advances for Ocean Modeling

Progress in ocean modeling in the future will depend on more detailed field studies of ocean variability. Such studies will establish the data base for the formulation of mixing by small-scale motions that must be included in the circulation model. Information on large-scale variability will provide a means for verifying the predictions of the models. The best way to satisfy this requirement appears to be the different arrays of automated buoys that have been proposed as part of the International Decade of Ocean Exploration Program. Coarse arrays covering entire ocean basins, as well as detailed arrays for limited areas, will be required.

Another technical requirement for ocean modeling is also common to a great many other scientific activities. This is the steady development of speed in electronic computers and the steady decrease in unit cost of calculations.

The Nature of Present Urgency

Numerical models and the IDOE Program: Numerical models of currents have now reached a point where they can be of very great value in the planning of observational studies and the analysis of the data collected at sea. For example, the models can be used in a diagnostic mode as well as in a predictive mode. This is particularly true of the buoy networks proposed as part of the IDOE. In order to do this more oceanographers will have to be trained to use the numerical models, and more support will be required to carry out the computations. This action will have to be taken quickly if numerical models are to have much significance in IDOE programs.

Perhaps the most urgent problem at the present time is the impact of human activity on the global environment. Other effects may turn out to be more important, but the best-documented factor is the rising CO_2 content of the atmosphere due to the burning of fossil fuels since the beginning of the Industrial Revolution. As pointed out by Roger Revelle and others, a large fraction of the added CO_2 is taken up by the oceans. However, little is known concerning the details of this buffering effect of the ocean and how long it will continue to be effective. The ability of the ocean to take up CO_2 depends very much on how rapidly surface waters are mixed with deeper water. More detailed studies of geochemical evidence and numerical modeling are essential to get an understanding of this process. A start in numerical modeling of tracer distributions in the ocean has been made by Veronis and Kuo at Yale University and in Holland at the Geophysical Fluid Dynamics Laboratory of NOAA.

Another urgent task is to make an assessment of the effect of CO_2 and particulate matter in the atmosphere on climate. Present knowledge of climate does not allow reliable, quantitative predictions of the effect of changing the "greenhouse effect" due to CO_2 or the screening out of direct radiation by particulate matter. All estimates that have been published so far have been based on highly simplified models, treating only the radiational aspects of climate. No climate calculation is complete without taking into account the *circulation of both the atmosphere and the ocean.* Some preliminary climatic calculations have been carried out with combined numerical models of the ocean and atmosphere. However, greater effort is required to develop more refined ocean models before these climatic calculations will be reliable enough to be the basis for public policy decisions on pollution control.

Time Scale of Significant Advances
Published papers on three-dimensional ocean circulation models have begun to appear only during the last five years, so that rapid development should continue for at least another five years along present lines. In five years, ocean models should have reached about the same level of development as the most advanced atmospheric numerical models today. Within five years' time adequate support for numerical modeling should insure that at least the *feasibility* of applications to small- and large-scale pollution studies, long-range weather forecasting, and hydrographic data analysis are well established. Another five years will probably be required to work out standard procedures to use numerical ocean circulation models in these applications on a normal basis.

Special Recommendation: International Agreements on Large-Scale Climate Modification Experiments
This recommendation properly belongs with the discussion of

atmospheric numerical models, but since numerical models of the ocean also play an important role in climatic research it will be included here. Through observational programs and numerical modeling, the mechanisms of climate are rapidly becoming better understood. Space technology is also being rapidly developed which will give man the capability of lifting huge payloads into the outer fringes of the atmosphere and near space. Thus we are rapidly approaching the time when *feasible* large-scale climatic experiments may be proposed. They may possibly be suggested as countermeasures to the inadvertent modification of climate caused by global pollution.

In any case, this whole area is one of immense danger in view of the possibility of uncontrolled experimentation by any one nation. For this reason it is recommended that the evolution of a climate modification technology should be *anticipated* by international agreements banning unilateral experiments in large-scale climate control.

Part VII Some Implications of Change

Though the primary focus of SCEP was on the scientific evidence which is available or must be generated to determine the true nature of critical, long-term global problems, one group of participants addressed the set of questions that must ultimately be resolved if changes in the status quo are to be effected and if remedial action is to be taken. This Work Group on the Implications of Change was chaired by Professor Milton Katz of the Harvard Law School and included lawyers, economists, social scientists, engineers from several industries, and government officials. The summary of the report of that Work Group is reprinted here from the SCEP Report as the first paper in this series.

The next paper is a short piece written for SCEP participants by Mr. Richard Carpenter, who had been a member of the Steering Committee that planned SCEP. It is a succinct bit of advice to all those who would attempt to synthesize scientific and technical material with a view toward raising the level of public discussion on important issues. Mr. Carpenter writes from the experience of six years of policy analysis in the Legislative Reference Service of the Library of Congress.

Effective management for environmental quality will require more scientific knowledge, but it will also require conceptual frameworks that make policy options, alternatives, and implications explicit. The paper by Dr. Spofford outlines such a framework for residuals management. In it he enumerates the various alternatives for achieving environmental quality by reducing in the environment the residuals generated from man's activities. He also discusses some of the types of incentives that may be necessary to assure that external costs of pollution are internalized at some point or points in the market or political processes.

In order to explore some of the general types of problems involved in devising technical, social, and political action to cope with environmental pollution, Dr. Brown prepared a paper during SCEP in which he considered three specific cases: phosphates, heavy metals (in water), and DDT-like materials. For each pollutant he has outlined the nature of the pollutant, the possible control technologies, the social alternatives, and the implications of pollution controls. His paper is neither comprehensive nor detailed, but it does provide some idea of the complex technical issues that will have to be addressed.

Part IV of this volume contains the SCEP Task Force report on pesticides in the oceans which documents some of the deleterious ecological effects of DDT. It should be recognized, however, that DDT is used because it provides significant health and agricultural services for man. The SCEP Report recommended "a drastic reduction in the use of DDT as soon as possible *and* that subsidies be furnished to developing countries to enable them to afford to use nonpersistent but more expensive pesticides as well as other pest control techniques." The paper by Dr. Rita Taubenfeld outlined for SCEP some of the social, policy, and economic trade-offs that will eventually have to be confronted if that recommendation is to be implemented.

29
Implications of Change and Remedial Action
Summary of SCEP Report

Introduction

The expansion and refinement of our knowledge and understanding are the necessary conditions for effective change in the present state of environmental management. However, these are not sufficient conditions. Even after optimal improvements have been made in our knowledge concerning the nature of key pollutants, their effects, their sources, their rates of accumulation, the routes along which they travel, and their final reservoirs, the questions will remain of how to apply our knowledge constructively and how to cope with the collateral consequences. As a practical matter, questions of environmental management will have to be faced before we have all the appropriate scientific and technical data, and this further complicates efforts of change or of remedial action.

In examining a wide range of specific problems at this Study, we have identified several aspects that are common to most of them and to many other critical environmental problems. These implications of change and remedial action are briefly discussed now.

Establishing New Priorities

Earlier in our history, the prevailing value system assigned an overriding priority to the first-order effects of applied science and technology: the goods and services produced. We took the side effects—pollution—in stride. A shift in values appears to be under way that assigns a much higher priority than before to the control of the side effects. This does not necessarily imply a reduced interest in production and consumption. When the implications of remedial action and the choices that must be made become clear, there may be second thoughts, confusion, and feelings of frustration.

In the effort to arrive at an optimal balance in specific situations, something will have to give. But the old routine assumption that it is the environment that must give has become intolerable.

Reprinted from Study of Critical Environmental Problems (SCEP), 1970. *Man's Impact on the Global Environment* (Cambridge, Massachusetts: The M.I.T. Press), pp. 32–36.

This assumption must be rejected in favor of an optimal balance to be reached from a point of departure in affixing the responsibilities for pollution.

Affixing Responsibilities

As a point of departure for taking action, we recommend a principle of presumptive "source" responsibility. While remedial measures can be attempted on the routes along which pollutants spread or in the reservoirs in which they accumulate, we believe that these measures should be generally taken at the "sources," which we define broadly to include (1) sources or the points in the processes of production, distribution, and consumption, at which the pollutant is generated, for example, factories, power plants, stockyards, bus lines; (2) protosources or earlier points that set the conditions leading to the emission of pollutants at a later stage, for example, the manufacturers of automobiles that emit pollutants when driven by motorists, or the brewers of beer sold in nonreturnable cans that are tossed aside by the consumer; and (3) secondary sources or points along the routes where pollutants are concentrated before moving on to the reservoirs, for example, sewage treatment plants or solid waste disposal centers.

The principle does not connote any element of blame or censure, nor is it intended to foreclose a judgment concerning where the financial costs of correction should ultimately be borne. It is intended, however, to indicate a point of departure for analysis and action. It rests, in part, on the basis that, if something goes wrong, it should be traced to its origin and corrected in terms of its cause; in part on a hypothesis that the source, protosource, or secondary source will typically be in the best position to take corrective measures, whether alone or with help from others; and in part on the view that the remedies available, the criteria for choice among them, and the implications of remedial action can best be appraised at the sources as here defined.

Accepting the Costs

Remedial changes will ordinarily involve financial costs, and the costs may be large in relation to the scale of the source enterprise. If the source enterprise can neither absorb the cost nor pass it on, it will be necessary to face a choice among failure of the enterprise, continuance of the pollution, or financial assistance out of public revenues. The initial change may have consequences reaching past the source enterprise to its employees, its suppliers, and its customers and beyond in widening waves of change that may engulf deep-rooted patterns of economic and social behavior. Our society is familiar with far-reaching readjustments caused by technological innovation or organizational change in the past. Comparable readjustments may be required by changes instituted to control pollution.

Assessing the Available Means for Action
The means available within the political process and legal system to encompass remedial changes include taxes designed as incentives, stimuli, or pressures, regulations, typically involving a statute, an administrative agency, and supplementary action through the courts; common-law remedies in the courts, incrementally adjusted to contemporary needs; governmental financing of research and assistance to facilitate costly adjustments to desired changes; and governmental operations, civilian and military. Governmental action in its own house can have a dual importance: in itself and as a model for others to follow.

Stimulating Effective Actions
The political, legal, and market processes of our society are profoundly affected by the nature and quantity of information available and the manner in which the information is infused into them. It is neither necessary nor feasible to postpone recommendations for action until scientific certainty can be achieved. The political process is accustomed to decisions in the face of uncertainty on the basis of a preponderance of the evidence or substantial probabilities or a reasonable consensus of informed judgment.

Thus, it is not enough for scientists and technologists to expand and refine their knowledge. They must also present their knowledge in a manner that clearly differentiates fact, assertion, and opinion and facilitates the task of relating the data to the possibilities of corrective action. But if such information is to be used, the Congress and state legislative bodies must be provided with instrumentalities and qualified staff to enable them more effectively to sort out and utilize the input of data, proposals, complaints, and suggestions that will flow into them in increasing volumes from all sectors of our society.

Developing New Professionals

In addition to general public education, we stress the special importance of some changes in scientific, technical, and professional education and training. A sensitivity to the relations between the processes of production, distribution, and consumption, on the one hand, and the processes of pollution, on the other, and a disposition to explore all the potentialities of technology and organization in the search for an optimal balance should be incorporated into their training. This applies to economists, lawyers, and social scientists as well as to scientists and engineers. Individual contributions may be undramatic now, but over time they will be critical.

Cooperating with Other Nations

Although many problems are global in nature, the solutions to these problems will generally require national as well as international action. Typically, remedial measures within one nation will need support from parallel actions within other nations. Frequently, collaborative international action will be required. The prospects for such cooperation are best for programs of collection and analysis of data. International cooperation on monitoring may also increase the likelihood of smooth relations should a global program ever demand strict international regulation or control of pollution-producing activities.

In the foreseeable future the advanced industrial societies will probably have to carry the major burden of remedial action. Developing nations are understandably concerned far more with economic growth and material progress than with second-order effects of technology. Similar attitudes were prevalent in the early stages of growth of present industrialized nations.

The challenges of international cooperation and collaboration in the critically important environmental areas studied by SCEP will be before the United Nations Conference on the Human Environment in 1972. We hope that this Report will provide useful inputs to that Conference and that the Study model furnished by SCEP will be applied to other critical problems of the environment.

Expectations of the Decision Maker

Richard A. Carpenter

Human beings make decisions constantly regardless of the adequacy of information and, to an extent, regardless of the penalty for being wrong. The basis for this Summer Study, however, is that leaders in society want to make better decisions (that is, optimum for human progress) and that science can provide approximate truths as a basis.

Decisions of individuals (freedom) can become a tyranny on the collective welfare as in the impacts of population on the world commons. Democratic political processes seek the proper trade-off between the common good and individual liberty. As technology and population increase, good collective decisions become more important, but that does not mean that they can or should all be made by political bodies. Through education and leadership the locus of decision can still often remain with the individual. Thus the results of the Summer Study are directed at both the citizen and public officials.

Both individuals and institutions have a limit as to the number of issues they can consider at any one time.

The conscious attention to a problem is only one of three mechanisms—the other two being (1) a sort of autonomic nervous system of society (ecosystem) where choices are continually made for us without thought and (2) the delegation of decision making to an established process such as the marketplace. The setting of priorities for consideration is necessary, but the stream of events interferes. (Witness the disappearance of the "environment" from the newspapers after the Cambodian affair—eight days past Earth Day.) Thus, the first expectation is that the Study will order the critical problems of the global environment as to priority. Although beyond the scope of the Study, it should be recognized that these problems will be viewed by the decision maker in a larger context, including local and regional environmental problems, world economic competition, national defense, human health care, and so on. Therefore, the "criticality" of the top-ranked global environmental problems must be commensurate with these other insistent demands for attention.

Prepared during SCEP.

It must be made clear why conscious decisions are necessary: why the conventional economic system is not taking care of the problem; why the common law cannot be relied upon to sort out the equities for all concerned; why the decisions of a prudent individual are inadequate to achieve the common good; why presently constituted management institutions are failing; why local or regional treatment cannot suffice.

Beyond this demonstration that the problem is worthy of special consideration, most of the following criteria should also be met in order for a particular problem to be ranked in high priority:

1. Man-made sources are important relative to natural sources or background values.
2. The effect is irreversible or very difficult (costly) to reverse.
3. The effect would have great economic damage.
4. Emissions or causes are mainly from the developed countries or from technology supplied by them.
5. Emissions are susceptible to control with present technology.
6. Effects are imminent.

The second expectation is that substantial uncertainties exist and that many problems can neither be dismissed (as local or unimportant) nor given high priority on the basis of present knowledge. This is acceptable if a productive course of further study is outlined and if speculation is avoided.

Extrapolations and "if-then" scenarios are perhaps useful in arousing public interest, but they confound the decision-making process. The decision maker is used to incomplete knowledge, in fact this is always the case. The scientist is used to hypothesis and experimentation. As long as the presentation of scientific advice is careful to separate out the "do know," "don't know," and "could know," the communication will be beneficial to the politician. A pro-and-con format may be useful. Another device is the admitted weighting of evidence (on the basis of peer judgment) in order to make a definitive statement—with dissent and its reasons placed in a footnote. Consensus should not be attempted in uncertain areas, and the Study should not be wishy-washy about what it does not know.

The third expectation is for a listing of alternative responses to the priority problems: what can we do about it. These actions are usually suggested by the factual description of the problem. They are augmented by a study of impacts, costs, side effects, feedback mechanisms, and so on. Scientific advice to the political process stops short of advocacy even though one course of action may be obvious to the scientist. We are seeking to reserve decision making to politically responsible officials, not to establish a technocracy or to lapse into a plebiscite of partially informed citizens on every issue that arises. Even in these complex technical matters the ability of the politician to integrate economic, social, raw political, and human intuitional inputs is valuable and should be guarded. As a citizen, the Study participant may opt for one or another solution, but the Study results should be an objective analysis of remedial alternatives as complete as possible.

The fourth expectation is recognition of the motivations of human beings which constrict the implementation of decisions. Only the archetypal ivory-towered scientist would pursue his work without consideration of human nature. The interpretation of the facts about global environmental problems should take due regard of the weakness of the flesh despite the willingness of the spirit.

For example, corrective actions will occur much more easily if they can be seen to coincide with selfish motives such as direct human health effects, welfare of offspring (but not very many generations—"What has the future ever done for me?"), damage to property, threat to livelihood, or disruption of the status quo. In contrast, appeals will go little heeded when based on goodwill toward man, voluntary reduction of standard of living to preserve the commons, stewardship of the affluent (noblesse oblige), or subjective aesthetic values.

There are possibilities for clever merging of baser human desires with actions for the long-term common good. For example, ecological principles will be obeyed whether we do it willingly or not. It can be shown that a high-productivity environment is indeed a high-quality environment. Another approach is the attitude toward the developing countries. We may very well make enormous investments and outright grants in these societies in order to preserve world order. Therefore, it should be practical to underwrite whatever concessions we ask them to make to abate pollution. A third example would stress the prudence of knowing more about global systems (research and monitoring) whether or not critical problems exist, simply so we can maximize the economic exploitation of resources and the environment over a long period of time.

The decision maker can be greatly helped by the purposeful seeking out of those facts and interpretations that strengthen implementation of desired courses of action by enlisting human nature. In fact, it would appear that people will do the right thing even at some personal inconvenience if some reinforcing of their ethical armament is provided by pointing out practicality and prudence. This does not suggest any distortion of the scientific information but only that effort be directed at revealing the relevance of good environmental management to personal health and welfare. And since the ecologists seem to be on to something, that ought to be easy enough to do.

31
Residuals Management   Walter O. Spofford, Jr.

During the past few years, we have become more and more cognizant of the fact that the discharge of certain residuals from the production and consumption activities of man may cause major changes in the behavior of global systems: physical, chemical, and biological. Consequently, we (society) now find ourselves faced with the problem of choosing a rational course of action on these matters. The process of deciding on a reasonable course of action involves, broadly, three functions: (1) estimating the changes in global systems (and the related subsequent damages to both present and future generations) associated with continuing our present policy of discharging various amounts of certain residuals into the environment; (2) estimating the implications (including social costs) of changing our present way of doing things and thereby reducing the quantities of these residuals discharged to the environment; and (3) given information on the benefits (that is, reduction in damages) to society for reducing the quantities of residuals discharged to the environment (as given in function 1), and given the costs to society of reducing the quantities discharged (as given in function 2), choosing those levels of environmental quality that we are willing to pay for.

Information on expected changes that may occur in natural global systems must come from the natural scientists, for example, meteorologists, oceanographers, and ecologists. Implications of changes (of the natural system, as well as social and economic changes associated with reducing the discharge rates of certain residuals) will be evaluated by social scientists, for example, economists and sociologists. Given this information, the choice among alternative courses of action will ultimately be made by the political decision process.

In the area of residuals management, it is the job of the economist to estimate the costs to society (that is, various ramifications throughout society, including, where applicable, the need for major social change) of adopting various alternatives for reducing residuals in the environment. His job is not to choose the "best" alternative or best set of alternatives for dealing with the residuals-environmental quality problem. Rather, it is his job only to provide decision makers (or more properly, the decision process) with

Prepared during SCEP and expanded after SCEP.

as many facts and best estimates as possible on which to make decisions. (This is not meant to imply that in some instances he should not at least suggest what he considers the most effective means of handling specific problems.) This paper outlines the residual management framework used by economists for these analyses.

Residuals are generated at virtually every stage in the production and consumption of goods and services. These residuals are of two basic types: material and energy. Most material residuals are readily transformed from one state to another (for example, solids to gases, as in combustion of municipal refuse or sewage sludge; gases to liquids, as in the wet scrubbing of gaseous effluent streams containing SO_2; gases to solids, as in the electrostatic precipitation of particles, and so on) in production, consumption, and residuals modification (treatment) activities. These state changes, however, frequently involve energy inputs and energy residuals. In the context of residuals management it is helpful to remember that residuals can be changed in form, but they do not disappear. The total weight of the material is still with us in one form or another. Hence, solids are sometimes referred to as the irreducible limiting form of residuals. By the application of appropriate equipment and energy, all undesirable substances can be removed from water and air streams (except CO_2, which might be harmful in the long run). And as the restrictions on the discharge of material residuals to our air and water resources become more stringent, we can expect to be faced with the problem of handling greater and greater quantities of solid wastes. Both air and water pollution may be reduced or even eventually eliminated with the application of the requisite resources, but we cannot eliminate the solid-waste problem without essentially total material recovery and reuse. We include this discussion here to emphasize the importance of considering all forms of energy and material residuals within a single analytical framework.

The discharges of sufficient quantities of residuals to air, water, and land environments (certain aspects of which may be considered common property resources) can, and generally do, impose external damages on other users of the environment. That is, within the normal function of the private market system, these damages are external to the operations of the discharger because there is no mechanism by which he is charged for the damages imposed on others—damages that are associated with his use of these common-property resources, specifically, the assimilative capacity of the environment. The extent of these externalities represents the failure of the private market system to allocate air, water, and land resources efficiently among all users. Stated differently, in terms of the many uses of the environment, the private market optimum does not correspond to the social optimum.

The ability of the natural environment to assimilate residuals is an extremely valuable resource. Complete elimination of all residuals discharges to the environment would be an extremely costly procedure. Indeed, as we have already pointed out, it would require complete recycling of *all* residual materials. But on the other hand, if no price or other use restriction were put upon the use of the assimilative capacity of the environment, it will be used too much, as we have already experienced in a number of cases. This in fact is what the "pollution" problem is all about.

Because the private market system fails to include all the relevant costs, governmental action is required. Types of incentives that have been used by federal, state, and local agencies in an attempt to compensate for the failure of individual units to consider externalities are (1) *legislative-executive,* including direct regulation, quality standards, and license requirements; (2) *litigation,* including compensation for damages as well as restrictions on operations and/or location; and (3) *economic,* including effluent charges or taxes, subsidies, and tax incentives, including accelerated depreciation allowances. Whatever the policy instrument(s), however, the intent is to achieve a more efficient allocation of resources from the standpoint of society as a whole.

Damages may be referred to broadly as the direct and indirect impacts on man stemming from deterioration in the quality of the various environments. These damages may assume many forms. Although in an economic sense all damages are conceptually direct, it is useful, at least for discussion purposes, to consider the following kinds of damages: (1) increased *private costs,* for example, increased costs of water treatment, medical treatment, and the cleaning of clothes, automobiles, and buildings; and (2) increased *nonmarket costs,* such as decreased aesthetic character of water, air, and land, and certain costs stemming from the impacts of residuals (for example, persistent pesticides and heavy metals) on flora and fauna, such as on food chains.

The quantities of material residuals that are ultimately discharged in, or returned to, the environment may be reduced by: (1) decreasing the quantities of residuals generated in production and consumption activities and (2) increasing levels of material recovery and reuse, such as materials recycling, by-product production, and product reutilization.

In the first case, the generation of residuals may be diminished in production and consumption activities by changing the nature of the product output; reducing the levels of production of a particular product through regulation or relative price change (tax); using substitute raw material inputs (or input mix); employing different technologies of production and consumption; and increasing the useful lifetime of goods, such as automobiles, buildings, machinery, and other durables.

The extent of material recovery and reuse is a function of the combination of private market prices and public and private constraints and incentives. The economic feasibility of materials recovery and reuse depends on the availability of a consistent supply of residuals of a specified quality and quantity, processing and/or reprocessing technology, product output specification, and the availability of markets for the secondary materials. Given a market capacity to utilize various quantities of secondary materials (together with the necessary institutions), *relative prices* of factor inputs are probably the most effective means of influencing the flows of materials to and from production and consumption activities. There is probably no recovery operation or industry that could not be expanded if the market price of the output were higher. It is quite common for recovered output to expand during periods of shortage of the competitive virgin raw material (the most notable example was the period during World War II).

In the past, prices have changed significantly over time, depending upon current supply and demand. What is not economically feasible today may be tomorrow (and the reverse, as in the case of steel scrap and the increased utilization of the basic oxygen furnace in steelmaking during the 1950s and 1960s). Today's prices, for example, generally do not reflect all the social costs associated with residuals generation and disposal (although with the current emphasis on air- and water-quality standards, prices are starting to include at least a part of these costs). In many cases, no costs are imposed on the producer for his use of the assimilative capacity of the environment as a factor input to his production process. If a fee were imposed on residuals discharged to the environment(s), relative prices of factor inputs to production would shift and process changes and/or increased materials recovery and reuse would likely be stimulated. Imposing this "effluent charge" would tend to induce alternative combinations of raw material inputs, production processes, types of product outputs, material recovery and reuse, and residuals treatment.

If the costs of reprocessing material residuals for inputs to production and consumption activities were decreased either through (1) changes in specifications of final product outputs that would reduce subsequent reprocessing costs, or even upgrade the residual as a secondary material (for example, elimination of the multimetal cans and substitution of aluminum electrical wire for copper in automobiles) or through (2) the development of more efficient reprocessing technology, relative prices would shift in a direction that would make material recovery and reuse more attractive. Similarly, if increased utilization of secondary materials could be found through either (1) new product uses of these residuals or (2) changes in production processes for final products that could process secondary material inputs along with other raw material inputs, relative prices would shift in a direction favorable to increased recovery and reuse. Technological innovation represents probably the most important alternative for reducing the unit costs of both residuals treatment and materials recovery and reuse and for expanding the opportunities for utilizing secondary materials.

Given this rather brief introduction to the general problem of the struggle between our ever increasing material and energy wants on the one hand and a decent quality environment on the other, we might ask ourselves how we propose to resolve this apparent conflict. That is, what kinds of alternatives are available to us for reducing the direct and indirect effects of pollution on man, and what sets of policy instruments might a regional environmental quality management agency invoke for effecting various changes.

At this point, it might be useful to examine one such regional management framework that has been proposed for this purpose by Resources for the Future, Inc. (Russell and Spofford, forthcoming). As shown in Figure 31.1, they have found it useful to think of the problem in terms of flows of residuals from human production and consumption activities (reflecting decisions about production mixes, levels, and methods, residuals modification, and recycle alternatives, and so forth) through the transforming environment (in which dilution, transportation, decay, and other processes may take place) to the receptors including man and the plants, animals, and inanimate objects of economic, aesthetic, or other interest to him (possibly after the application of final protective measures such as air filtration or water chlorination). In the overall system, these flows of residuals and the resulting damages give rise to corresponding flows of information from the large and diffuse groups of receptors to such generalized response organizations as conservation groups, city governments, and state legislatures. The information ultimately reaches whatever more specialized environmental quality management agencies exist and have jurisdiction, and these in turn take whatever actions seem both desirable and politically feasible. Such actions may include limiting discharges, requiring certain levels of treatment (or building treatment plants), imposing effluent charges, subsidizing recycling operations, modifying the assimilative capacity of the environment, or installing collective final protective measures. Whatever actions such agencies take will modify the ultimate damages and hence the content of the information being fed back into the response organization. Presumably, at some point the management agency will judge that the costs of action to reduce residuals concentrations reaching receptors will no longer be justified by the resulting decreases in damages (all these adjusted for the equity and political influence factors). At such a point, we can say that, at least temporarily, "a solution" to the residuals management problem has been reached.

Figure 31.1 is, at best, a generalized schematic diagram of a residuals management system and is not meant to include all the details of a system as complex as that required for residuals-environmental quality management. But it is useful for discussing the range of possibilities for reducing the damaging direct and indi-

483 Residuals Management

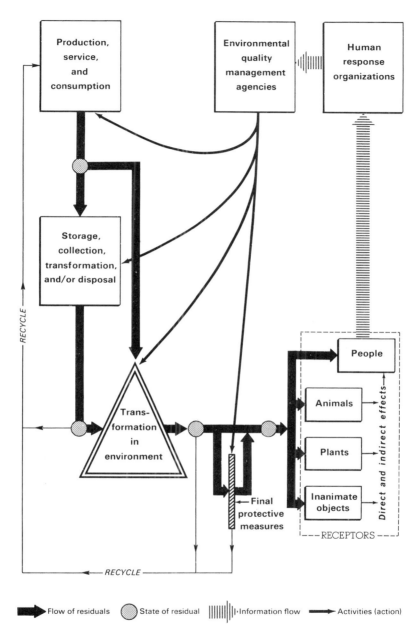

Figure 31.1 Schematic diagram—residuals management system
Source: Resources for the Future, Inc. 1968.

rect effects of certain residuals in the environment, as well as for discussing the various policy instruments available to a management agency and how each might be used for inducing the desired changes in activities.

Although the design, as well as the establishment, of effective institutions is necessary if we ever hope to be able to allocate our air, water, and land resources equitably, a detailed discussion of this, as important as it is, is not within the purview of this paper.

For the remainder of this section, we shall concentrate on (1) means for reducing the damaging effects of residuals discharges, and (2) policy instruments available to a management agency for inducing change(s). The reason for reemphasizing this is to point out that in many cases there is a variety of alternatives or combination of alternatives for coping with the residuals problem. The number of available alternatives, of course, varies among the residuals to be controlled and the nature of the activity involved.

There are various options available for improving environmental quality; we find it convenient to classify these alternatives in the following order:

1. *Reduction in the generation of residuals at the source* (that is, at the production or consumption activity). Alternatives within this category are (1) changing the nature of the product output through a change in product specification, for example, the change from high-phosphate to low-phosphate (or no-phosphate) detergents; (2) reducing the levels of production of a particular product through regulation or relative price change (for example, by imposing a tax), for example, restrictions on the use of persistent pesticides (among others, DDT) and the proposed tax on the lead in gasoline; (3) using substitute raw material inputs (or input mix), such as a switch from high- to low-sulfur fuel; (4) employing different technologies of production and consumption, such as processes that convert energy more efficiently, or replacement of the internal-combustion engine with an external-combustion engine; and (5) increasing the durability of products, thereby extending their useful lifetimes.

2. *Residuals modification processes.* This category includes the storage, collection, transformation, and/or alternative methods of disposal of residuals after generation, including possibilities for

materials recovery and reuse. The objective here is to modify the form of the residual in order to increase the potential for reuse or to decrease the impact on the environment upon discharge. It should be noted here that, although the form of a particular material residual may be modified, the total weight to be handled and disposed of in some manner—either through reuse or discharge to the environment—is not reduced and in some cases it is actually increased because of the material additions in the treatment process (for example, the addition of limestone in the wet limestone process for removing sulfur oxides from stack gases of power plants).

3. *Modifications to the environment which improve its assimilative capacity.* At the present time, alternatives in this category are few. However, two good examples in the water quality field are (1) instream aeration and (2) low flow augmentation.

4. *Discharge operations and/or structures that make better use of the assimilative capacity of the environment.* In this group we include the following activities: (1) temporary storage facilities so that residuals discharges may be scheduled to conform more closely with time varying assimilative capacities, and (2) effluent distribution, for example, use of higher stacks, and dispersing the locations of discharge points to make better use of the mixing and/or diluting characteristics of the environment.

5. *Final protective measures.* This category involves means of protecting the receptor(s) from the degraded environment, for example, water treatment facilities, sound proofing and air conditioning of buildings, and careful preparation of contaminated foods.

6. *Separation of residuals and receptors.* Distance alone in many cases is enough to protect receptors from insults to the environment. Zoning regulations have been used for this purpose and have proved, in some cases, an effective means of ensuring this separation.

We now discuss the various policy instruments available to society as incentives for effecting change. Again, we find it convenient to identify these in the following order:

1. *Legislative-executive.* Alternatives within this category are (1) direct regulation; (2) input, process efficiency, effluent, and ambi-

ent quality *standards;* and (3) licenses, for example, the Atomic Energy Commission's control over the use of radioactive materials. These are usually enforced by court action and/or fines.

2. *Litigation.* There are basically two types of litigation used in environmental quality matters: (1) lawsuits for the purpose of compensating for damages and (2) prevention of an industry from locating in a particular area or closing down industries that do not comply with regulations (or standards).

3. *Economic.* Alternatives in this category are (1) effluent charges on the kinds and amounts of residuals discharged to the environment; (2) direct subsidies, for example, the federal involvement in the construction of municipal sewage treatment facilities; (3) tax incentives, depreciation allowances, and so on; and (4) user charges, that is, charges to industry for using a collective treatment activity.

4. *Education.* This might be used mainly for inducing major social change through alteration of people's preferences and sets of values. Changes in relative values feed directly into the economic system through the concept of "willingness to pay" and into the political system through the pressures applied to a representative by his (or her) constituency.

One point we have not mentioned up to now, but must not omit, is how much it will cost society to improve the quality of the environment and who will pay for what. In the first case, society through the political process will decide what levels of environmental quality it is willing to pay for. And society as a whole will ultimately pay for this improved quality. In the second case, the question of who pays for what (that is, the old distribution question in economics), again, will ultimately be decided by the political process.

What does the preceding discussion suggest to us about the implications to society (private costs as well as other nonmarket costs) of controlling the rates of discharge of the "critical" residuals such as carbon dioxide, particles, heavy metals, oils, pesticides, and radioactive nuclides? First, it tells us that there is a variety of available alternative means for coping with environmental quality problems relating to these residuals. The number of options, however, depends on the particular residual considered. This is important for us to keep in mind, as frequently possible management alternatives are overlooked. For example, the traditional sanitary engineer usually thinks only of "end of pipe" treatment. Second, it points out the strong interrelationship among components of the overall system, especially among the states of residuals (liquids, gases, and solids) and between material and energy residuals. Remedying an individual situation may only cause a problem elsewhere in the system. Because we are dealing with such a complex interrelated system, a plea is made here for additional effort into the development of quantitative approaches, that is, mathematical decision models, to aid in the preparation of alternatives upon which decisions regarding overall social welfare might be made. Third, the approach to the critical environmental problem, which in some cases may involve drastic social change, should be extremely imaginative, and all possible technologically feasible alternatives should be given at least a cursory review. Alternatives should not be excluded solely on our present-day value judgments as to their economic or political feasibility. Fourth, decisions involving social costs and benefits, which usually involve a complex set of value choices (especially where the allocation of a common property resource is involved), will be made by a political decision-making process and not by physical scientists, engineers, and economists outside the political arena. The job of scientists and engineers is to provide decision makers with the best-available information upon which to make decisions regarding a particular issue, including environmental problems. Last, our decision makers may not choose the alternative some of the physical scientists, engineers, or economists would consider a "rational" choice. But in terms of the decision maker's very complex objectives, it most likely will be his most rational choice. The point to be made here is that whether we (as scientists and "rational" beings) like it or not, people, as a society, will somehow determine their own destiny based on an implicit evaluation (whether or not they are "properly informed") of the social costs and benefits accruing both during their own expected lifetime as well as during future generations. An example of this apparent "irrational behavior" is the cigarette problem, where, *given* the statistical association between smoking and health hazards, so many people continue to smoke.

For some of the critical residuals, various alternatives of the direct cost variety and which do not involve major changes in the patterns of people's lives are available; for example, sulfur dioxide, particles, and heavy metals. But for such residuals as carbon dioxide, heat, and radioactive nuclides, the number of options are limited, and those that are available involve high social costs possibly even requiring, in some cases, vast social change. In all of these cases, extensive analyses of the factors involved should be undertaken so that these residuals-environmental problems can be resolved.

Acknowledgments
Preliminary discussions with Blair T. Bower of Resources for the Future, Inc. were extremely helpful to the author. In addition, helpful comments on the manuscript were received from Blair Bower and Allen V. Kneese.

References
Resources for the Future, Inc., 1968. *Annual Report* (December), p. 44. Figure 31.1 reprinted with permission.

Russell, C. S., and Spofford, W. O. A quantitative framework for residuals management decisions, *Environmental Quality Analysis: Research Studies in the Social Sciences,* edited by A. V. Kneese and B. T. Bower (Washington, D.C.: Resources for the Future, Inc., in press).

Phosphates, Heavy Metals, and DDT: Pollution Control Costs and Implications

John F. Brown, Jr.

Phosphates

The Pollutant

Phosphate is frequently the limiting nutrient for the growth of algae in lakes, rivers, and estuaries; hence, increasing the aquatic phosphate level over that naturally present will frequently accelerate the rate of eutrophication. Total U.S. phosphate discharges into waterways are probably three to four times what they were fifty years ago, and eutrophication of some lakes and rivers has progressed to the point where it is evident even to the layman. United States farms and cities emit over 1 million tons of phosphate annually. In the Lake Erie basin, the estimated breakdown is rural runoff, 17 percent; urban runoff, 7 percent; industrial discharges, 4 percent; discharges into sewers, 72 percent (Federal Water Pollution Control Administration [FWPCA], 1968). The latter figure probably represents about 15 percent from human wastes and 57 percent from phosphate detergents. This pattern is probably reasonably characteristic for the nation as a whole, except for the unknown, but possibly large, contribution made by feedlot runoffs (FWPCA, 1970).

Control Technology

A number of existing and foreseeable control techniques are possible:

1. Reduction in agricultural emissions through better soil conservation practice (to limit surface erosion), discontinuance of fertilizer- or manure-spreading on frozen fields, and provision of effluent treatment or waste recycling for feedlot and chicken farm wastes. Such procedures would be very helpful in many local situations (FWPCA, 1970). Nationwide, they would remove only a small part of the phosphate discharges.

2. Application of phosphate-removal techniques to industrial discharges. Presumably, most industrial phosphate discharges could be treated by processes similar to those developed for municipal wastes; however, it would appear that only a few percent of the national emissions would be thus eliminated.

Prepared during SCEP.

3. Reduction or elimination of the phosphate in detergents. Presuming that nonbiotoxic, biodegradable agents having suitable detergency could be found, their use could remove the largest single source of aquatic phosphate. Currently, NTA (nitrilotriacetic acid) is the leading candidate; it is stated to be not quite as active a polyphosphate as a detergent, and its biological effects are not known in detail.

4. Removal of phosphate from sewage. Phosphate can be precipitated from either primary or secondary treatment effluents using lime (80 to 90 percent removal) or iron or aluminum salts (90 to 95 percent removal). Treatment costs run about 5 cents per 1,000 gallons (American Chemical Society, 1969). It can also be removed (> 90 percent) in the activated sludge process for secondary treatment, provided the sludge is precipitated rapidly and not subjected to the usual anaerobic digestion, which releases most of the nutrients back into the effluent.

Social Alternatives

There appear to be several ways by which the various sorts of control technology could be put in place:

1. Continued education of farmers by the U.S. Department of Agriculture. This might somewhat reduce the runoff losses and drainage from improperly managed feedlots.

2. Require registration and inspection of feedlots by state health departments. This is already being done in Kansas.

3. Establishment of water quality and emission standards for phosphate, and their enforcement as regards industry, farmers, and municipalities. In the absence of incentives, the enforcement of compliance by the municipalities, which are the major sources, is likely to be very slow.

4. Outlaw the use of polyphosphate in detergents. This is currently a politically popular approach that might solve half the phosphate problem. If applied immediately, it would result in reduced cleanliness of our dishes and clothes, particularly those washed in automatic dishwashers and washing machines.

5. Tax the use of phosphate in detergent mixtures and use the proceeds to pay municipalities for removing the phosphate from their sewage. On a break-even basis, a tax of about 5¢/lb on the detergent should suffice. If higher, the municipalities might be able to perform phosphate removal for profit. This would simul-

taneously provide incentive to the municipality to commence phosphate removal, the water treatment industry to find cheaper materials or equipment for phosphate removal, and the detergent industry to find inexpensive but effective substitutes for polyphosphates.

Implications of Change

The benefits of phosphate removal from waterways would be improved aesthetic and recreational quality in many lakes, rivers, and possibly estuaries, probably partially offset by lower fish production in some areas. The major cost would be in the price of detergents, which would occur whether the polyphosphate were taxed or replaced. If not taxed, then additional tax revenues would have to be collected to pay for the treatment cost ($1 to $2 per capita annually). Minor costs, also paid by taxes, would be required for the agricultural education work; establishing, monitoring, and enforcing the water quality standards, and so on.

Heavy Metals

The Pollutants

The heavy metals that pose problems when discharged into the aquatic environment include mercury, lead, nickel, cadmium, manganese, chromium, copper, and zinc. Although the last four of these are normal and essential trace elements in living organisms, all eight can be highly toxic at higher levels. They are capable of killing the normal aquatic biota of rivers and estuaries and are frequently implicated in poisoning the biomasses in secondary sewage treatment plants. In addition, mercury may be converted by bottom mud microorganisms into the very toxic substance methyl mercury. This behaves as a nonbiodegradable cumulative poison that concentrates upon being passed up the biological food chain (like DDT), so that predator species (notably fish) may contain manyfold higher concentrations than the waters in which they live (Jensen and Jernelov, 1969).

All of the heavy metals are known to have airborne as well as waterborne routes into the environment. However, in all cases of actual aquatic life poisoning or bioaccumulation reported to date, the evidence indicates that the initial discharge was via a sewer or waterway.

Lead, copper, and zinc may enter waste waters as a result of the corrosion of pipes, but in amounts rarely troublesome any more. The more usual route is via discharges from (generally) small technical operations, involving both large and small industries and, occasionally, scientific laboratories as well. Mercury is used in a host of manufacturing and research operations that ultimately result in discharge of aqueous wastes (Chemical Engineering News, 1970). Chromium, nickel, cadmium, and copper are widely used in electroplating, a highly fragmented industry with many small operators having few inhibitions about disposing of their waste solutions in the easiest way possible. A typical sewage analysis (from the Los Angeles County Sanitation District) indicated the following concentrations of heavy metals (in parts per million): chromium, 0.61; lead, 0.17; cadmium, 0.08; copper, 0.48; nickel, 0.24; zinc, 0.50; and manganese, 0.09 (Brooks, 1969).

Control Technology

The basic chemical techniques for removing heavy metal ions by precipitation or ion exchange have been long known and involve only inexpensive chemical operations. For example, effluents from mercury cell chlor-alkali plants (the major source of mercury discharges) have been completely freed of mercury by using a settling basin and recycle system (Chemical Engineering News, 1970). The more usual problems are that any form of effluent control is a bother in a small operation, and the operator himself may know neither the amount of heavy metal in his effluent nor of its toxicity. This is probably just as true for the director of the scientific laboratory discharging mercury as it is for the operator of the small plating shop discharging chromium, cadmium, and nickel.

At any event, heavy metals do have to be removed as close to their point of origin as possible because of the dilutions involved, removal at a central sewage treatment plant is impractical.

Social Alternatives

Several types of social action offer hope for reducing the levels of heavy metal pollution:

1. Publicize the problem. Knowledge of the toxicity of the heavy metals under discussion is surprisingly limited; lead is the only one of the group for which the U.S. Food and Drug Administration (FDA) has set a food content standard. Although the

volatility of metallic mercury and the toxicity of its vapor have been long known, neither the magnitude of mercury discharge into the aquatic environment nor the hazard posed by its biological conversion to methyl mercury was recognized until very recently. Many—perhaps most—polluters are willing to change their practices once they know that a real problem exists.

2. Establishment of water quality standards. Particularly in a situation where individual treatment of many small discharges is required, the question of standards becomes vexing. Does every small laboratory or industrial discharge require treatment? How complete should the treatment be? No treatment can ever be 100 percent complete, nor does it need to be, since all of the heavy metals have finite abundances in nature anyway. (In seawater, the normal heavy metal concentrations run between 10^{-1} and 10^{-4} of the values for Los Angeles sewage mentioned earlier [Brooks, 1969].) Public agreement on emission standards is required if the polluter is to know where further treatment is required, the water treatment industry is to know what improved treatment materials and equipment will find use, and the law enforcement agent when his intervention is required.

3. Water quality surveys and monitoring. Pollution by trace amounts of heavy metals, unlike that by massive amounts of organic wastes or particles, is not immediately apparent to the casual observer. Knowledge that a problem exists is an essential prerequisite to either technical or legal action.

4. Registration of users of heavy metal salts and mercury. If all purchases of key chemical intermediates likely to be used in polluting operations were registered, it would be much easier for the local water quality management authorities to locate the potential trouble spots.

5. Enforcement of water quality standards. The validity of the public's interest in water quality will have little credence unless the relevant legislation is at least occasionally enforced by one means or another.

Implications of Change

Control of heavy metal discharges into waterways would get us away from the problems of poisonous mercury-containing fish, copper-containing green oysters, and erratic performance in sewage treatment plants and would generally improve the quality of our rivers and estuaries. The costs of such pollution controls would not appear large, and there are probably few, if any, laboratory or business operations that would be shut down on their account. There would, of course, be a cost to the taxpayer for the governmental actions required to establish, monitor, and enforce the water quality standards.

DDT

The Pollutants

The story on DDT, the other "hard" insecticides, and one class of nonbiocidal persistent chlorinated hydrocarbons, the polychlorinated biphenyls (PCBs), needs little further elaboration at this point. DDT, the type species, during its quarter-century of use, has saved millions of tons of agricultural production from insect attack and millions of lives of men and livestock from malaria, typhus, and other insect-carried diseases. At the same time, much has translocated to lakes and oceans, accumulated in the biological food chain, and reduced the populations of many species of plankton, crustaceans, mollusks, fish, and birds. The other chlorinated insecticides, introduced to deal with DDT-resistant insects, and the PCBs, used industrially as noninflammable fluids and plasticizers, have similar effects in the aquatic environment.

DDT is used in agriculture and public health work because it is persistent at the point of application (infrequent application required), effective against a considerable range of insects, of very low toxicity to men and animals, and cheap (17¢ per pound). The alternative pesticides, both "hard" and "soft," suffer by comparison to varying degrees on every one of these scores. The PCBs as constituents of the commercial chlorinated biphenyl mixtures (Arochlors, and so on) are also cheap chemicals (15¢ per pound) and less expensive than most other organic fluids and plasticizers (though not than the hydrocarbon oils that they were originally introduced to replace in order to achieve noninflammability).

Control Technology

There are six general strategies that may be used to reduce environmental pollution by DDT and other chlorinated hydrocarbons (American Chemical Society, 1969):

1. Minimize pesticide use. Many studies have indicated that the amounts of DDT used in agriculture and forestry, and the frequency of application, are often excessive. The cost of DDT is so low that the user is strongly motivated to spray first and question the necessity later.

2. Improve application equipment. Reports that as much as half the insecticide sprayed on a field winds up in the atmosphere indicate ample room for the improvement of applicator design.

3. Improve the formulation. The physical persistence of the pesticide at the point of application, where its insecticidal activity is needed, and where its rate of chemical and biochemical detoxification is probably the greatest, can be markedly altered by careful formulation (for example, the selection of the oil or resin used to "glue" the pesticide to the substrate).

4. Use a nonpersistent substitute. The aldrin group of chlorinated insecticides appears somewhat less persistent than DDT, and the carbamates and organophosphorus derivatives very much less so. All of these are, however, considerably more expensive to use and are also more toxic to humans. Ideally, research should be done to develop low-cost, nonpersistent, nontoxic insecticides for general use. This is slow and expensive, however, and there is no great assurance that the next pesticide discovered will be any better in this combination of characteristics than the 900 already registered for use.

5. Use nonchemical methods of pest control. A variety of biological techniques, involving such approaches as bacterial and viral diseases, use of insect pheromones (feeding or sex attractants) as baits, encouragement of predator species, and planting of sacrificial crops have been developed for controlling specific insect pests. Considerable research is continuing in this area.

6. Eradicate the pest. The extreme of the foregoing approach is to aim at total eradication of the insect species. In recent years, the Mediterranean fruit fly has been eradicated in Florida, and the screwworm fly throughout most of the country. It has been estimated that the eradication of three species—the boll weevil, the bollworm, and the codling moth—could reduce U.S. insecticide usage by 40 percent.

Social Alternatives

1. Publicize the problem. The gains to be achieved by voluntary actions should not be overlooked. In the years between Rachel Carson's book *The Silent Spring* and the formal U.S. Department of Agriculture restrictions on many uses of DDT, U.S. consumption of DDT dropped in half. Within a year of the time that the PCB problem became known, the sole U.S. producer, Monsanto, announced plans for progressive discontinuance of sales and introduction of nonpersistent substitutes.

2. Promote research on more efficient application and usage techniques.

3. Promote research on nonchemical alternatives. In both cases, government funding (for example, by the Department of Agriculture and the states) of the work would be required, since the incentives for private industry to do the work are not apparent.

4. Promote research on nonpersistent pesticides. In this area, virtually complete funding and performance of the work by industry would be expected. Effective stimulation of such research could be achieved by progressive taxation or banning of the persistent insecticides.

5. Tax the persistent pesticides in proportion to their demonstrated ecological effects. This would be the most direct route of internalizing the external costs of use.

6. Ban the manufacture and/or use of all toxic chemicals found to persist in the environment. This alternative, often heard today, just might be politically possible in the United States and certain other Western nations. It would not be acceptable in most other countries.

7. Require that all new commercial chemicals be cleared for environmental compatibility before use. This proposal visualizes environmental clearance procedures analogous to those now practiced in the food and drug area.

Implications of Change

The benefits of environmental conservation, preservation of fisheries and birds of prey, and so on, have often been stressed. A few of the problems associated with some of the social alternatives merit mention, however.

First, eliminating the use of DDT and other persistent chlorinated hydrocarbons could undoubtedly be accomplished without seriously injuring those involved in most agricultural and industrial applications. There are some very real hard-core problem areas, however. The most important of these involve use in underdeveloped countries for disease control and crop yield maintenance and use on the U.S. cotton crop. Replacement of DDT there would add perhaps 3 to 5 percent to the cost of production, or 1¢/lb to the final product, which is a commodity already in a severe price squeeze. Some protection for the U.S. cotton farmer might be required if a total DDT ban were to be politically acceptable.

Second, banning the use of DDT in the United States will be of limited value as long as the bulk of world production is used in other countries. In view of the magnitude of the current social dependence upon DDT in many areas, it would appear that the most reasonable target for the environmentalist would be a progressive reduction in DDT use, achieved by gradual replacement, increased sophistication in usage and application techniques, biological controls, improved farming practices, and so on, rather than an immediate ban.

Third, the requirement of FDA-type clearance on all commercial chemicals would multiply by manyfold the cost of industrial research and probably bring technical progress to an almost complete halt in the chemical industry. In view of the facts that 99 percent of the known industrial chemicals do not accumulate (as mercury and DDT do), that most are produced only in small volume, that the mechanisms of biological conversion and accumulation are unpredictable, and that man-made effects within the oceanic environment are most unlikely to develop suddenly, it would appear that control after the demonstration of a problem rather than before would be the more realistic course of action.

References

American Chemical Society, 1969. *Cleaning Our Environment. The Chemical Basis for Action* (Washington, D.C.: The American Chemical Society).

Brooks, N. H., 1969. *Some Data on Municipal Waste Discharges to the Pacific Ocean in Los Angeles Area,* Preliminary report from the W. M. Keck Laboratory of Hydraulics and Water Resources, California Institute of Technology, Pasadena, California.

Chemical Engineering News, 1970. Mercury stirs more pollution concern, *Chemical and Engineering News* (June 22), pp. 36–37.

Federal Water Pollution Control Administration (FWPCA), 1968. *Lake Erie Report* (Washington, D.C.: U.S. Department of Interior).

FWPCA, 1970. *The Economics of Clean Water, Vol. II. Animal Wastes Profile* (Washington, D.C.: U.S. Department of Interior).

Jensen, S., and Jernelov, A., 1969. Biological methylation of mercury in aquatic organisms, *Nature, 223:* 753.

33
DDT: The United States and the Developing Countries
Rita F. Taubenfeld

Introduction

Dichlorodiphenyltrichloroethane, the forerunner of DDT, was synthesized in 1874. When its insecticidal properties became known in the 1930s, DDT became the first of a series of chlorinated hydrocarbons that have come to be known as the "persistent," "residual," "long-lasting," "long-lived," or "hard" pesticides. They "have contributed tremendously to preventing human disease and to increasing the production of food and fiber" (U.S. Department of Agriculture, 1969). They have allowed the opening of new lands, previously infested by malaria and the tsetse fly. Because of DDT, "World War II is said to be the first in history where more soldiers died from bullets than from typhus, a louse-borne disease" (Keller, 1970).

There is also evidence of the buildup in man, beast, vegetation, and in the atmosphere and the seas of the persistent chlorinated hydrocarbons. DDT is known normally to have a relatively long decaying time. Apparently this buildup of DDT in the environment is worldwide and has a tendency to concentration in some living organisms; often the higher the link in the food chain, the greater the concentration.

It is important to note, however, that there is dispute in the scientific community as to the implications of accumulations of DDT in the tissues of living organisms and in general as to the dangers to the ecology and to men of the use of DDT. The recent authoritative study undertaken for the U.S. Secretary of Health, Education, and Welfare (often referred to as the Mrak Report after the name of the chairman of the commission) carefully sifted almost all available evidence on the effect of pesticides on the environment and on man and reached the same conclusions as did the SCEP Report with regard to effects on nontarget organisms other than man (Report of the Secretary's Commission on Pesticides and Their Relationship to Environmental Health, 1969; SCEP, 1970).

Persistent chlorinated hydrocarbons (of which group DDT is a member) . . . are causing serious damage to certain birds, fish

Prepared during SCEP and expanded after SCEP. The original paper was shortened considerably by the editors for inclusion in this volume.

and other nontarget species. . . . Some of these species are useful to man for food or recreation, some are essential to the biological systems of which he is a part, and some merit special attention because they are already endangered.

However, for the purposes of this paper and in the calculus of the developing countries, the most important and debatable question remains: What is the direct effect of the growing pool of persistent pesticides in the environment on man? On this important question, after warning on the difficulties of reaching conclusions on the basis of available information, the *Report of the Secretary's Commission on Pesticides* (1969) concluded:

It appears, however, that present levels of exposure to DDT among the general population have not produced any observable adverse effects in controlled studies on volunteers. . . . These findings acquire greater force when combined with observations on other groups, such as occupationally-exposed persons.

On the basis of the present knowledge, the only unequivocal consequence of long-term exposure to persistent pesticides, at the levels encountered by the general population, is the acquisition of residues in tissues and body fluids. No reliable study has revealed a causal association between the presence of these residues and human disease. . . .

While there is no evidence to indicate that pesticides presently in use actually cause carcinogenic or teratogenic effects in man, nevertheless, the fact that some pesticides cause these effects in experimental mammals indicates cause for concern and careful evaluation.

From the point of view of the current needs of the developing countries, some special points in favor of DDT should also be noted. Defenders of DDT have pointed to its relative safety in application, compared even to some of the less-persistent pesticides. There has been, for example, "no documented instance in which human deaths have resulted from the proper application" (Abelson, 1969). "DDT as applied has not caused any side effects among domestic animals" or in the wildlife of the countries participating in the widespread World Health Organization (WHO) malaria eradication campaign. It proved safe for both the spraymen and the affected population, and this covers a sample of 130,000 spraymen and 535 million people at the peak of the campaign (National Communicable Disease Center [NCDC], 1969; World Health Organization [WHO], 1969).

It is not the intent of this paper to come to any definite conclusions on the conflicting evidence or the conflicting interpretations of the evidence of impact of DDT on living organisms and, most important, on man himself. Nor is the intent to present supporting data on either side of these questions. For this discussion of policy alternatives, it is necessary only to note that there is serious dispute on these issues among well-qualified, concerned investigators. This was clearly identified in the commission cited earlier.

The present policy toward the use of DDT in the developing states can best be understood and evaluated with the following circumstances in mind: in any "rational" cost-benefit calculations on the uses of DDT, the direct dangers from DDT to man remain unknown and unproved even in 1970; and the seriousness of the implicit dangers to the present ecological balance is not fully understood and not easy to quantify as to value to human society. On the other hand, the past and present benefits to human communities from the use of DDT seem more quantifiable, in principle. Some of the major historical components of these benefits, particularly as they remain relevant to the points of view of the developing states on the appropriate current uses of DDT, are presented now.

DDT and the United States

The Regulation of DDT in the United States

Over the years there have been growing pressures against the use of DDT from many environmentalists and ecologists. In addition, pesticide tolerance levels have been established by the U.S. government for fruits, vegetables, meat, fish, and other products. Partly as a result of fears of breeching these tolerances Michigan, Wisconsin, and Arizona have banned the use of DDT for many purposes, while other states have restricted its use.

Continuing popular pressures and the growing general interest in ecology and the environment suggest that further intensification of government efforts to control the use of persistent pesticides and especially the use of DDT can be expected.

The Federal Insecticide, Fungicide, and Rodenticide Act requires that all pesticides be registered before they can be marketed in interstate commerce in the United States. Under this act the Secretary of Agriculture is responsible for the registration and regulation of such pesticide products (U.S. Department of Agriculture, 1969). In the future the registration of pesticides will be part of the duties of the Environmental Protection Agency (EPA). In addition, in November 1969, the U.S. Department of Agriculture issued a statement concerning DDT regulations that outlined several uses for which DDT could no longer be employed and noted that other uses were being considered for cancellation (U.S. Department of Agriculture, 1969).

Although it has been estimated that about 14 million pounds, or 35 percent of the total DDT used in this country, is manufactured for these currently banned purposes, these policies probably did not greatly affect the major U.S. DDT users, the growers of cotton and some growers of corn (*Science,* 1969). Then in January 1971, the U.S. Court of Appeals for the District of Columbia ordered the administrator of the EPA "To issue immediately notices of cancellation of all uses of DDT" and to "determine whether DDT was an imminent hazard to public health" (*New York Times,* 1969). A more comprehensive U.S. ban on the use of DDT seems likely.

Of course, the U.S. government itself uses pesticides. In 1969 the U.S. Department of Agriculture pest control operations were reviewed by use. Some persistent pesticides, which include DDT, were replaced by less-persistent ones in some cooperative federal-state programs. The department issued a policy statement on October 23, 1969, which states the interesting compromise policy which the department had worked out to apply to its own pesticide applications:

Persistent pesticides will not be used in Department pest control programs when an effective nonresidual method of control is available. When persistent pesticides are necessary to combat pests, they will be used in minimal effective frequencies. (U.S. Department of Agriculture, 1969.)

As this policy statement suggests, the fact is that for some very important uses there is no genuinely adequate replacement for the hard pesticides, including DDT.

The Production and Distribution of U.S. DDT

It is reported that the U.S. Department of Agriculture has estimated that there would be a 25 to 30 percent drop in U.S. food production if pesticides were banned (Kramer, 1969). No such ban is presently foreseen.

United States production of pesticides continued its ten-year increase in 1968 to meet the rising consumer demand both at home and abroad. However, in 1968 DDT represented only 24 percent of the U.S. production of insecticides, fumigants, and rodenticides and 12 percent of the U.S. production of all types of pesticides. Furthermore, domestic use represented only about 26 percent of U.S. DDT production in 1968 (U.S. Department of Agriculture, 1969).

But though on a relative decline, DDT is not disappearing as an important pesticide. For while overall demand for DDT, at home and in some areas abroad, appears to be downward, the volume produced and quantity exported in 1968 were both up from the previous year (U.S. Department of Agriculture, 1969). "DDT continued in 1968 to be the insecticide in most use worldwide judging from production and export data" (U.S. Department of Agriculture, 1969). (See Table 33.1.) Nearly 80 percent of the U.S. production of DDT in 1968 went for export. Of this, about 58 percent was shipped in the 75 percent wettable powder

Table 33.1 DDT: United States Exports of Formulations Containing 75 Percent or More DDT, by Country of Destination, 1965–1968

Country of Destination	1965 Pounds	1966 Pounds	1967 Pounds	1968 Pounds
India	17,443,172	5,762,441	11,111,055	31,670,776
Pakistan	9,873,921	16,421,149	1,693,430	8,121,152
Brazil	3,509,742	3,700,132	3,661,431	3,713,703
Philippines	1,271,401	560,117	407,519	2,794,384
Ceylon	0	276,600	11,610	2,280,552
Iran	4,263,150	3,269,190	1,344,900	2,155,684
South Vietnam	1,773,190	299,926	0	1,287,648
Guatemala	280,000	147,778	1,274,250	1,259,100
Mexico	2,509,771	2,563,078	2,259,424	1,199,823
Thailand	6,855,468	1,833,769	8,607,605	981,251
Afghanistan	214,150	1,387,520	3,750	773,250
Iraq	930,075	1,402,676	1,916,450	702,300
Colombia	777,000	864,631	345,667	694,800
Ecuador	886,148	223,315	39,250	691,280
Haiti	0	110,000	247,381	563,480
Other	5,127,202	6,325,640	9,056,774	4,337,196
Total	55,714,390	45,147,962	41,980,496	63,226,379

Source: U.S. Department of Agriculture, 1969

formulation presumably going mostly for the control of malaria mosquitoes. Three countries, India, Pakistan, and Brazil, together received more than two-thirds of this formulation (U.S. Department of Agriculture, 1969).

Widespread use of DDT on crops in the United States began in the late 1940s. Other organochlorine insecticides were developed following the introduction of DDT. Total usage of the more popular of these insecticides, including DDT, peaked in 1959 at 156 million pounds but came down to 71 million pounds in 1968 (U.S. Department of Agriculture, 1969). The U.S. Department of Agriculture (1969) credits this decline to three factors: "(1) increased insect resistance to insecticides, (2) the ability of less persistent insecticides which are efficient in insect control, and (3) the concern about the risks from use of persistent insecticides."

DDT and the Developing Countries

DDT and Agriculture

Most developing countries use little or no pesticides on the majority of crops (President's Science Advisory Council [PSAC], 1967). Data on the losses caused by pests in foreign countries "are largely unavailable but indications are that the overall loss in developing countries is greater than in the United States. According to FAO estimates, preharvest losses to food and industrial crops due to insects and diseases in 11 developing countries of the Near East Region are equivalent to 23 percent of the crop production. . . . The average losses in potato yields . . . in Chile have been estimated at about 23 percent per year" (PSAC, 1967). Even in the USSR, annual crop losses in recent years due to pests apparently amount to 20 percent of total production. "Reportedly economic benefits from chemical control amount to 10 to 20 times the cost, but less than 50 percent of the demand for pesticides is being satisfied." By 1970 seven times the 1963 output of plant protection chemical is expected (PSAC, 1967).

DDT and related compounds account for well over half the insecticides used for crop protection in poor lands. The Food and Agriculture Organization estimates that without DDT-like compounds, 50 percent of the cotton production in less-developed countries would be chewed. . . .

In India, where insects eat 15 percent to 30 percent of all farm crops each year, the government and U.N. consultants are working to increase crop land protected by pesticides—mainly DDT—to 20 percent of the total from 10 percent. That measure alone, it's expected, will bring India an extra 1.4 million tons of rice, 100,000 tons of peanuts, 65,000 tons of Sorghum, 250,000 tons of sugar, 46,000 tons of corn, and 200,000 tons of potatoes every year." (*Wall Street Journal,* 1970.)

Despite the development of biological and other nonchemical methods of pest control, "ravaging insects such as the desert locust, army worms, boll weevil, codling moth, seed weevil, and borers attacking sugarcane, rice, maize, cotton, fruit and many other crops can be controlled presently only by use of (chemical) pesticides" (PSAC, 1967).

The amounts of DDT used in agriculture in 1967 in several countries and the world are given in Table 33.2.

While DDT has been selling for about 17¢ a pound, alternative chemicals often cost as much as a dollar a pound, and more (U.S. Department of Agriculture, 1969). Many of these are less long lasting than DDT and must be reapplied several times for one application of the latter. This also implies increased expenses of storage, transport, and labor in application.

It has been argued that "American farmers, who currently spend only 5 percent of their total operating expenses on insecticides, might find it relatively easy to adjust to costlier chemicals." But in countries like Mexico, if pesticides are used at all, they would tend to make up a quarter or more of the total outlay of farmers. In such instances substitution of less long lasting, eco-

Table 33.2 DDT Used in Agriculture, 1967
(metric quintals* of active ingredient)

United States	182,602†
United Arab Republic	35,650
India	28,710
Poland	27,080
Italy	18,211
World‡	336,042

Source: Food and Agriculture Organization, 1968.
* 1 metric quintal = 220.5 U.S. pounds.
† U.S. figure includes all uses. These are believed to be mainly agricultural.
‡ Does not include mainland China.

logically preferable pesticides at the farmer's expense appears financially impossible *(Wall Street Journal,* 1970).

Crop yields by country are significantly correlated with pesticide usage (PSAC, 1967). In the context of widespread hunger it seems clear for the present at least that there is the need for more, not less, use of pesticides or for adequate substitutes for pesticides in the developing countries. DDT is preferred by them for agriculture because of its safety, broad effectiveness, and most important, its cheapness. If it is desired by the United States and others that more expensive methods of control be used by them, the developing states can be expected to ask: "Who is to bear the burdens of the extra costs?"

DDT and Health Programs

"In 1968, 75 million lbs of DDT produced in the U.S.A. went towards public health," according to a statement prepared by Vector Biology and Control (WHO, 1969).

Almost 90 million pounds of the technical DDT per annum is employed in the campaign against malaria. This remains its most important health use. DDT has also been used as a mosquito lavicide in water and as an aerosol fog against adult mosquitoes. These latter uses have apparently been largely discontinued in the United States and are being supplanted in Asia and Africa against the vectors of hemorrhagic dengue and yellow fever due to the development of resistance by the vectors to DDT. Another important use of DDT is in the control of the tsetse fly that transmits sleeping sickness to man and nagara to cattle. DDT is also still used against vectors of bubonic plague, epidemic typhus, and relapsing fever, though in some areas resistance is reported and other insecticides are being substituted. DDT still remains unsurpassed for control of Phlebotomus sandflies vectors of leishmanissis (WHO, 1969). It is also reported to give significant protection against yellow fever. It is useful in the United States against vectors of eastern equine encephalitis, and it has many other public health uses.

Despite some problems of resistance, DDT remains the insecticide of choice in combating malaria. The inception of resistance by some anopheline spurred a worldwide search for substitutes, in which many thousands of chemicals have been screened without yielding any that compete in safety, versatility, cost, or residual effectiveness in malaria eradication programs (WHO, 1970). The safety criteria applied in this screening are reported to be quite stringent, and it has been suggested that some of the earlier compounds now in use such as DLD might not have passed such standards.

The U.S. National Communicable Disease Center (NCDC) has noted that as of August 1968 more than 1,300 chemicals have been studied. However, only two compounds—malathion and arprocarb (Baygon)—show any real promise for even limited operational use, and both have limitations (NCDC, 1969). A third promising compound, fenetrothion, is still being tested. The NCDC Health Services estimate that "neither malathion nor arprocarb is as long-lasting as DDT, both are much more expensive, and arprocarb is much more toxic both for spraymen applying it and for the occupants of the treated houses" (NCDC, 1969). In addition, they are bulkier, and handling and storage costs are at least 50 percent higher.

Despite their various drawbacks, however, relatively safe, effective biodegradable alternatives to DDT that "could replace the chlorinated hydrocarbons in controlling almost every species of public health importance" exist (Wright and Wurster, 1970). It will be noted that the principle defect of these alternatives to DDT is their high cost. At present this is an overwhelmingly important defect. We return to appraise alternatives to DDT later. The point is that currently the worldwide malaria eradication program is dependent on financial support from the United States, United Nations International Children's Emergency Fund (UNICEF), Pan American Health Organization (PAHO), WHO, and some other developed countries such as France, the USSR, and Great Britain. United States foreign assistance, including assistance for malaria eradication has been steadily declining. The latter was approximately $40 million several years ago, $30 million in the fiscal year 1969, and it is expected to decline to $25 million in the fiscal year 1970. Shortage of funds is perhaps the most common limiting factor in the overall global malaria effort currently employing low-cost DDT.

We shall briefly review this program and its achievement to date (NCDC, 1969). The global malaria eradication program was established after the postwar success of most of the advanced countries of the world in malarious areas, including the United States. Before the program it was "estimated that each year malaria was contracted by 300 to 400 million persons and that it killed between three and four million of these" (Wright and Wurster, 1970). Since that time, WHO, PAHO, and UNICEF have substantially assisted most of the 116 countries originally classed as malarious. Eradication has been claimed in 36, mostly technologically advanced states such as the United States and Western Europe or island countries; 53 countries are currently engaged in national malaria eradication campaigns; 27 countries are carrying out significant large-scale malaria control efforts, and 30 countries, mostly in Africa, are still without definite antimalaria projects. It has been estimated that in the first eight years of these programs, 5 million lives were saved and 100 million illnesses prevented (WHO, 1969). By 1968, 78.1 percent of the populations in the originally malarious states, or 1,353 million people, were reported as totally protected from malaria because of these programs (NCDC, 1969).

It is repeatedly claimed by health authorities that this program has been achieved with a minimum of pollution to the environment. The malaria eradication campaign is limited to the interior of houses, a situation that is claimed tends to minimize contamination of the environment.

The WHO reports, nevertheless, that

the general attitude and feeling of WHO toward the use of DDT is at present agonizingly ambivalent. On the one hand it is proud of its amazing record. . . . On the other hand WHO is still pressing its search for new compounds with the view of finding some to validate as DDT substitutes. It has investigated the possibilities of biological control since 1959 and has not given up although the outlook appears so unpromising. . . . In short, WHO has been working towards a progressive transfer away from DDT in public health operations.

But it hopes to do so without jeopardizing large numbers of human lives. Without overwhelming evidence of "adverse effects on man, his domestic animals and wildlife," WHO has little grounds to persuade members to abandon a chemical the discontinuation of which would result in thousands of deaths and millions of illnesses.

Although the malaria program calls for the eventual discontinuation of insecticide spraying, evidence as to the likely order of magnitude of the effect of abandoning vigilance against malaria prematurely is persuasive. When spraying was discontinued in Ceylon in 1964, there had been only seventeen cases of malaria reported the year before. By 1968 after a combination of administrative and operational deficiencies "and lack of funds and insecticides and a series of unusual meteorological conditions" more than a million cases occurred in Ceylon (Wright and Wurster, 1970). Another nationwide eradication program, based on DDT house spraying, was required.

The NCDC points out that the United States is the principal source of the DDT used in malaria eradication. If the United States were to ban the production and export of DDT, "and if the malarious countries wished to continue the use of DDT in their programs," the NCDC feels that "it is unlikely that the remaining production facilities in the rest of the world could meet the demands without much plant expansion which would require time . . . during which the malaria programs would suffer serious setbacks" (NCDC, 1969). Furthermore, if this happened it is feared that the United States would not permit the use of U.S. funds, now so important to the eradication program, for purchase of DDT from other sources, thus entailing a greatly diminished program.

These dark predictions may not be at all likely. A scenario that foresaw another country—perhaps France, the Soviet Union, or Communist China, or a group of countries—coming forth with $25 million worth of aid, or even three times that, with which the developing states could acquire DDT denied by a withdrawal of United States support for the malaria eradication program seems at least a possibility. This might simply imply perhaps some delay while new sources of DDT were developed—the Japanese chemical industry is reportedly making great strides in its DDT production—and a great blow to the image of the United States with a corresponding propaganda coup to the donor(s).

Policy Alternatives Open to the United States

Introduction

Obviously, the U.S. government has several policy options it might seek to pursue if it wishes to reduce or eliminate any harm to the environment from the use of DDT on a worldwide basis and if it is to seek to satisfy the needs of the developing states for the kind of effective cheap protection from crop damage and disease offered by DDT as well. It could seek to encourage or subsidize the creation of preferable chemical substitutes for DDT or subsidize a search for making DDT and other hard pesticides less hazardous. The United States could also seek to encourage or subsidize the discovery of other alternative avenues to the control of pests which would prove to be adequate substitutes for pesticides, such as those sought by biological techniques. Some combination of these approaches is likely, for at present each of these approaches seems to offer promising possibilities. In the interim, the United States could immediately subsidize the use of existent expensive biodegradable alternatives to DDT in the developing countries insofar as they are likely to be equally effective.

In addition, the United States could pursue and urge others to pursue a policy generally suggested by the recent U.S. domestic policy of minimizing the contamination of the environment by pesticides in all other ways and restricting their use to essentials, including within this category the principal uses of the developing states.

Alternative Insecticides as an Interim Measure

In contrast to a DDT cost of about $15.4 million, the insecticide cost of switching to presently available alternative insecticides for a typical (recent) year in which a relatively low estimate of 40,000 metric tons of DDT were used in antimalarial programs could range from something over $30 million per annum for a switch to other hard pesticides to $408 million if Baygon, the presently preferred alternative to DDT, were used (see Table 33.3). These calculations were based, wherever possible, on prices paid for health shipments during the period 1968–1970.

These estimates should be used guardedly, as they are probably high. The increased demand for any substitute that would occur as a result of a switch from DDT could result in a scale of production that, if combined wih competition, could lower the costs. Baygon (Arprocarb, OMS 33), the present substitute of choice is a patented product and hence relatively secure from competition. Fenitrothion, however, is not, and if approved for large-scale use, may go down to two or three times the price of

Table 33.3 Comparative Costs of Insecticides and Estimated Total Costs of Switching from DDT for Antimalarial Programs

Insecticide	Approximate Duration of Residual Effect (months)	Cost per Square Meter— 6 Months Treatment	Comparative Cost: DDT = 1	Total Insecticide Cost (current prices) for Equivalent to a 40,000 Metric Ton DDT Program (millions)
DDT (OMS 16)	6	$0.001026	1	$15.4
BCH (gamma-BCH-OMS 17)	3	0.001998	1.95	30.03
Dieldrin (OMS 18)	4–6	0.002280	2.2	33.8
Malathion (OMS 1)	3	0.00696	6.8	104.72
Malathion (OMS 1)	2	0.010440	10.1	155.54
Fenitrothion (OMS 43)*	3	0.017600	17.15	264.11
Baygon (Arprocarb, OMS 33)	3	0.0272	26.5	408.10

Sources: For OMS numbers, see WHO (Organisation Mondiale de la Santé [OMS]), *Evaluation of Insecticides for Vector Control, Compounds Evaluated in 1960–1967*, Part I. Doc WHO/VBC/68.66.
* OMS, (Organisation Mondiale de la Santé) number placed first, since this chemical is still being tested and is not now considered operational by the WHO for antimalarial programs.

DDT or even less. As a modest counterbalance, the use of less-persistent pesticides will require additional costs for repeated sprayings. A rough calculation based on WHO experience suggests that perhaps $3.5 to $7 million per year should be anticipated in additional local expenses for spraying costs alone. These calculations do not include the implied additional moving or storing costs.

The most difficult problem in calculating costs of switching from DDT is that any alternative overall strategy against malaria would probably have to utilize several of these alternative formulations, depending on the local biological, epidemiological, ecological, and other circumstances. Strong points and defects vary for each substitute, depending on such factors as geographic region, vector resistance, and local conditions such as the composition of building materials, meteorological conditions (WHO, 1970). Several of the cheapest alternatives are other hard pesticides. Widespread use of Dieldrin and BCH is expected to lead to rapid vector resistance. As for the other compounds, we do not know fully the biological or ecological implications of widespread use of any of these DDT substitutes nor how quickly vector resistance would develop if they were used massively.

At present, world health authorities repeatedly stress that for inexpensive, reliable, safe, widespread, broad, long-lasting effectiveness DDT remains at present the "insecticide of choice" in antimalarial programs. It seems fair to say that it is generally used except in those cases where for one reason or another it has not proved effective. Then one of the other of the alternatives listed in our chart is tried as a substitute; normally apparently the cheapest substitute that will work is employed. Often this is another hard pesticide. For less than one-half billion dollars, the currently best substitute for DDT, OMS 33, could be utilized at current prices in a somewhat expanded worldwide antimalarial program. Another possibly significant substitute for DDT not covered in our chart must be mentioned. The practicality and safety of a biodegradable form of coated DDT is now being studied by a private firm under contract to the U.S. Department of Interior. The utility and cost of this product remain unclear, but the concept appears very promising.

Biological Controls

The amount of DDT necessary to eradicate all target organisms is many times greater than that required to eliminate 90 percent of the pests. In order to eliminate all target organisms without endangering the environment a method of "integrated pest control" is proposed. This requires the combination of pesticides with biological and other related controls.

Biological control of insects may take many forms. Among these are the introduction of a parasite or predator, which will reduce the population of the target insect, the introduction of large numbers of sterile insects of that species to reduce the population of the next generation, the introduction of a fatal disease through the target population, the use of sexual attractants to lure insects into a chemical or mechanical death trap, and the development of crop strains that are resistant to insect damage.

Outstanding examples of successful pest control by use of biological techniques have been reported. The milky white disease was successfully introduced into populations of Japanese beetles. The use of male-sterility techniques has apparently controlled the screwworm in the southern United States. The cooperation of Mexico in this project is reported. The cottony-cushion scale has been controlled in the United States for more than seventy years by the Vedalia beetle introduced from Australia. Over 110 cases of successful or partially successful biological control of different insect pests have been reported in sixty countries (Kramer, 1969; PSAC, 1967). In general, biological control methods "reduce pest population more slowly, and they are almost invariably less dependable than chemical pesticides" (Report of the Secretary's Commission on Pesticides, 1969). Furthermore, as noted, they too must be introduced with great caution. The Department of Agriculture is reported to require some evidence that imported predators, for example, will not change hosts or that the ecological balance will not be deleteriously changed.

The fundamental conceptual problems for resolving challenges of biological pest control are still to be sought. It is reported that the current stress of the Department of Agriculture is on the development of biological control, and the entomology research division budget of the Agricultural Research Service of the department has tripled in the last decade, but university interest has remained meager. Industrial research has also not been enthusiastic in this area, though the opportunities to seek useful chemosterilants and other hormones clearly exist. Generally, the narrow market for such chemicals will not support such private research. It probably has to be undertaken as a public service with public support.

The Economics of Chemical Innovation of Alternative Insecticides and DDT

Estimates are that it costs a company between $3 and $5 million and takes five to seven years of work to develop and register a new compound (*Wall Street Journal*, 1970), or $2.5 to $6 million, and that it takes six to ninety-six months to develop and market a new pesticide (Kramer, 1969). United States patent protection lasts only seventeen years, so companies must concentrate on products that can provide a return of research costs plus a profit in less time than that. Some feel that pesticides do not fall into this category. In fact, it is likely that current registration procedures that already require expensive trials will become more cumbersome. The likelihood is that costs of innovation will rise while profits on individual compounds may tend to fall, due to the increasing stress of government standards on the production of chemicals with a very specific range of effectiveness which therefore are likely to have a narrow market.

In any case, one reads threats that some firms may leave the pesticide chemical business. The firms in question are generally subsidiaries of conglomerates, which can give up one line without going out of business. One also hears repeatedly that research has been stultified because of the institutional setting in which a profitable outcome is difficult to predict, given the intrinsic risks and the man-made hurdles. As one example, the Olin Corporation announced that it would halt production of DDT as of June 30, 1970, and would shut down its Huntsville, Alabama, plant. The company reportedly produced above 20 percent of the DDT manufactured in the United States (*New York Times*, 1969).

However, contrasting evidence also exists. For example, Eli Lilly is reported to have made an estimated $40 million on Triflurin in 1966. At times, then, innovation in insecticides that yields patents and high prices pays well. The forty firms engaged in industrial research in chemical pesticides are reportedly investing something over $60 million annually on research and development, a relatively modest figure for such socially important research. They are reportedly concentrating on producing softer pesticides, especially organophosphates and carbamates because of the difficulties of gaining acceptance for other hard pesticides. The organophosphates, however—the family from which it is expected most substitutes must be taken in the immediate future—are close cousins to the original nerve gases and are widely considered to be possible environmental hazards as well. In all, then, one survey has concluded: "measured in terms of rate of introduction of a new product or sales growth, the industrial research effort is a success. But measured by the standards of environmentalists it is not" (Kramer, 1969).

We can add that, measured in terms of the generation of new compounds to be tested for public health utility by the WHO screening program, the current worldwide research effort is definitely flagging. The WHO sources report that, for whatever reasons, manufacturers are in fact presenting many fewer new compounds for WHO testing than before. At one time, for example, the program had fifteen compounds in stage four of testing. Now it has only one, and no more than three or four are expected by the end of 1970. Without new compounds to test, obviously, new chemical alternatives are unlikely to be discovered. Furthermore, as the experience with OMS 33 suggests the current U.S. institutional and patent system as well as that of other Western countries may imply that even if effective, safe new pesticides were found by chemical industry effort, motivated primarily for private profit, they would not be cheap enough for many years to be useful to either the agriculture or health programs of the developing countries.

The Policy Implications

We have already seen that research in DDT improvements and substitutes is likely to produce social benefits "external" to the private profits received by the chemical companies that would normally be expected to generate relevant research in response to the incentives of such private profits. Under the circumstances, it is now traditional to note that without an adequate socially subsidized program the resultant research effort is likely to be suboptimal from the social point of view. Traditionally, unless the market could be manipulated to "internalize the externality," such a case would call for an explicit political decision as to the value to society of the activity that generates such "externalities" and the level at which it should be socially supported, followed by decisive government action to achieve these policy decisions. In such a case it becomes important for the relevant decision makers, in this case, ultimately the President, or the President and Congress, to come to an informed, rational decision on the issue, in this case, on the extent of the dangers of the present use of DDT and the value to the United States of replacing DDT on a worldwide basis by environmentally preferable disease and pest control technique, considering the needs of other countries and their financial capacities and preferences. The rough time horizon that can be accepted by the decision makers for these achievements should also be sought. As the *Report of the Secretary's Commission on Pesticides* (1969) comments: "If the crisis with respect to the use of pesticide chemicals for vector control is to be overcome, a large increase in research is mandatory." The first decision must be on the question: Is it worth it to us, given the costs and risks?

This accomplished, it would be rational to call on the executive branch of the U.S. government to take the necessary steps to design, promote, and execute an adequately subsidized long-run research program and an adequate set of interim programs of education, technical assistance, and financial assistance to domestic users and to other states so that, for example, expensive already-existent substitutes to DDT would be utilized optimally given their costs and implied benefits. Such an increased research effort could well be expected to combine greater government research grants to universities as well as more aggressive in-house research programs in the government laboratories and in the NCDC and elsewhere in interested departments of government. Probably also it would call for subsidies to generate desired, private industrial chemical research, perhaps in the form of generous cost-plus contracts. Government-assisted private research could be conducted with assurances attached that the chemicals produced would be sufficiently reasonably priced to be potentially useful for public health programs, perhaps by providing for the free licensing of resultant patents.

The expressed interest of PAHO and more important of the WHO in such programs also suggests the possibility of a more vigorous, more heavily subsidized international research effort focused on "integrated" control of vector diseases that would involve the developed and the developing states together in the early stages of a more aggressive effort to solve what should be viewed as *mutual* environmental and public health problems.

In addition to its chemical evaluation program, the WHO already has a program seeking to innovate in and to evaluate the promise of various types of biological control of disease vectors. This research is due to be expanded. But it is felt by world health authorities that it could very creatively absorb a much higher budget than is presently and prospectively to be devoted to it by the WHO. Current estimates are that at the present pace "it may be as long as ten to 15 years before any of these procedures can possibly be used operationally" (Wright and Wurster, 1970).

In addition to a greatly expanded domestic and international public health research effort, similar vigorous domestic and international efforts for innovations in biological and chemical and other forms of integrated control of crop damages could be organized by the Department of Agriculture and other relevant administrative departments in the U.S. government and internationally by or with the cooperation and participation of the FAO.

The bitter words of spokesmen for the developing countries sum up the value choice the world political community faces in the DDT issue: "Do we save the lives of birds or the lives of human beings?" This and similar questions were aired in a conference of the FAO and the WHO in Rome in late 1969, in which officials from 121 member countries were called together in part to listen to the case for DDT. The classical political way out of such dilemmas is to minimize confrontation—that is, to do both if possible.

To achieve great strides in this direction, it seems likely that a much greater U.S. and international effort at scientific and technical innovation will be necessary. A much more significant U.S. and international program of subsidizing substitution of the use of environmentally preferable alternative to hard pesticides by the developing states in the interim may also be the most rational course for all.

Under present circumstances there is no adequate substitute for DDT for the developing states. "Reports from authorities interviewed indicated that it was likely that malaria programs would gradually be discontinued if they were forced to use substitutes for DDT" (*Report of the Secretary's Committee on Pesticides*, 1969). This would probably imply "the affliction of hundreds of millions of cases of malaria and millions of deaths from it within the next decade" (NCDC, 1969). In this situation the issue of the worldwide use of DDT is clearly not an issue for Americans alone to decide. If there is a danger of growing accumulations of DDT in the oceans, for example, and if this should seriously suggest that continued use of present types of DDT might become a grave, worldwide ecological danger, the appropriate, potentially effective democratic approach to such a threatened common disaster is to seek a common effort to avert it. Indeed, the search for such a common effort is likely to be the only potentially effective approach to all genuinely global pollution problems.

Acknowledgments

The following persons were very helpful in providing information, data, insight, and comments on versions of this paper: Philip Kearny, Roy F. Fritz, Howard Taubenfeld, Dale Jenkins, Edward Goldberg, James W. Wright, G. Sambasivan, H. Rafatjah, Terry Schaich, and innumerable participants of SCEP.

References

Abelson, Philip H., 1969. Persistent pesticides, *Science, 164:* 633.

Keller, Euglina, 1970. The DDT story, *Chemistry* (February).

Kramer, Joel R., 1969. Pesticide research: industry, USDA pursue different paths, *Science, 166:* 1383.

National Communicable Disease Center (NCDC), 1969. *DDT in Malaria Control and Eradication,* Health Services and Mental Health Administration, Department of Health, Education, and Welfare.

New York Times, July 14, 1969; January 8, 1971.

President's Science Advisory Council, 1967. *The World Food Problem* (Washington, D.C.: U.S. Government Printing Office).

Report of the Secretary's Commission on Pesticides and Their Relationship to Environmental Health, 1969 (Washington, D.C.: U.S. Department of Health, Education, and Welfare, U.S. Government Printing Office).

Science, 166: November 28, 1969.

Study of Critical Environmental Problems, 1970. *Man's Impact on the Global Environment* (Cambridge, Massachusetts: The M.I.T. Press).

U.S. Department of Agriculture, 1969. *The Pesticide Review.*

Wall Street Journal, February 16, 1970. DDT dilemma: poor countries insist pesticide is essential despite its dangers.

World Health Organization, 1969. The present place of DDT in world operations for public health, addendum V, *The Biological Impact of Pesticides,* paper presented at Oregon State University Symposium, Vector Biology and Control.

World Health Organization, 1970a. *Executive Board 55 Session,* Part II (January).

World Health Organization, 1970b. The significance of the organochlorine insecticide in the control of vector of public health importance. (Mimeo).

Wright, James W., and Wurster, Charles F., 1970. Good words and bad for DDT, *Smithsonian, 1.*

Name Index

Abelson, Phillip H., 500
Acree, F., Jr., 280, 282
Adam, N. K., 340
Adams, D. F., 118
Addy, C. E., 286
Aitken, T. H. G., 159
Albert, R. E., 87
Alexander, L. T., 137, 138
Allee, W. C., 186
Altshuller, A. P., 325
Anderson, D. W., 286, 287, 288
Anderson, Franklin K., 118
Anderson, Henry W., 244
Anderson, J., 283, 284
Applegate, H. G., 118
Arnason, Geirmundur, 1, 377–391, 429, 430–447
Ayres, R. U., 328
Azevedo, J. A., Jr., 156

Backus, R., 306, 341
Baer, L., 438
Bailey, T. E., 279
Bailey, W. A., 110
Baker, D. G., 184, 186, 188
Baker, W. L., 221
Baldwin, F. M., 149
Balsi, Gianetto, 121
Banks, J. E., 438
Barghoorn, E. S., 184
Barker, R. J., 157
Barrett, Thomas W., 106
Barter, G. W., 220
Barth, D. S., 88, 94
Baum, W. A., 206
Beal, M. L., Jr., 280
Beerstecher, Ernest, 309
Beilmann, A. P., 213
Bender, F. W., 119
Bentz, W. W., 236, 237
Berbee, J. G., 220
Beroza, M., 280
Berry, C. R., 119, 124, 219
Bien, G. S., 448, 457
Billings, W. D., 184, 186
Birkhoff, G., 433
Bitman, J., 288
Bjerknes, V., 448
Blackman, N. R., 284
Blumer, A., 298, 309, 311
Blus, L. J., 287
Bodine, B. R., 438
Boicourt, W. C., 439, 441
Bolin, B., 134, 457
Bonneli, J. A., 89

Bonner, J., 134
Bormann, F. H., 2, 33–46, 54
Bowen, H. J. M., 6
Bowman, F. C., 280
Boyce, J. S., Jr., 221
Brady, D. K., 442
Brandt, C. S., 105
Braunshweig, S., 140
Brennan, E. G., 107, 108, 139
Brean, H., 206
Brenner, L. G., 213
Brewer, R. F., 107
Britt, C. S., 232, 236
Broadbent, L. 221
Brown, A. W. A., 148
Brown, C. L., 258
Brown, George W., 247, 253
Brown, John F., Jr., 467, 489–498
Brown, R. P., 252
Bryan, Kirk, 429, 436, 448, 460–465
Bryan, R. J., 102
Buell, J. H., 208
Buell, M. F., 51
Burbank, Stephen, 2
Burwell, R. E., 231, 232, 235
Busch, K. A., 88
Butler, Philip A., 148, 281, 283, 284, 285, 287

Cadle, Richard D., 325, 337
Cain, S. A., 186
Cantlon, J. E., 51
Carpenter, J. H., 440
Carpenter, Richard, 467, 473–476
Carraker, John R., 234
Carroll, Robert, 90
Carruthers, 309
Carter, H. H., 440
Carter, Robert L., 234
Carver, T. C., 337
Cecil, H. C., 288
Changnon, S. A., 140
Chapman, D. W., 247
Cherniak, I., 102
Chichester, C. O., 149
Citron, Robert, 325, 326, 339, 392–428
Clark, J., 220
Clark, R., 258
Clements, F. E., 189
Cobb, Fields W., Jr., 119, 220
Cochrane, E., 313, 314
Cole, L. C., 136
Collyer, E., 161
Conney, A. H., 288
Cope, O. B., 148
Corbett, J., 282
Cordone, Alma J., 246

Corino, Edward R., 297–318
Costonis, A. C., 106, 119
Cottam, Clarence, 275, 276
Coulter, M. C., 288
Cox, M. D., 462
Craig, H., 448, 457
Crocker, W., 107
Crompton, R., 147
Cronin, L. E., 322
Cronquist, A., 184
Cross, D. J., 158
Curry, J. A., 245
Cusumano, Robert D., 123

Daines, R. H., 106, 107, 108, 139
Darley, E. F., 102, 105, 107
Daubenmire, R., 208
Davidson, C. W., 105
Davis, 162
Davis, B. N. K., 157
Davis, F. D., 288, 309
Davis, Harry C., 283
Davis, J. J., 138
DeBach, P., 148
Delay, W. H., 442
Demb, Ada, xi
Denning, R. E., 257
DeWit, C. T., 186, 189
Dochinger, L. S., 119
Dole, R. B., 257
Dortignac, E. J., 138
Dragoun, F. J., 234, 235
Dreibelbis, F. R., 235
Dubois, K. P., 148
Dugger, W. M., Jr., 108
Duke, T. W., 284, 285
Duncan, W. G., 186
Dunning, J. A., 105, 108
Dyrness, C. T., 244

Easton, F. M., 105
Edelberg, Seymour, 325
Edinger, J. E., 442
Edwards, R. W., 160, 164
Edwards, W. M., 231, 232, 235, 236, 237
Einstein, H. A., 431
Ekman, Walfrid, 436, 437, 448
El-Sharkawy, M. A., 185, 186
Elton, C. S., 165
Emery, K. O., 257
Eschner, A. R., 245
Ewing, Gifford, 325

Feder, W. A., 104
Feltz, H. R., 336

Fisher, G., 438
Fiske, W. F., 144
Flaccus, Edward, 244
Fleming, R. H., 280
Fogel, M. M., 135
Fowler, D. L., 135
Frankenberg, T. T., 121
Freeman, J. C., Jr., 438
French, M. C., 289
Fries, G. F., 288

Galston, A. W., 134
Galtsoff, P. S., 312
Garrett, W. D., 341
Garrison, J. M., 438
Gates, D. M., 184, 186
Genelly, R. E., 148
Georghiou, G. P., 148
Gessel, Stanley P., 240
Geyer, J. C., 442
Gilbertson, C. B., 232, 235
Gilmour, J. W., 215
Gleason, H. A., 184
Glymph, L. M., 138
Goddard, R. E., 212
Goldberg, Edward D., xi, xii, 1, 227, 261–274, 304, 326, 371–376, 377–391
Goldberg, I., 343
Goldsbrough, G. R., 448
Good, R. D'O., 184
Gordon, A. G., 52
Gorham, E., 52, 372
Graves, W. L., 442
Greenfield, R. S., 436
Gress, F., 287, 288, 289
Grill, E. V., 443
Gross, M. Grant, 1, 227, 252–260, 371–376, 377–391
Grosso, J. J., 103

Haagen-Smit, A. J., 102, 107
Hagestead, 199
Hall, W. C., 105
Hampson, G. R., 310
Hannum, J. R., 279
Hansen, D. V., 437
Hansen W., 438
Hanshaw, Bruce B., 1
Hanson, W. C., 138
Harder, J. A., 431
Hardy, E. P., Jr., 137
Harleman, D. R. F., 443
Harner, Frances, 118
Harrold, Lloyd L., 227, 230–239

Harvey, G. W., 341
Hasler, A. D., 320
Hasler, Arthur, 1
Havens, J. M., 206
Hayes, G. L., 208
Heagle, A. S., 104, 124
Heath, R. G., 286
Heck, W. W., 105, 108, 124
Hedgcock, G. C., 120, 139
Heggestad, H. E., 99, 101–115
Hellmers, Henry, 192, 194, 195, 196
Hendrix, J. W., 118
Henry, A. W., 218
Hepting, George H., 116–129, 139, 203–226
Herman, S. G., 287
Hesketh, J. D., 185, 186
Hewitt, E. J., 89
Hewitt, W. B., 103
Hibben, C. R., 102, 104
Hickey, J. J., 286, 287, 288
Hiesey, W. M., 185
Higer, Aaron L., 280
Higgins, Elmer, 275, 276
Hill, A. C., 102, 104, 106, 108
Hill, Austin Bradford, 85
Hilmon, J. B., 1, 25
Hindawi, I. J., 105, 108
Hitchcock, A. E., 106, 107
Hitchcock, S. W., 165
Hobbs, P. V., 140
Hodges, C. S., 217
Hollister, H. L., 137, 138
Holloway, J. T., 205, 209, 223
Holly, E. R., Jr., 443
Holme, Robert A., 310
Holmes, D. C., 90
Holmes, Robert W., 310
Holswade, W., 280
Holt, R. F., 231, 232, 235
Hoover, Marvin D., 246
Horn, M. H., 306, 341
Horton, Robert J., 80–97
Hovermale, J. B., 436
Howard, L. O., 144
Hueter, F. Gordon, 80–97
Hulett, H. R., 61
Hull, H. M., 103
Humphrey, H. B., 203
Hunt, E. G., 165
Hursh, C. R., 120, 126
Hurst, C. K., 253
Hutchinson, G. L., 232

Ide, F. R., 165
Idso, Sherwood B., 184–191

Jacobs, W. C., 448
Jacobson, J. S., 106
James, Richard D., 194, 196
Jaske, R. T., 442
Jeffries, D. J., 157, 287, 289
Jehl, J., 287
Jelsnianski, C. P., 438
Jenkins, Dale W., 1, 325, 351
Jensen, S., 286, 491
Jernelov, A., 491
Johnson, M. W., 280
Johnston, W. R., 231–232
Jones, Eustace W., 124
Jung, G. H., 438
Junge, C. E., 207

Kearny, Philip C., 1
Keeling, D. C., 25, 134, 331, 333
Keen, F. P., 208
Kehoe, R. A., 87, 88
Keller, Euglina, 499
Keller, F. J., 244
Kelley, Don W., 246
Kellogg, William W., xi, 429
Kendrick, J. B., 102
Kent, R. E., 440
Ketchum, Bostwick H., 2, 59–79, 227, 297–318
Keulegan, G. H., 432
Kimmey, J. W., 216
Kirven, M. N., 287
Kleuter, H. H., 110
Klingman, D. L., 137
Kneese, A. V., 328
Koch, R. B., 288
Koeman, J. H., 90, 287
Koisch, Maj. Gen. F. P., 253
Koukol, Jane, 108
Kovner, Jacob L., 241
Kozlowski, T. T., 208
Kramer, Joel R., 502, 513, 514
Kramer, P. J., 208
Krantz, W. C., 286
Kratzer, P. A., 140
Kreitzen, J. F., 286
Krizek, D. T., 110
Krygier, James T., 247
Kuenen, D. J., 149, 160, 161
Kuhlman, E. G., 217
Kuo, Han-Hsiung, 448, 464

Lacasse, Norman L., 122
Laird, M., 167
Landsberg, H. E., 140, 206, 223
Lane, C. E., 312
Larson, Karl R., 192–202

Last, Fred, 200
Laverton, S., 166
Leaf, Charles, 246
Leake, C. P., 207
Leaphart, C. D., 219
Leendertse, J. J., 438
Lehenbaur, R. A., 184
Lehman, Jules, 325
Lennon, G. W., 438
Leone, I. H., 107, 108, 139
Levno, Al, 247
Lichtenstein, E. P., 90, 280
Lieth, H., 134
Lind, C. J., 107
Lindner, C. P., 433
Ling, Y. Y., 257
Linzon, S. N., 118, 121
Littlefield, N., 108
Love, Gory J., 80–97
Lowe, J. I., 284, 285
Lowman, Frank G., 1, 377–391
Ludwig, F. L., 88
Lull, Howard W., 227, 240–251
Lundegardh, H., 184, 186
Lutmer, R. F., 88
Lynn, Ronald J., 449

McCormick, F., 51
MacDonald, J., 220
McGuinness, J. L., 235
Machta, Jon, 25
Machta, Lester, 25
McHugh, J. L., 322
McKee, Herbert C., 122
Mackin, J. G., 311, 312
MacLeish, Archibald, 39
McManus, D. A., 257
McMillan, 196
Mahan, J. J., 135
Manabe, S., 436
Manigold, D. B., 279
Manning, W. J., 104
Margalef, R., 131
Marks, Jonathan, 325
Marshall, R., 209
Martin, H., 148
Masch, F. D., 431
Mason, H. L., 189
Mather, J. R., 249
Matsumura, Fumio, 288
Matthews, William H., x, xi, xii, 429
Meade, R. H., 252
Menhinick, E. F., 162, 164, 166
Menser, H. A., 103, 107, 124
Menzel, D. W., 281, 283, 284, 304
Metcalf, R. L., 148

Meyer, R. E., 136
Middleton, J. T., 101, 102, 103, 105, 107
Mihursky, Joseph, 74
Miksche, Jerome, P., 199
Miller G. L., 51
Miller, P. R., 119
Miller, R. G., 88
Milner, H. W., 185
Mitchell, J. W., 137
Mitchell, Wilfred C., 245
Monteith, J. L., 186
Montgomery, M. L., 136
Montolla, Paolo, 121
Moore, N. W., 99, 144–172
Morton, H. L., 136
Moss, R. A., 186
Mudd, J. B., 108
Muir, R. C., 161
Mulhern, B. M., 287
Müller, P., 148
Munk, W. H., 448, 457
Murray, W. S., 331, 337
Muskie, Edmund S., 40

Nash, C. E., 73
Nash, R. G., 280
Negherbon, W. O., 148
Newberg, E. A., 436
Newill, Vaun A., 2, 80–97
Nicholas, D. J. D., 89
Nicholas, C. W., 105
Nilson, R., 86, 89
Nimmo, D. R., 284, 285
Nixon, Richard M., 40, 44
Noble, W. M., 124
Norris, L. A., 135, 136
North, W. J., 310
Nozaki, K., 87

Oda, A., 312
Odum, E. P., 47, 137
Okubo, A., 440
Olson, Jerry S., 1, 25
O'Regan, W. G., 213
Osburn, W. S., 138
Ostrom, Carl E., 192–202

Paine, L. A., 213
Pales, J. C., 331, 333
Palmer, H. E., 138
Palmer, R. L., 108
Paltridge, G. W., 186
Pandolfo, J. P., 442
Parameter, J. R., 119
Parizek, R. R., 249

Park, T., 186
Parker, A. K., 219
Parkin, D. W., 372
Parmalee, D. W., 249
Pase, C. P., 135
Patil, K. C., 288
Peakall, D. B., 288, 289
Pearce, J. B., 258
Pearson, G. A., 213
Pechanec, J. F., 135
Peirson, H. B., 121
Pendleton, J. W., 185
Perkins, J., 104
Perring, F. H., 152
Peterle, T. J., 280
Peterson, E. K., 134
Philips, D. R., 372
Pierce, R. S., 248
Pires, E. G., 105
Plass, G. N., 207, 208
Polunin, N., 189, 190
Porter, R. D., 286
Post, A., 160 161
Pratt, 288
Preddy, W S., 443
Prestt, I., 157, 160, 287
Pritchard, D. W., 430, 437, 440

Radke, L. F., 140
Rakestraw, N. W., 448, 457
Randtke, A., 283, 284
Ratcliffe, D. A., 157, 165, 286
Rattray, M., Jr., 437
Raup, H. M., 212
Redmond, D. R., 218
Reichel, W. L., 286
Reichle, Henry, 325
Reid, Joseph L., 1, 377–391, 429
Reid, R. O., 438
Reinhart, K. G., 245, 248
Reitemeier, R. F., 137, 138
Renzetti, N. A., 102
Revelle, Roger, 25, 297–318, 463
Richards, B. L., 103
Ricker, W. E., 322
Riley, G. A., 443
Ripper, W. E., 148
Ripperton, L. A., 119, 219
Risebrough, R. W., 90, 280, 287, 288, 289
Robbins, C. S., 165
Robinson, E., 88
Robinson, E. D., 136
Robinson, G. D., xi, 325, 429
Robinson, Thomas W., 242
Rohlich, G., 320

Romanovsky, J. C., 102
Ronningen, T. S., 135
Rossiter, J. R., 438
Rothacher, Jack, 247
Rudd, R. L., 147, 148, 156
Rudolph, Thomas D., 192–202
Rusden, P. L., 220
Russell, C. S., 482
Russell, R. J., 204, 206
Ryther, John H., 62, 64

Sand, F. F., 337
Sanders, H. L., 310
Sandstrom, J. W., 448
Sass, J., 311
Satterstrom, C., 107
Sawhney, B. L., 138
Schaeffer, Milner B., 62
Scheffer, T. C., 120, 139
Schimper, A. F. W., 189
Schneider, F., 165
Schreiber, R. W., 287
Schroeder, H. A., 89, 90
Schutze, J. A., 279
Schwalm, H. W., 102
Scott, D., 184
Scurfield, G., 122, 139
Seaders, J., 442
Seba, D., 313, 314
Seliskar, C. E., 119
Sheets, T. J., 136
Shepard, H. H., 135
Shuman, F. G., 436
Shwartz, H. L., 289
Shy, Carl M., 92
Sibley, F. C., 287, 288
Siggers, P. V., 217
Silen, R. R., 212
Simmons, H. B., 432, 433
Simmons, J. H., 90
Sinclair, W. A., 119
Sisler, F. D., 309
Smith, D. D., 230, 252, 253
Smith, Dixie R., 130–143
Smith, Frederick E., xi, xii, 1, 25
Smith, J. E., 312
Snider, G. R., 344
Sopper, William E., 241
Southern, H. N., 153
Southwood, T. R. E., 158
Souze, G., 311
Spann, J. W., 286
Spofford, Walter O., 25, 467, 477–488
Spooner, G. M., 312
Spooner, M. F., 312
Squillace, A. E., 212

Stabler, H., 257
Stage, A. R., 219
Stark, R. W., 119
Stephens, E. P., 104
Stern, V. M., 148
Stewart, 309
Stewart, B. A., 232
Stewart, R. W., 449
Stickel, W. H., 157, 278
Stoller, Robert, 2
Stommel, Henry, 443, 449, 457
Storey, H. C., 138
Street, J. C., 90
Striffler, W. David, 246
Suess, H. E., 448, 457
Sullivan, R. A. L., 372
Svansson, A., 438
Sverdrup, H. V., 280, 448
Swank, W. T., 241

Takano, K., 437
Tarrant, K. B., 280
Tarzwell, Clarence, 1, 247, 312
Tatton, J. O. G., 90, 280
Taubenfeld, Rita, 468, 499–518
Taylor, A. W., 231, 232
Taylor, O. C., 102, 103, 104, 105, 119
Teal, J. M., 306, 341
Ten Noever de Brauw, M. C., 90
Tepper, Morris, 325
Thomas, Adrian W., 234
Thomas, G. W., 135
Thomas, M. D., 119
Thompson, D'A., 184
Thomson, C. R., 103
Thornthwaite, C. W., 248
Thornton, N. C., 107
Timmons, D. R., 231, 232, 235
Tingey, D., 102, 104
Toba, Y., 378
Toole, E. R., 219
Trapido, H., 159
Treshow, M., 108, 118
Trimble, G. R., 245
Triplett, G. B., 235
True, R. P., 221
Tschirley, F. H., 55
Tsuchiya, K., 90

Ukeles, Ravenna, 283
Urie, Dean, 241

Van Arsdel, E. P., 216
Verduin, J., 319, 322
Veronis, Klaus, 448, 464
Viets, F. G., 232

Vieweg, F., 140
Vinton, W. H., 89
Vogenberger, R. A., 245
Voight, G. K., 192
Vollenweider, 320
Vos, R. H. de, 90

Wadleigh, Cecil H., 138, 139, 232, 233, 236
Wagener, W. W., 214, 215, 216
Waggoner, P. E., 136, 186, 203
Waldichuk, M. W., 257
Walker, C. H., 157
Walker, J. T., 102
Walters, S. M., 152
Ward, E. W. B., 218
Wark, J. W., 244
Warren, G. F., 136
Wasser, George L., 123
Watkins, 309
Watson, D. G., 138
Watt, K. E. F., 131
Way, 162
Webber, B., 443
Weidensaul, Craig T., 122
Weimeyer, S. N., 286
Weisser, D., 123
Welander, Pierre, 448
Wellington, W. G., 221
Wenger, Karl F., 192–202
Wenk, Edward, 2, 297–318
Went, F. W., 103, 123
Wentzel, K. F., 124
Willis, A. J., 159, 162
Wilson, A. J., Jr., 281, 284, 285
Wilson, Carroll L., x
Wilson, J. W., 184
Wilson, P. W., 107
Windom, H. L., 389
Winter, S. W., 184
Wischmeier, W. H., 230
Wolman, A., 67, 74, 281
Wood, F. A., 106, 108
Woods, F. W., 214
Woodwell, George M., 2, 47–55, 130, 131, 199, 200, 289
Wooldridge, David D., 240
Worthington, L. V., 449
Wright, James W., 507, 508, 509, 516
Wright, L. A., 124
Wurster, Charles F., 57, 283, 507, 508, 509, 516
Wyrtki, George, 448, 449

Yarwood, C. E., 203
Yates, M. L., 280

Yemm, E. W., 159, 162
Youker, R. E., 235

Zimmerman, P. W., 106, 107
Zobel, B. J., 212
ZoBell, C. E., 304, 306, 307, 308, 309, 311, 312
Zon, R., 208, 212

Subject Index

Abyssal circulation, 448
Accumulators, 355, 360, 362, 367
Acetone, in oceans, 267
ACMRR. *See* Advisory Committee on Marine Resources Research
"Advances in Pest Control Research," 148
Advisory Committee on Marine Resources Research, 426
Africa, food for, 35
Agrichemicals, transport of, 237
Agricultural Research Service, 230, 243
Agriculture, DDT and, 504
  modern, 35
  production in, 7
  research and development in, 36
  swidden, 41
  yields, 8
Agung effect, 180
Air Quality Criteria Documents, 82, 85
Alaskan North Slope, oil production in, 305, 315
Albemarle Sound, 431
Aldehydes, 102
Aldrin, 21
Aldrin-Toxaphene Group, 278, 279
Alfalfa, 106, 107
Algae, 64. *See also* Eutrophication
  of sewage plants, 57
Aluminum, in oceans, 272
American Chemical Society, 194
American Petroleum Institute, 301
Ammonia, 107
Anaconda, smelters at, 120
Andrews Experimental Forest, 244
Anemia, 90
Animals, predatory, 365. *See also* Predators
Antarctic Bottom Water, 456
Antarctic Circumpolar Current, 449, 457
Aphids, as virus vectors, 221
Appalachia, soft-coal-burning in, 127
  tree disease in, 216
Apricots, 106
Arborvitae, 118
Army Corps of Engineers, 431, 433
  Committee on Tidal Hydraulics of, 436
Arprocarb (Baygon), 507
Arsenic, 110
Atmosphere, composition of, 14
  interaction with ocean of, 448–449, 450, 457–458, 464

Atomic bomb, radioactivity from, 40. *See also* Radioactivity
Atomic Energy Commission, 486
Attrition, problem of, 12
Autecology, 144

Baltic sea, 42
Bamboo, 54
Barium, in sea water, 265
Barley, 106
Baygon, 507
Bays. *See also* Coastal waters; Estuaries
  filling of, 342
BCH, 511
Bear oak, 51, 54
Beetles, bark, 119
Begonias, as monitors, 356
Beltsville, Md., 199
Benthic organisms, 270, 321
Benzanthene, 309
Benzene, 309
Benzophrene, 309
Beryllium, 110
Bioassay, organisms for, 359
Biocide, 150, 353
Biological concentrators, 156
Biological oxygen demand (BOD), 67
Biomass, 19
Biosphere, 4, 40
Birds, air-fouled, 308
  as biological indicators, 364
  in Britain, 162
  chlorinated hydrocarbons and, 286–288
  fish-eating, 90, 276, 277
  insecticides and, 165
  marine, 266–267
  predatory, 57, 64, 157, 165, 364
Blister rust, 216, 217
Bluefish, 343
Botrytis, 104
British Insecticide and Fungicide Council, 148
British Weed Control Council, 148
Brookhaven National Laboratory, 48, 50, 52
Bureau of Commercial Fisheries, 285
Bureau of Criteria and Standards, 80, 88
Bus stop disease, 122
Butyraldehyde, in oceans, 267
Buzzards, insecticides and, 154

Cadmium, 110, 373, 374, 491
  hypertension and, 89–90
Calcium, 85

CALCOFI, 370
California, DDT residues in, 285
  photochemical oxidants in, 109
  pollution in, 119, 120–121
California Current, 449
Canadian Northern Archipelago, oil production in, 305, 315
Carbon, in atmospheric dusts, 372
  organic, 26, 27
Carbon cycle, 25–27
Carbon dioxide, 19
  in atmosphere, 41, 42, 175, 176, 268
  evaluation of, 328
  from fossil fuels, 175–177
  global temperatures and, 207
  increase in, 192–193, 197
  light intensity and, 186
  ocean systems and, 268
  plants and, 110
  range ecosystems and, 134–135
  SSTs and, 181
Carbon monoxide, 357
  in atmosphere, 268
  automobile emission of, 38, 102, 120, 127, 264
  tropospheric, 182
Carcinogens, in marine environment, 309
Carnivores. *See also* Predators
  pollutants and, 56
Celery, 106
Cesium-137, 137, 138, 269
Chaparral, 242
Chemical production data, public access to, 292–293
Chesapeake Bay, 430
  waste disposal sites in, 256
Chesapeake Bay Institute, 440
Chesapeake River, sediment load of, 257
Chestnut blight, 214
Chicago, pollution studies in, 92, 93
Chlorine, accidental spills of, 106–107
Chlorophyll. *See also* Photosynthesis
  cadmium and, 89
  in marine waters, 68–69
Chlorosis, 104
Chromatography, gas, 338
Chromium, 110, 491
CIG. *See* Comité International de Géophysique
Cigarette problem, 487
Circulation, of atmosphere and ocean, 450, 457–458, 464
  geostrophic, 448
  surface, 449

Citrus crops, 126
  smog and, 118
Cladonia cristatella, 52
Clean Air Act, 80, 83
Climate, changing of, 204–208
  disease vectors and, 220–222
  effect of CO_2 on, 176
  fluctuations in, 174
  regulation of, 14
  space experiments and, 465
  surface changes and, 182–183
Clinical studies, of pollutant effects, 91
Clouds, cirrus, 179
  role of, 178
Coal. *See also* Fossil fuel
  burning of, 127, 372
Coastal waters, eutrophication and, 321
  excess heat of, 74
  overfertilization of, 68
  pollution of, 59
  power plants and, 71
  waste disposal into, 252–260
Cobalt, 86
Columbia River, sediment load of, 257
Combustion, fossil-fuel, 19. *See also* Fossil fuel
Comité International de Géophysique (CIG), 426
Committee on Space Research, of ICSU, 426
Community Air Pollution Effects Surveillance, 92
Comptonia peregrina, 52, 53
Computers, for environmental modeling, 441. *See also* Models
Conifers, 117, 197. *See also* Forests; Pines; Trees
  destruction of, 127
  erosion and, 246
  photosynthesis and, 193–194
  in Tokyo, 199
Conifer-spruce budworm system, 11
Connecticut River, sediment from, 257
Conservation, water, 230
Consumption, 78
  pollution and, 472, 478
Continental shelf, sampling of, 385
  waste deposits on, 258
Contrails, 179
Control, pollution, 94, 323, 470–471, 489–498
  necessity for, 323

Cooperative Investigations of the Caribbean and Adjacent Regions (CIGAR), 393
Cooperative Investigations of the Northern Part of the Eastern-Central Atlantic (CINECA), 413
Cooperative Study of Kuroshio and Adjacent Regions (CSK), 393
Copper, 86, 491
Copper Hill, 126
  smelters at, 119, 120
Coriolis force, 434
Corn, 106
COSPAR. *See* Committee on Space Research
Cotton, 105, 106
Cottonwood, 201
Council on Environmental Quality, 258
Crab, 364. *See also* Shellfish
Crab grass, 52
Crops, chronic injury to, 99
  "cool" weather, 190
Crustaceans. *See also* Shellfish
  chlorinated hydrocarbons and, 284
  sampling of, 384
CSK. *See* Cooperative Study of the Kuroshio and Adjacent Regions

Data, availability of, 336
DDD, 361
DDE, 286
DDT, x, 10, 21, 64, 148
  agriculture and, 504
  in antimalarial programs, 510, 511
  atmospheric transport of, 280–281
  biological controls for, 512
  in birds, 364
  circulation of, 42
  control of, 494
  cost of, 495, 506
  defenders of, 500
  developing countries and, 497, 499–518
  distribution of, 502–504
  effects of, 167
  egg production and, 161, 267
  emission of, 374
  food chain and, 375
  and health programs, 506
  in humans, 358
  in marine environment, 281, 282
  in oceans, 57, 266, 517
  predatory birds and, 91
  production of, 278, 279, 502–504
  regulation of, 501–502
  social alternatives to, 496, 507
  in soil, 360
  spread of, 41
  in subsurface flow, 233
  transport of, 279
  United States exports of, 503
  warnings about, 275
Decision maker, 473–476, 477
Deforestation, effects of, 28. *See also* Forests
Demand, ecological, 8
  environmental, 15
Department of Agriculture, U.S., xi, 99, 134, 173, 490, 496, 497, 502, 513, 517
Department of Transportation, U.S., 300
Deserts, cold, 133
  spreading of, 342
  warm, 133
Detectors, 354, 360, 362
Detergents, phosphates in, 323, 490
Developing countries, 33
  DDT and, 499–518
Diatoms, 368, 369
Dieldrin, 280, 313, 511
Digitaria sanguinalis (crab grass), 52
Dilution, 65
Disease, climate change and, 213–215
  forest, 203
  fungus, 215, 223
  threshold, 214
  tree, 222
Dispersants, 299
Disposal, dredged waste, 258
  of residuals, 484
  waste, 253–257
Distribution, pollution and, 472
Diversity, effects of pesticides on, 162
  pollution and, 131
  species, 368
Division of Health Effects Research (DHER), 80, 81, 85, 91, 93
Donora, Pa., 119
Dover sole, 73
Dredging, 227, 252, 257, 258
  in Canada, 253
  regulation of, 253
Drosophila melanogaster, 360
Dry-cleaning solvents, 374
Dustiness, 41. *See also* Turbidity
Dutch Elm disease, 123, 214, 220–221

Eagles, bald, 287
  sea, 286

Earth, nonrenewable resources of, 61
Earth slides, 244
Earthworms, 157
Easterlies, polar, 378
Ecological Society of America, 2
Ecological stations, 352
Ecology, ecological growth, 23–24
  education and, 24–25
  population, 145 (*see also*
    Population)
  theory of, 56
Economic and Social Council
  (ECOSOC), of U.N., 426
Economy, growth of, 43
Ecosystems, 57
  attrition of, 11–12
  effects of pesticides on, 161
  range, 130–143
  regulation of, 11
Education, re: environment, 486
Effluents, from sewage treatment
  plants, 253
Eggshells, thinning of, 286, 287, 288
Ekman transport, 448
Electric power, demand for, 71
Elements, mobilization of, 6
Elm, fluoride sensitivity of, 118
Elm phloem necrosis, 221. *See also*
  Dutch elm disease
Endangered Ecosystem Monitoring
  Programs, 393
Engine, external-combustion, 484
  internal-combustion, 102, 120, 127,
    264
Environment, aquatic, excess heat
  in, 72
  polluted, 65
  degradation of, 308
  demands upon, 7
  holocoenotic, 186
  pesticides in, 168 (*see also*
    Insecticides; Pesticides)
Environmental Health Surveillance
  and Monitoring Programmes, 393
Environmental Protection Agency
  (EPA), 502
Environmental Science Services Administration (ESSA), 461. *See also*
  National Oceanic and Atmospheric Administration
Enzymes, cadmium and, 89
Epidemiology, 82, 83
Epilobium angustifolium (fire weed),
  53
Erosion. *See also* Runoff
  potential for, 244

  in range ecosystems, 138
  sediment and, 243–244
Eskimos, accumulations of cesium-137
  in, 138
Estuaries, bar-built, 430
  coastal-plain, 430
  destruction of, 42
  eutrophication of, 219, 321
  filling of, 342
  fjords, 431
  modeling of, 429, 430–447
  monitoring of, 332, 338
  multiple uses of, 75, 76, 77
  nutrient discharges to, 322
  overfertilization of, 68
  pollution of, 59, 69
  power plants on, 71
  resuspended wastes and, 258
  sampling of, 385
  waste disposal in, 253, 254
Ethylene, 105
  detection of, 356
Eutrophication, 54, 64, 231. *See also*
  Lake Erie
  fish and, 363
  phosphorus and, 319
Evergreens, 116. *See also* Conifers;
  Forests
Evolution, biological, 56
Exhaust, automobile, 102, 120, 127,
  264

FAGS. *See* Federation of Astronomical and Geophysical Services
Falcons, 140. *See also* Birds,
  predatory
Famine, 42, 60
  predictions of, 34
FAO. *See* Food and Agriculture
  Organization
Far East, food for, 35
Farming, conservation, 235. *See also*
  Agriculture
Federal Aviation Agency, 180
Federal Insecticide, Fungicide, and
  Rodenticide Act, 501
Federal Water Pollution Control
  Administration (FWPCA), 63, 72,
  320–321, 322, 489
Federal Water Quality Administration, 301
Federation of Astronomical and
  Geophysical Services, 426
Fenetrothion, 507
Feral pigeon, insecticides and, 154
Fernow Experimental Forest, 245, 246

530 Index

Fertilization. *See also* Eutrophication
  excessive, 68
  increased use of, 34, 39
Field trials, 146
Filter, carbon, 102, 126
  forest as, 248
Fire weed, 53
Firs, 124. *See also* Conifers; Forests; Trees
  Douglas, 195, 213, 214, 217
Fish, anadromous, 64
  chlorinated hydrocarbons and, 41, 285
  DDT and, 281, 282
  flying, 385
  herbicides and, 55
  landing of, 382
  mosquito, 358
  oil pollution and, 266, 307
  pesticides and, 273, 363
  radioactivity in, 273
  sampling of, 384
Fisheries, 13, 382
  demise of, 57
  productivity of, 383
Fish farming, in British Isles, 73
Floods, 205, 242–243
  control of, 14
Fluorides, 106, 110
  trees and, 116, 117, 118, 124
  vegetation and, 101
Food, increase in, 8
  shortages of, 35
Food and Agriculture Organization (FAO), of United Nations, 5, 326, 504
Food chain, 47, 110, 353, 365, 367
  hydrocarbons and, 312
  radionuclides in, 140
Forests, 5
  air pollution in, 125, 126–128
  atmospheric conditions and, 192–202
  carbon dioxide absorption and, 207
  climax, 205, 208–213, 222
  destruction of, 127
  effect of climate on, 203–226
  erosion in, 246
  as filter, 248
  fires, 242
  as flood source, 242
  hydrologic cycle and, 227
  monitoring of, 200–201, 332
  oak-pine, 50, 51, 54
  pesticides and, 24
  pollutants in, 199, 200
  radioactivity and, 199, 200
  rainfall and, 198, 240
  runoff from, 240–251
  sediment and, 244–246
  in Southeast Asia, 55
  sunlight in, 193
  temperature and, 213
Forest Service, U.S., 173
Fossil fuel, burning of, 371, 372
  carbon dioxide from, 175–177
  oceans and, 262, 268
Fraser Experimental Watershed, 245
Fuel, low-sulfur, 484. *See also* Fossil fuel
Fungus, canker, 215
  disease-bearing, 214
  and forest diseases, 203
  effects of ozone on, 104
Fusiform rust, 216
FWPCA. *See* Federal Water Pollution Control Administration

GARP. *See* Global Atmospheric Research Program
Gasoline, lead in, 88, 89
  tax on lead in, 484
Gaylussacia, 51
Gaylussacia baccata, 50
Geophysical Fluid Dynamics Laboratory, of ESSA, 436, 462, 464
Glacier, Franz Joseph, 205
Glaciers, 204, 205, 273
  in Alaska, 206
  sediment samples from, 389–390
Gladiolus, 106
Global Atmospheric Research Program (GARP), 370, 413
Global Network for Environmental Monitoring (GNEM), 393, 422
Global Network for Monitoring the Biosphere (MABNET), 393, 422, 427
Global Telecommunication System (GTS), of World Weather Watch, 426
Grapes, 106
Grasses, tufted, 159
Grasslands, monitoring of, 332
  temperate, 132
Great crested grebe, insecticides and, 154
GNEM. *See* Global Network for Environmental Monitoring
GNP, growth of, 23, 39, 44
Great Smoky Mountains, 123
Greenhouse effect, 194, 464
Green plants, production of, 36–37

Green Revolution, 34, 35, 43. *See also* Developing countries
Gross Domestic Product, 23
Growth, retardation of, 104
  uncontrolled, 75
GTS. *See* Global Telecommunication System

Habitats, pesticides and, 158
Hardwoods, 197. *See also* Trees
Health, effects of pollution on, 80–97
  indicators of, 93
Heat, excess, 74 (*see also* Pollution, thermal)
  pollution effects of, 71
  rangelands and, 136
Heavy metals, 491–492. *See also specific metals*
  control of, 492
  monitoring of, 335–340
  social alternatives to, 492
Herbicides, 166
  aerial spraying of, 248
  effects of, 53, 162
  forests and, 247–248
  partridges and, 158
  susceptibility to, 164
  in United States, 135
Herbivores, 55
Herbs, 53, 57
Herons, insecticides and, 154
Heterotrophs, pollutants and, 55
HMH. *See* Hydrometeorological Monitoring for Hydrological Purposes
Homo sapiens, 59. *See also* Human activity
Horse chestnut, 121
Hubbard Brook Forest, 54
Hudson River, 68
Hudson River Estuary, 70
Human activity, global ecological effects of, 9–12
  increase of, 7
Humans, as monitors, 359
Hurricanes, 330
Hydrocarbons, in atmosphere, 268, 302
  chlorinated, 277, 278–279
  dispersal of, 280
  long-term effects of, 289
  in marine environment, 275–296, 315
  monitoring of, 335
  physiological effects of, 288–289
  toxicity of, 283

emission of, 101–102, 374
global dispersion of, 375
low-boiling aromatic, 309
monitoring of, 369
petroleum, 228, 304 (*see also* Petroleum)
produced by marine plants, 304
removal from ocean of, 306
Hydrogen cyanide, 107
Hydrogen sulfide, 107
Hydrometeorological Monitoring for Hydrological Purposes (HMH), 393
Hypertension, cadmium and, 89, 90

IABO. *See* International Association of Biological Oceanography
IAEA. *See* International Atomic Energy Agency
IAPSO. *See* International Association for the Physical Sciences of the Ocean
IBP. *See* International Biological Program
Ice. *See also* Glaciers
  polar, x, 41
  sea, 207
ICE. *See* International Center for the Environment
Ice Age, 174, 204, 207
ICSM. *See* International Cooperative Studies of the Mediterranean
IDOE. *See* International Decade of Ocean Exploration
IGOSS. *See* Integrated Global Ocean Station System
IGU. *See* International Geographical Union
IHD. *See* International Hydrological Decade
India, Green Revolution in, 34, 35
  pesticide use in, 505–506
Indicators, 354, 360, 362
  mollusks as, 361
Industry, environment and, 75
Insecticides, 238. *See also* Pesticides
  alternatives to, 510
  control of, 494
  costs of, 513–515
  in food chains, 168
  organochlorine, 154, 164, 504
  in runoff, 234, 238
  secondary poisoning and, 157
  social alternatives to, 496
  susceptibility to, 164

Integrated Global Ocean Station System (IGOSS), 370, 413
Intergovernmental Maritime Consultative Organization (IMCO), 426
Intergovernmental Oceanographic Commission (IOC), 426
International Association of Biological Oceanography (IABO), 426
International Association of Meteorology and Atmospheric Physics, 330
International Association for the Physical Sciences of the Ocean (IAPSO), 426
International Atomic Energy Agency (IAEA), 426
International Biological Program (IBP), 27, 393, 426
International Center for the Environment (ICE), 393, 422, 426
International control. *See also* Control
 for tankers, 317
 necessity for, 315
International Cooperative Studies of the Mediterranean (ICSM), 413, 426
International Council for the Exploration of the Sea (ICES), 426
International Council of Scientific Unions (ICSU), **426**
International Decade of Ocean Exploration (IDOE), 426, 463
International Geographical Union (IGU), 426
International Hydrological Decade (IHD), 393, 426
International Indian Oceanographic Program, 370
International Union of Biological Sciences (IUBS), 427
International Union for the Conservation of Nature (IUCN), 427
International Union of Geodesy and Geophysics (IUGG), 427
International Union of Geological Sciences (IUGS), 427
Inter-Union Commission on Solar-Terrestrial Physics (IUCSTP), 427
Iron, 86
Iron oxide, in sea water, 271–272
Irradiation, chronic, 49. *See also* Radiation
Irrigation, sprinkler, 248–249

Japan, fishing fleets of, 36
 pesticides in, 8
 protein available in, 62
Japanese Meteorological Agency, 461
Jet aircraft, cirrus clouds from, 179

Keithia Thujina, 214
Kestrels, 160. *See also* Birds, predatory
Keynesian economics, 43
Kuroshio Current, 449

Lake Erie, 39, 42, 369, 489
 eutrophication of, 320
Lakes, eutrophication of, 319, 320
 herbicides and, 135
Land, agricultural, 35
 forest, 5
 uses of, 5
Landslides, causes of, 244
Latin America, food supplies for, 35
Law and order, 33
Lead, 86–89, 110, 491
 in atmosphere, 373
 monitoring of, 335
 in oceans, 262–265
Leaves, abscission of, 104
 absorption of pollutants by, 112
Lichens, 53
 as monitors, 124
 radioactivity and, 138
Lindane, 233
Lidar, 178, 182
Litigation, 486
"Load on Top" (LOT), 300
Lobsters, 310, 364. *See also* Shellfish
 kerosene and, 313
Logging, 123, 243, 244, 245, 246
 in Ozarks, 213
London, tree growth in, 200
Long Island Sound, waste disposal in, 256
Long-term and Expanded Program of Oceanic Exploration and Research (LEPOR), 413
Los Angeles. *See also* Smog
 pollution control in, 40
 pollution studies in, 92
Lupinus, 184

MABNET. *See* Global Network for Monitoring the Biosphere
Mackerel, California, 63, 286
Magnesium, 85
Maize, photosynthesis and, 186, 187
Malaria, 507
 DDT and, 506, 517

Malaria (continued)
  eradication of, 508, 509
Malathion, 507
Malnutrition, in U.S., 62
Mammals, as biological indicators, 364–365
  migratory, 365
Man, activity of, 4–7 (see also Human activity)
  chemical invasion of ocean by, 261–274
Man and the Biosphere (MAB), 413
Manganese, 491
Manganese oxide, in sea water, 271–272
Maple, fluoride sensitivity of, 118
  red, 52
  sap streak of, 214
Marine organisms, metabolic processes of, 20
  temperature and, 63
Megalopolis, 127
Mercury, 491. See also Heavy metals
  in atmosphere, 373
  increase in, 263
  monitoring of, 335
  in oceans, 261, 262–265
Metals. See also Heavy metals
  emissions at earth's surface, 373
  monitoring of, 369
  trace, 86
Meteorological Services to Air Transport (MSAT), 393
Methyl ethyl ketone, in oceans, 267
Meuse Valley, in France, 119
Microcoulometry, 338
Mimosa wilt, 214
Mimulus cardinalis, photosynthesis in, 185
Minimata disease, 262, 264
Mining, 6
Mississippi River, DDT in, 279
  mouth of, 430
  sediment load of, 257, 259
Mistletoe, 215
Mites, tetranychid, 161
Mixing layers, in ocean, 378
Models, of atmosphere-ocean system, 429
  dynamic (mathematical), 431, 432, 436–439, 443–444
  electric analog, 431
  of environmental processes, 429
  hydraulic, 431, 432–434
  hydrodynamic, 460–465
  mathematical, 432, 434–436, 443–444
  one-dimensional, 438–439
  simulation, 435
  thermal pollution, 441, 444
  three-dimensional, 439–440
  two-dimensional, 438–439
  water quality, 441, 443, 444
  of watershed hydrologic processes, 237
  of waves, 460
Moisture, changes in, 218
Mollusks, 364. See also Shellfish
  and chlorinated pesticides, 285
  as indicators, 361
  oil pollution and, 307
  sampling of, 384
Monitoring, 173
  biological, 331–333, 351
  chemical, 330–331, 353
  of chlorinated hydrocarbons, 290–291
  concept of, 327–328
  economic and statistical, 328–330
  of estuaries, 323
  global, 112, 345, 377
  international, 330, 332, 351, 366, 397–428
  of marine environment, 369
  modern technology and, 333–335
  of oil, 340–342
  operational programs of, 393, 394
  physical, 330–331
  planned programs of, 413, 414, 422
  recommendations for, 329–330, 331, 335, 339, 342, 344, 346–348
  of runoff water, 236–237
  stations for, 366, 370
  of surface changes, 342
  techniques for, 328, 343–344
  of water quality, 493
Monitoring of Basic Environmental Climate (MBEC), 393
Monitoring Sun-Earth Environment (MONSEE), 393
Monitors. See also Indicators
  begonias as, 356
  bioassay, 354
  trees as, 116
Monks Wood Experimental Station, 148
Mosquito. See also Malaria
  DDT and, 159
Mosses, 53. See also Lichens
  radioactivity and, 138
Mount Agung, volcano eruption of, 180
Mrak Report, 499

Mulberry, fluoride sensitivity of, 118
Mullite, in fly-ash, 372
Mutagens, control of, 48

Napthalene, 309
National Academy of Sciences, 81, 268, 344
  Committee on Oceanography of, 228, 229
National Air Pollution Control Administration (NAPCA), 80, 88, 329, 330, 337
National Air Sampling Network, 337
National Communicable Disease Center (NCDC), 507
National Environmental Policy Act, 80
National Insurance Act, of United Kingdom, 89
National Oceanic and Atmospheric Administration (NOAA), 461, 462
Nature Conservancy, 147
New England, climatic changes in, 212
New Jersey, ozone damage in, 127
Newton rings, 341
New York, climatic changes in, 212
  pollution control in, 40
  pollution studies in, 92, 93
  waste disposal sites in, 255, 256
Nickel, 110, 491
  in atmosphere, 373
Nitrates, increase in, 8
  in watershed, 238
  water transport of, 232
Nitrogen, 231
Nitrogen dioxide, 105
Nitrogen oxides, 19, 102
  in atmosphere, 268
  emission of, 102
  trees and, 116, 123
North Equatorial Current, 449
Norway, fishing fleets of, 36
Norwegian Sea, 450, 453
N-P-K fertilizers, on range lands, 136–137
Nuclear energy, 57–58
Nuclear explosions, 273
Nuclear power, 71
Nutrients, in estuaries, 322
  forestry and, 248
  monitoring of, 344
  pollution and, 131
  recycling of, 323
  in runoff, 234
  in subsurface flow, 232–233
  in surface flow, 231–232
  in world's water, 344
Nutrition, 111

Oak, red, 52. *See also* Trees
Oak wilt, 217, 221
Ocean, Atlantic, 454
  atmosphere and, 448–449, 450, 457–458, 464
  carbon dioxide taken up by, 463
  chemical invasion of, 261–274
  circulation patterns of, 448–459
  currents of, 380 (*see also specific currents*)
  depths of, 451
  hydrostatic stability in, 450, 452
  Indian, 455, 462
  lead in, 265
  low stability regions in, 453
  mean state of, 461
  mixing in, 455–457
  modeling of, 461, 462, 463 (*see also* Models)
  Pacific, 456
  petroleum in, 261
  resources of, 36
  river discharge to, 6
  sampling of, 380
  sediment in, 389–390
Octopus, 364
Oil. *See also* Peroleum
  base load of, 311
  burning of, 316
  Kuwait, 309
  monitoring of, 340–342
  in oceans, 316
  offshore wells, 299, 300
  world production of, 297
Oil spills, 63, 297, 298–300, 302, 310–311
Ontario, smelters in, 120
Orchids, injury to, 105
Organic chemicals. *See also specific kinds*
  emission of, 374
  halogenated, 375
Oxidants, monitoring of, 111
  photochemical, 101–105
Oxidation, of oil, 315
  of petroleum hydrocarbons, 306–307
Oxygen, atmospheric, 182
Oxygenation, by trees, 125
Oysters, in Great South Bay, 64
  as indicators, 361
  larvae of, 10
  "tainting" of, 311

Ozone, 102, 110
  biologic indicators of, 102
  detection of, 355
  monitoring of, 111
  trees and, 116
  white pine and, 119

Pakistan, Green Revolution in, 34
Pamlico Sound, 431
PAN. *See* Peroxyacetyl nitrate
Pan American Health Organization (PANO), 507, 508, 516
PAPA, 370
Particles, in atmosphere, 177–178, 372
  monitoring of, 178
  rangelands and, 139–140
  residence time of, 373
  stratospheric, 182
Pathogenesis, 218
Pathogen-pollutant interactions, 111
Pathogens, weather and, 216
PCBs. *See* Polychlorinated biophenyls
Pelican. *See also* Birds, predatory
  eggs of, 287, 288
Pennsylvania, pollution damage in, 122
Pennsylvania State University, 249
Peregrine falcon, 91
  insecticides and, 154
Peroxyacetyl nitrate (PAN), 102, 104, 118
  trees and, 116
Pest control, 12–13
Pesticides, 8
  addiction to, 24
  avian predators and, 147
  in birds, 358
  chemical, 514
  definition of, 149
  dependence on, 34
  development of, 513
  ecology and, 145
  effects of, 146, 150, 151, 163, 499
    delayed, 156
    ecological, 99
    on single species, 155
  eggshells and, 140
  government use of, 502
  halogenated hydrocarbon, 228
  monitoring of, 331
  in oceans, 261, 266
  predatory animals and, 56, 363
  public interest in, 145
  in subsurface flow, 232–233
  in surface flow, 231–232
  on rangelands, 135–136
  restrictions on, 484
  symposia on, 148–149
  "Pesticides and the Living Landscape," 147
Pestilence, 60
Petrochemicals, 21
  plants, 297, 298, 301, 316
Petroleum. *See also* Oil
  emission of, 374
  in marine environment, 266, 297–318, 375
  submarine reservoirs of, 305
Pheasants, DDT and, 156
  insecticides and, 154
Philippines, Green Revolution in, 34
Phosphates, control of, 489–490
  increase in, 8
  social alternatives to, 490
  in water, 68–69, 489
Phosphorus, 231, 238, 320
  eutrophication and, 319–324
  in Hudson River, 69
  production data for, 319
  in runoff, 320
  washoff of, 234
  in watershed, 238
Photosynthesis, 53, 57, 186
  calculations of, 188
  carbon dioxide and, 192
  and chlorinated hydrocarbons, 283
  daily net, 187
  respiration and, 130
  sunlight and, 193
  temperature and, 184, 185, 196
Phynanthrene, 309
Phytophtora cinnamomi, 222
Phytophthora root rot, 214
Phytoplankton, 21, 64. *See also* Plankton
  phosphate content and, 69
Phytotoxicants, 119
Picea glauca, 199
Pine pitch canker, 214
Pines. *See also* Conifers; Trees
  air pollution and, 126
  Austrian, 124
  infection of, 216
  ponderosa, 117, 119, 208, 210, 213, 215
  Southern, 118
  stone, 121
  white, 106, 118, 121, 124, 211, 219
    ozone damage to, 215, 219
Pinus banksiana, 199
Pinus palustris, 51
Pinus rigida, 50

Pinus strobus, 53
Pitch streak, 220
Plaice, 73
Plankton, 17. See also Phytoplankton
  chlorinated hydrocarbons and, 283
  DDT and, 281
  sampling of, 385, 386
Plants, 56. See also Vegetation
  air pollution and, 101–115
  as biological indicators, 365
  as biological monitors, 351
  environmental factors affecting, 108
  pollutants and, 55
  as sensors, 368
  temperature and, 189
Pleistocene, 174
Poisoning, secondary, 156
Pole blight, 219, 223
Pollination, insect, 13
Pollutants, artificial, 130
  in atmosphere, 371
    ecosystem and, 139
    effects of, 108
    forests and, 199
    sources of, 379
  base-line sampling program for, 377–391
  effects of, 48, 91
  increase of, 127
  in marine environment, 277
  movements of, 101
  natural, 130
  predators and, 10, 365
  primary, 109
  runoff as source of, 210–239
  sampling stations for, 379
  secondary, 109
  sensitivity to, 333
  surveillance of, 22
  transport paths of, 273
Pollution, x, 39
  air, 94–95
    bioassay for, 124
    effects of, 99
    pathways of, 95
  alternatives to, 72, 487
  attitudes toward, 469
  biological effects of, 63–65
  chronic, 55, 64
  control of, 94, 323, 470–471, 489–498
  direction of, 57
  domestic, 65
  effects of, 56
    on health, 80–97
  environmental, 16
  estuarine, 227
  hidden costs of, 65, 66, 72
  international aspects of, 43
  needed research on, 110–112
  oil, 228, 303, 305–306, 307 (*see also* Petroleum)
    biological effects of, 315
  sediment, 233, 234
  social costs of, 478
  sources of, 470
  sulfur dioxide, 105
  theories on, 130–133
  thermal, 71, 74, 183, 344
  urban air, 80
Pollution Project Groups, 83, 84
Polychlorinated biphenyls (PCBs), dispersal of, 280
  emission of, 374
  in marine biosphere, 267–268
  production of, 278
  transport of, 279
Polygonum cilinode, 53
Population, control of, 144
  growth of, 23
  increase of, 7
  national policy and, 43
  overpopulation, 35
  politics and, 33
  technology and, 37, 44
  world, 59–61
Portugal, protein in, 62
Potassium, 85
  eutrophication and, 320
Potentiation, 168
Power, nuclear, 71
  thermal, 121
Precipitation, atmospheric particles and, 177
  CO_2 and, 193
  in forests, 241
  in ninth century, 204
  temperature and, 197
  urban-produced increases in, 140
Predators, 10, 164, 365
  avian, 57, 64, 157, 364
  damage to, 17
  DDT and, 41
  reduction of, 160–161
  selective impairment of, 9–11
President's Science Advisory Committee (PSAC), 5, 134, 192, 197, 200, 504, 506, 513
Production, pollution and, 472, 478
Protective measures, 485
Protein, deficiency of, 62
  sea as source of, 62
Prunes, 106

Public Health Service, U.S., 249, 337
Puget Sound, waste disposal sites in, 256

Quercus rubra, 52

Radial growth, of wood, 210, 211
Radiation, effects of, 47
  ecological, 48–49, 50, 51
  forests and, 199
  solar, 174, 178, 194, 330
  water temperatures and, 247
Radioactivity, 42
  DNA content and, 199
  marine biosphere and, 269
  in range system, 137
Radionuclides, 48
Radishes, 107
Rain. *See also* Precipitation
  "misty," 139
  sampling of, 339
  sediment samples of, 389, 390
Range, arctic, 138
  chaparral, 138
  environmental conditions for, 132–133
  locations of, 132–133
  in New Mexico, 138
Rats, as monitors, 356–357
Recycling, 7, 67, 78, 94
Red cedar, 214
Red heart, 214
Redwoods, 195
Refineries, 297, 298, 301, 302, 316.
  See also Oil; Petroleum
Refuse Acts, 253
Regional Plan Association, 259
Reprocessing, 479
Research Natural Areas (RNA), 200, 201
Residence time, of ocean pollutants, 272
Residuals, management of, 477–488
Resources, natural, 37
  consumption of, 38
Rhine Valley, forests in, 199
Rhisomatous species, 159
Rivers, 273. *See also* Estuaries; Runoff
  elements in, 20
  pollution of, 39
  sampling of, 385
RNA. *See* Research Natural Areas
Rubus, 52
Ruhr Valley, 127
Runoff. *See also* Sediment; Sewage
  mean phosphorus content of, 319
  reduction of, 241
  transport of, 231–236

Salinity, 445
  precipitation and, 453
Salmon, 246
  Pacific, 323
Sambucus pubens, 53
Sampling. *See also* Monitoring
  of marine environment, 18, 381–385
  stations, 385
San Francisco, mercury concentrations in, 263
San Francisco Bay, benthic organisms in, 270
  smog in, 262
Santa Barbara, oil spills in, 266, 299, 303, 310, 316
Saprogenesis, 218
Sargasso Sea, 305–306, 450
Satellites, unmanned earth, 334
  weather, 330
Savanna, tropical, 132
Savannah River Laboratory, 51
Scattering, optical, 182
Scavengers, in marine environment, 290
SCEP. *See* Study of Critical Environmental Problems
Science, 2
Scientific Committee on Antarctic Research (SCAR), of ICSU, 427
Scientific Committee on Ocean Research (SCOR), of ICSU, 427
Scrub, hawthorn, 162
  wild, 167
Scrub stage, 166, 167
Sea. *See also* Ocean
  food from, 61–63
  oil pollution in, 300–304
Seabirds, DDT and, 281. *See also* Birds; DDT
Seabrook Farms, 248
Seaports, 59. *See also* Estuaries
Secretary's Commission on Pesticides, 500, 515
Sedge, 51
Sediment, aquatic life and, 246
  carried by streamflow, 243
  deep-sea, 390
  erosion and, 243–244
  from forested watersheds, 243
  pollution in, 338
  riverborne, 252, 257
  samples of, 387
  in surface waters, 138

Sensors, biological, 353
  satellite-type, 343
Sentinels, 354
Sewage, disposal of, 66, 67
  domestic, 320
  heavy metals and, 492
  phosphates in, 490
  treatment plants, 74
Shag, insecticides and, 154
Shellfish. *See also specific kinds*
  harvests of, 364
  oil spills and, 310–311, 312
  pesticides and, 358, 363
Shoaling, 433, 444
Shrimp, 342, 364
  as bioreagents, 357
Shrubs, 53, 57
  herbs and, 131
  on range, 135
Silicon tetrafluoride, 118
Silver iodide, in atmosphere, 200
Slash burning, 247
Slicks, 375–376. *See also* Oil; Petroleum
  natural, 303
Smelters, in British Columbia, 120
  forests and, 126
  in Tennessee, 120
Smog, London-type, 118, 121
  Los Angeles-type, 118, 119, 126
  photochemical, 102, 105, 118, 120, 121, 124
  trees and, 125
Smoke, in atmosphere, 268
Smoky Mountains, 54
Snow, "dusty," 139
Snowmelt, 231, 243
Sodium, 85
Soil, 201
  erosion of, 42
  formation of, 14
  monitoring of, 338
  nutrient transport by, 235
  of range ecosystem, 135
  retention of, 14
  in runoff, 234
  stability of, 245
Soil Conservation Service, 230
Sorghum, photosynthesis in, 186, 187, 188, 189
Soviet Hydrometeorological Service, 462
Soviet Union, politics of, 43
  population control of, 37
Sparrow-hawk, 160
  insecticides and, 154

Special Committee on the International Biological Program (SCIBP), 427
Species, substitution of, 159
Spruce, 124
Squid, 364
Standard, living, 23
  water quality, 21, 22, 493
Starvation, mass, 62
Steam plants, nuclear-powered, 122
Stratosphere, monitoring of, 181
  water vapor in, 182
Streams. *See also* Runoff
  herbicides and, 135
Strontium-90, 138, 269
  in United States, 137
Study of Critical Environmental Problems (SCEP), x, xi, 1, 173, 301
  Task Force of, 275–296
  Work Group on Ecological Effects, 1, 2, 4–32, 319
  Work Group on Monitoring, 327–330
Study of Man's Impact on Climate (SMIC), 429
Succession, pesticides and, 166
Sulfur, 374
  in atmosphere, 372
Sulfur dioxide–ozone ratio, 107
Sulfur oxides, 121
  in atmosphere, **268**
  removal of, 122–123
  sulfur dioxide, 19, 105–106, 107
  trees and, 116, 117, 119, 124
  vegetation and, 101
Sunlight, CO_2 and, 193
Supersonic transports (SSTs), 179–181
Sverdrup transport, 448
Sweetgum blight, 223
Symbiosis, 218
Synecology, 144

Tampico, 310, 311
Tankers, 297, 298, 299, 300, 305. *See also* Oil; Petroleum
TDE, application of, 160
  effects of, 165
Technology, growth of, 39
  monitoring, 333–335
  population and, 33
  restrictions on, 57
Temperature. *See also* Pollution, thermal
  agriculture and, 184–191
  biological processes and, 73

Temperature (continued)
  changes in, 194–196, 219
  global, 195
  photosynthesis and, 184
  in polar regions, 205
  precipitation and, 197
Tetraethyl lead, 264
Texas Parks and Wildlife Department, 285
Thorium, in oceans, 272
Thunderstorms, 174
  urban-produced, 140
Tides, storm, 461
Timber. See also Forests; Trees
  destruction of, 119
Tobacco, 106
  blue mold and, 217
  damage to, 103
  for monitoring, 356
  pollution studies and, 107
  weather fleck and, 219
Tolerance limit (TL_m), 357
Toluene, 309
Tomatoes, 106
Torrey Canyon, 298, 310, 316
Toxaphene, 21
Toxicity, cadmium, 90
  of pollutants, 20
Toxicology, 82, 140, 152
Trace metals. See also Metals
  safe levels for, 86
Tractors, increase in, 8
Transport mechanisms, lateral, 378
  vertical, 378
Trash. See also Wastes
  production of, 39
Trees. See also Forests
  adaptation of, 198, 212
  air pollution and, 99, 116–129, 219
  air purification by, 125
  chronic injury to, 121
  coniferous, 116 (see also Conifers)
  decrease of, 214
  future for, 122–126
  as monitors, 125
  PAN and, 116
  radiation and, 50
  rings, 18, 210, 211, 219, 222
  roots of, 218
  shrubs and, 131
  surface runoff and, 235
  temperature changes and, 220
  warm-climate species of, 211
Triflurin, 514
Troposphere, particles in, 177
Trout, 21, 246

sea, 285
Tsunami Warning System (TWS), 393
Tuna, 21, 385
Tundra, alpine, 133, 184
  arctic, 133
Tundra-lemming system, 11
Turbidity, of atmosphere, 41
Turkey, Green Revolution in, 34

Ungulates, 365
U.S.S.R., fishing fleets of, 36
United States, politics of, 43
  population control of, 37
United Nations, Economic and Social Council, 426
  Educational, Scientific and Cultural Organization (UNESCO), 427
  International Children's Emergency Fund (UNICEF), 507, 508
  World Health Organization, 500, 507, 508, 511, 516

Vaccinium, 50, 51, 53
Vanadium, 110
  in atmosphere, 373
Vegetation. See also Plants; Trees
  air pollution injury to, 108
  broad sclerophyll, 132
  change of type, 342
  climate and, 184
  effects of sulfur dioxide on, 106
  photochemical oxidant injury to, 101
  phreatophytic, 242
  pollution and, 99, 109
  woody, 125
Viet Nam, herbicides in, 53, 55
  war in, 42

Wall Street Journal, 506
War, 60
Wastes, arsenic-containing, 257
  disposal, sites for, 253–257
  domestic, 262
  dredged, 259
  globally distributed, 371–376
  industrial, 322, 323
  in marine environment, 269–270
  measurements of, 371
  riverborne, 321
  sources of, 252–253
Water. See also specific water bodies; Runoff
  resources of, 20
  temperature changes in, 57, 247
Waterways Experiment Station, 433
Waves, 460–461

540 Index

Weather. *See* Climate
Weather Bureau, U.S., National Meteorological Center of, 436
  Office of Climatology, 223
  storm warnings of, 461
Weddell Sea, 454
Weeds, reduction in, 158
WEI. *See* World Environmental Institute
Westerlies, 378, 390
West Wind Drift, of North Pacific, 449
Wetlands, destruction of, 42
Whales, DDT and, 281
Willow, fluoride sensitivity of, 118
Wind, 273
  ocean circulation and, 449, 450
  prevailing, 378
  sampling and, 377–381
  seasonal modification of, 378
  trade, 390
  westerly, 378, 390
Woods Hole Oceanographic Institution, 272, 306, 310
World Environmental Institute (WEI), 393, 422, 427
World Health Monitoring and Information Network, 413
World Health Organization, of U.N., 500, 507, 508, 511, 516
World Meteorological Organization (WMO), 330, 393
World Weather Watch (WWW), 393, 427
WMO. *See* World Meteorological Organization
WWW. *See* World Weather Watch

Xerophytic species, 197
Xylene, 309

Zea, 184
Zinc, 86, 491
Zoning, 485